FOREWORD

Topics the OECD/NEA Nuclear Science Committee (NSC) has addressed over the years in co-ordination with the Committee on the Safety of Nuclear Installations (CSNI) include, among others:

- Surveillance and diagnostics.
- In-core instrumentation and reactor core assessment.

A series of symposia and specialists meetings were held covering these topics, the most recent of which were:

- SMORN-VII: A Symposium on Nuclear Reactor Surveillance and Diagnostics, Avignon, France, 19-23 June 1995 (proceedings available at http://www.nea.fr/html/science/rsd/).
- Specialists Meeting on In-Core Instrumentation and Reactor Core Assessment, Mito, Japan 14-17 October 1996 (proceedings available at http://home.nea.fr/html/science/rsd/ic96/).

Both of these meetings addressed the issue of signal registration, its processing and interpretation for the purpose of surveillance, diagnosis and improved operation of nuclear power stations.

The ninth meeting of the OECD/NEA Nuclear Science Committee (NSC), held in June 1998, was partly devoted to an in-depth discussion of core monitoring. The NSC concluded that it would be useful to follow up this discussion with a workshop to discuss improvements and further development of the present monitoring systems and methodologies with the aim of enhancing their ability to handle modern fuel under present or foreseen operational strategies.

The Workshop on Core Monitoring for Commercial Reactors: Improvements in Systems and Methods (CoMoCoRe'99) took place from 4-5 October 1999 in Stockholm. It was jointly organised by Vattenfall AB, ABB Atom and the Swedish Nuclear Power Inspectorate.

The present workshop, although having a similar scope, is separate from the series of specialists meetings on In-Core Instrumentation and Reactor Core Assessment. The main objective was to discuss how instrumentation, methods and models used in core monitoring could be validated, or, if needed, improved and further developed to provide more reliable and/or detailed information on local power in the core and on other parameters indirectly affecting fuel duty as, e.g. the core decay ratio in a BWR. Another important objective was to show how the core monitoring system can be used to support reactor operation in normal and anticipated transient modes and to supply data used to derive initial key core parameters for transient and accident analysis.

Although the present methodology is adequate, it has some shortcomings when applied to more advanced fuel and core designs. Several possible improvements were also mentioned that would lead to smaller uncertainties in design and to better known uncertainties and margins in operation. Many of these improvements are based on having better models for fuel and core calculations.

These proceedings present the motivation for holding the workshop. Included are summaries of the discussions held, the papers presented in the different sessions and some conclusions.

This text is published under the responsibility of the Secretary-General of the OECD. The views expressed do not necessarily correspond to those of the national authorities concerned.

TABLE OF CONTENTS

EXECUTIVE SUMMARY

Introduction

The ninth meeting of the OECD/NEA Nuclear Science Committee (NSC), held in June 1998, was partly devoted to an in-depth discussion of core monitoring. In preparation for the discussion at the NSC meeting, a written report had been distributed to the committee members. This report (see Annex 1) gives an overview of requirements, system layouts and operational experience with regard to core monitoring for BWRs and PWRs. It also discusses improvements and further development of the present monitoring systems and methodologies that would enhance their ability to handle modern fuel under present or foreseen operational strategies. As a result of the discussion engendered by this report, the NSC concluded that a workshop would be a useful follow up.

CoMoCoRe'99, organised jointly by Vattenfall AB, ABB Atom and the Swedish Nuclear Power Inspectorate, took place from 4-5 October 1999 in Stockholm. The workshop was timely, as it dealt with an issue which should be addressed in parallel to the internationally ongoing discussion among authorities, utilities and vendors on how to deal with the rapid technical development and optimisation of nuclear fuel and its utilisation under new, more aggressive fuel management strategies.

Objectives

Although having a similar scope, CoMoCoRe'99 should be considered separate from the series of specialists meetings on In-Core Instrumentation and Reactor Core Assessment, the last of which was held in 1996. The main objective of the present workshop was to discuss how instrumentation, methods and models used in core monitoring could be validated, or, if needed, improved and further developed to provide more reliable and/or detailed information on local power in the core and on other parameters indirectly affecting fuel duty as, e.g. the core decay ratio in a BWR. Another important objective was to show how the core monitoring system can be used to support reactor operation in normal and anticipated transient modes and to supply data used to derive initial key core parameters for transient and accident analysis.

Technical programme and participation

Presentations were invited dealing with applications for all types of commercial LWRs, including VVER. Twenty-three papers were accepted for presentation, structured into four technical sessions (see Annex 2):

- Session I. Requirements on Core Monitoring Systems.
- Session II. Sensors, Signal Processing and Evaluation.
- Session III. Improved Core Models in Core Monitoring.
- Session IV. Improved Core Monitoring Systems, Design and Operating Experience.
- Session V. Discussion and Conclusions.

The workshop was attended by about 60 participants from 30 organisations representing 15 countries (see the *List of Participants*), and was concluded with a discussion, the highlights of which are presented in the following.

Summary of discussion

During the meeting a good overview of present efforts to improve the capabilities of the core monitoring (CM) systems of different commercial reactor types was provided.

Some general trends may be seen:

- The introduction of more detailed physics models in on-line calculations for both BWR and PWR.
- More wide-spread discussion on possible advantages of backfitting some PWR types with fixed in-core detectors.
- Methods to combine the information from on-line measurements and on-line calculations.

There is in fact a very rapid development of more advanced fuel design and methods of operating the core. Consequently, there is a need to reconsider how the core is monitored.

Regulatory perspectives

From the regulators perspective, core monitoring should not be regarded as an isolated issue, but as part of overall fuel cycle issues. For introduction of advanced core designs, surveillance systems must be part of the strategy. It is important to treat the whole core monitoring chain, starting with detectors and signal processing. Most countries have no formal requirement to license CM systems as they are not safety critical for the reactor protection system. Discussions with regulators regarding modifications to the CM system should however be encouraged without a need of formal approaches. As for physics models, 3-D best estimate methods are accepted today by regulators if they are accompanied by thorough and well founded analysis of uncertainties, in particular as they relate to advanced fuel and core design. Obviously, old methods should not be used for advanced cores. Changes in CM systems need to be well founded, and penalties and benefits should be discussed with regulators at an early stage, as should the question of how fuel safety limits and operation margins are set.

Methods, risks – benefits, operating margins

The balance of risk (compromised fuel integrity) and benefit (fuel performance) need to be considered simultaneously in all CM upgrades. Improved core monitoring has a potential economic benefit; better in-core instrumentation and physics evaluation methods improve accuracy. The operator may benefit from this by operating the reactor core closer to thermal limits (obtaining higher performance out of the fuel) as long as the remaining thermal margins are well understood and accounted for. This can be a difficult task, which is exemplified by the fact that pellet-clad interaction (PCI) failures, even with liner fuel, are still observed. This indicates that fuel pin powers have not been adequately estimated, nor the limiting cases well identified. When more heterogeneous fuel bundles are introduced it is important to be able to carry out more precise measurements and to improve accuracy in calculations. Crude methods, such as those using one and a half energy groups and spatial resolution only on the level of a whole fuel node, now need to be replaced by the methods representing the present state of the art for a better estimation of damage risks. Commercial codes that are currently available include full two-group energy representation and detailed pin by pin calculations. For BWRs more advanced methods for coupling of 3-D neutronics with thermal-hydraulics will soon emerge. When implementing these new models it is important that detailed, realistic experiments be used as a base for the validation.

It should be emphasised, however, that fuel reliability is today recognised as a goal in and of itself, balancing in a natural way the goal of higher fuel performance. Ensuring reliability also means that margins have to be left for the unexpected. The emerging deregulated electricity markets in Europe and the USA lead to extra economical pressure on operators and therefore also on vendors. It is important to maintain a proper balance between risk and balance under these circumstances.

New core monitoring systems should improve the view inside the reactor. Thus, a closer look should be given at the process signals from the reactor. Here, signal qualification is an important aspect that should be further addressed. Signals must be checked and validated to be useful. The role of measurements in CM is to reveal anomalies in the core for the purpose of taking action. A neural network kind of approach could be one method. How to best combine measurement and calculation is a big challenge to be tackled. Solving this challenge, though, should lead to a good industrial product.

In some countries (France, in particular) there are challenging demands on core follow operation for nuclear power plants. In PWRs this leads to an operation mode with so-called grey control rods, which affect peaking in the core. Additionally, PCI is a concern in Class II events. This means that present thermal margins are more or less used up. One way to improve operation flexibility in the future is to upgrade the present CM system using fixed in-core detectors together with full 3-D on-line simulation with short response time. However, a CM system should not be too complicated with regard to maintenance, interpretation, evaluation, etc. Confidence can be built up only under these circumstances. It is mainly the reactor operator that should run a system for continuous operation and not a group of engineers.

Some conclusions

- There is an ongoing development in physics models in the reactor physics, thermal-hydraulics and other related research communities. This will provide improved models that can be implemented in core monitoring (CM) systems.

- Signal validation is of prime importance in any CM system and could be a subject for further study in the framework of the NEA.

- More rigorous methods to combine information from measured and calculated data should also be very useful in future CM systems.

Workshop organisation

Chairman:	Tomas Lefvert, Vattenfall AB, Sweden
Organising committee:	Tomas Lefvert, Vattenfall AB, Electricity Generation
	Stig Andersson, ABB Atom AB, Sweden
	Oddbjörn Sandervåg, SKI, Sweden
	Enrico Sartori, Nuclear Energy Agency
Program committee members:	Tomas Lefvert,Vattenfall AB, Sweden
	Stig Andersson, ABB Atom AB, Sweden
	Oddbjörn Sandervåg, SKI, Sweden
	Herbert Finnemann, Siemens/KWU, Germany
	Alan Wells, Siemens Power Corporation, USA
	Richard Cacciapouti, Duke Engin. and Services, USA
	Daniel Janvier, EDF, France
	Öivind Berg, Halden, Norway
	Etsuro Saji, Toden Software, Japan
	Yoichiro Shimazu, Hokkaido University, Japan
	Rudolf Vespalec, NPP Dukovani, Czech Republic
	Moon Ghu Park, Korean Electric Power Corp., Korea
	Roy Olmstead, AECL, Canada
	Enrico Sartori, Nuclear Energy Agency

Annex 1

On-Line Core Monitoring for BWR, PWR Overview and
Comments on Operation Experience and Further Development

Presented by Tomas Lefvert at the Ninth NSC Meeting, June 1998

General requirements

Reference is made to the General Design Criteria put forward in US CFR 50 App. A, namely:

- GDC-10 Reactor Design
  The reaction core and associated coolant, control and protection systems shall be designed with appropriate margin to assure that specified acceptable fuel design limits are not exceeded during any condition of normal operation, including the effects of anticipated operational occurrences.

In order to show that these margins exist, and to take protective action if they become too small, the following criteria also apply:

- GDC-13 Instrumentation and Control
  Instrumentation shall be provided to monitor variables and systems over their anticipated ranges for normal operation, for anticipated operational occurrences, and for accident conditions as appropriate to assure adequate safety, including those variables and systems that can affect the fission process, the integrity of the reactor core, the reactor coolant pressure boundary, and the containment and its associated systems. Appropriate controls shall be provided to maintain these variables and systems within prescribed operating ranges.

- GDC-20 Protection System Functions
  The protection system shall be designed (1) to initiate automatically the operation of appropriate systems, including the reactivity control systems, to assure that specified acceptable fuel design limits are not exceeded as a result of anticipated operational occurrences, and (2) to sense accident conditions and to initiate operation of systems and components important to safety.

In the following, we will only address the system for monitoring the fission process in the core. This is one of the systems referred to in GDC-13 and it is often based on neutron flux detectors. However, there is also a connection to the protection system which often takes data from the monitoring system but treats them in another way to assure redundancy, operability in adverse conditions, etc.

There are also systems for core surveillance meaning that the fission process is checked periodically rather than continuously.

The operating domain of the reactor is defined through criteria in the technical specifications and constitutes a certain 2-D region in the power-core flow (BWR) or power-axial offset (PWR) space within, but well separated from, the various protection lines. If, through core monitoring or core surveillance, we find that the operating point falls outside the operating domain, the technical specifications defines the action to be taken, e.g. lowering reactor power. Thus, there is both an administrative and an automatic protection system in place.

System layouts

BWR

The BWR typically has a heterogeneous distribution of fuel and moderator over the core, both radially and axially. This is true at zero power due to the variation in enrichment in the fuel pins, the gadolinium absorber, the fuel channel and water gaps between assemblies different from the pin-to-pin gaps in the fuel assembly. At power there is also a 3-D distribution of void due to the two-phase flow. In addition, there is the local perturbation from the control rods used to control part of the cycle excess reactivity and to shape the power distribution.

Under such conditions it is difficult to reveal unwanted local power peaks unless the neutron detector is situated in the core. All BWRs have a similar detector layout with roughly one detector string per 4*4 fuel assemblies in the core. Each string typically has four fission detectors (local power range monitors – LPRMs) in different axial positions along the active height of the fuel assemblies. The LPRM readings are frequently compared with the results of 3-D on-line core simulation using nodal diffusion codes to solve the neutron transport equation. This is normally done in the adaptive mode, where the calculated power distribution is fitted to the measured data points before evaluating the margins to the various thermal limits on the fuel pin. Alternatively, the calculated results can be used directly to find the margins and the observed deviation between measured and calculated data used to define the uncertainty to be included in the margin.

The sensitivity of the LPRMs will vary with detector burn-up. Therefore they need to be calibrated regularly, typically every 2-4 weeks. The calibration is effectuated by a system of movable fission or gamma detectors (TIPs) that can be pushed through a tube adjacent to the LPRMs. Normalisation to absolute power is done by comparison with the heat balance of the plant.

Thus, the BWR core monitoring system consists of the LPRMs, the TIPs and their associated software for treating the detector signals, etc., plus the on-line core simulator. As a result, the error in determining the local power and thermal margin is affected both by certainties in the flux measurement and calibration, by model uncertainties in the simulator and by uncertainties in input data to the models, given by the deviation between the nominal and the actual core and fuel geometry and fuel isotopics. Theoretically, and since the BWR is operated with either quarter-core or half-core symmetry, the model uncertainty could be separated from the rest if measurements were taken in symmetrical positions. However, unlike PWRs, this is normally not the case in BWRs.

The BWR protection system against high power is based on LPRM signals. Signals from a number of LPRMs spread evenly over the core are added to form an average power range monitor (APRM) signal. Typically four different sets of LPRMs form four independent APRMs, the signals of which are compared with setpoints for reactor protection.

At shutdown and low power conditions other in-core detectors are used in the source and intermediate power range, respectively. Today, many BWRs use one set of detectors for the whole range from shutdown to low power, namely wide range neutron monitors (WRNMs). They are used to monitor sub-criticality during, e.g. core shuffling and control rod scram tests, and in the measurements of shutdown margin. They also protect against inadvertent local criticality using appropriate set points for doubling time and flux level. Modern designs of WRNM are sensitive enough to allow absolute reactivity measurements down to about 1% sub-criticality. In the low power range they provide a diversified overlap with the LPRMs.

PWR

Originally the PWR core had a fairly homogeneous layout of fuel and moderator. There are no fuel channels, there are uniform pin-to-pin water gaps over the whole core and there is always one-phase flow. Originally there were no burble absorbers and control rods were not used during full power operation. However, unlike the BWR case, the layout of the core monitoring system for PWRs tends to vary between reactor vendors. The dominating designers, Westinghouse and Framatome, do not use fixed in-core detectors but rely on large volume external detectors for core monitoring and a TIP system for calibration of the external detectors and for core surveillance. The TIP can run in the centre tube of about one-third of all assemblies. Considering that one PWR assembly is roughly equivalent to four BWR assemblies, we note that more core positions are measured in a PWR than in the BWR. Some of the positions are in symmetrical core positions allowing a separate analysis of model uncertainty in calculating TIP signals.

The external detectors are also used for overpowering protection. The monitoring and protection limits are defined in terms of the axial offset evaluated directly from the ex-core detectors, which integrate the thermal flux coming from the upper and lower part of the core periphery, respectively. The limits have been set to conservatively cover an envelope of possible core perturbations where local power remains within specified fuel design limits.

The TIP data, taken about every four weeks, are evaluated off-line with a core simulator and the results compared with technical specification limits with regard to departure from nucleate boiling (DNB) and LOCA-related power peak.

PWR designs from Siemens, the former Combustion Engineering and the former B&W all have fixed in-core detectors in addition to the TIP (reference) system and ex-core detectors. The Siemens PWR has prompt responding detectors while the other designs use rhodium, which gives a delayed power response unsuitable for use in a protection system. The use of in-core detectors (PDD) and the availability of a fast reference system (Aeroball Measurement System) in Siemens PWRs also permit a PDD signal validation in perturbed conditions. Thus PDDs register all power density changes caused by power density perturbation modes or by control rod misalignments (mode detection capability). *Applying a simple calibration procedure for the PDDs under reference conditions the maximum PDD signal directly indicates the peak power density.*

PWR core monitoring with fixed in-core detectors will allow a wider operating domain. If, in addition, the in-core detectors can be used in the protection system, the operating domain may be extended further allowing even higher flexibility in operation.

Ex-core detectors are used also for monitoring and protection during shutdown and start-up.

Finally, the latest Framatome designs make use of a new system of external detectors with six axial levels instead of two and the appropriate supporting software. This allows a better reproduction of core axial power at the core periphery, which is important considering the rather demanding mode of operation that the EDF follows with both frequency control and load-follow. These grid requirements have also prompted the EDF to generally introduce an operating mode where so-called grey rods, made of steel with no extra absorber, can reside in the core during operation in order to shape the power distribution.

Comments on operational experience

BWR

There are examples where fuel pin power has exceeded design limits and failed during normal reactor operation, without the core monitoring system giving the proper indications. The main causes for this have probably been off-nominal water gap geometry and/or less than anticipated model accuracy. As already mentioned, there is no way in a BWR to separate model uncertainty from the rest of the uncertainties when comparing measured and calculated power distributions. The cross-section homogenisation models for the fuel assembly can be tested, e.g. in critical experiments with fresh fuel and in irradiation experiments, e.g. for gadolinium burn-up. Moreover, lattice code intercomparison has been made by the NEA/NSC. However, the accuracy of the older nodal diffusion codes and thermal hydraulic models in calculating power distributions in an operating BWR are not known. Also, there is no way to know the true lattice geometry in the core or the true lateral position of a LPRM or TIP detector in the narrow water gap between assemblies. We can only observe the combined uncertainty from all these sources of error.

A prudent way to deal with these uncertainties is to introduce better models, try to avoid materials and mechanical design of fuel assemblies that could lead to distortion of nominal dimensions under irradiation and to have a fuel lattice and nuclear design that provides extra thermal margins in a given mode of operation.

Thus the modern advanced nodal core simulators with two full energy groups, discontinuity factors and pin power reconstruction for the 3-D neutron transport, are now coming into use in core monitoring and will reduce the model uncertainty. Also, the thermal hydraulic models and/or the empirical correlations they use could be improved in order to cope with the increasing design complexity of modern BWR fuel. It is likely that with the new neutronic models mentioned above the T&H modelling will be limiting the total model uncertainty of the BWR core monitoring system. The 10*10 lattice is becoming a BWR industry standard, at least in Europe and the USA. It provides more margin to pellet clad interaction (PCI) and generally lowers the absolute pin powers in the core which in turn gives more margin for inadvertent variations in inter-assembly water gaps and also less fission gas release. Moreover, BWR operators keep a close check on fuel channel dimensional changes. Channel bow has in the past caused fuel failures resulting from increased water gaps which then cause corner and peripheral fuel rods to obtain better moderation and therefore higher power.

PWR

The modern PWR core is becoming less homogeneous with the introduction of burnable absorbers and low-leakage loading patterns. These patterns also cause low power assemblies to reside in the core periphery, which affects the relative contribution of high and low power assemblies to the integrated signal in the external detectors. Recently, the assumption of uniform water gaps between assemblies has also been challenged with the manifestation of the assembly bow problem in many PWRs. However, we know of no reports on fuel failures due to rod overpower within the operating domain. This is good news, indicating that there was ample margin originally. We should also keep in mind, however, that re-licensing of transients and accidents and/or power uprating using less conservative methodology, have been performed recently by many operators, allowing operation with higher load factors on individual fuel rods.

In total, it is likely that this development has decreased the available margin for unexpected and new fuel related phenomena (not detectable during operation). We also must continue to expect the unexpected in future operation and allow for appropriate margins. Again, a prudent strategy would be to try to reinstate extra margin by using a similar approach as that described for BWR, but adapted to the specific demands of the PWR design being studied, e.g. stiffer skeleton to avoid assembly bow. In the assessment of uncertainties we are better off in the PWR since analysis of measured data in symmetrical core positions allows an estimate of the model uncertainty itself. Thus, a systematic analysis of TIP comparisons along these lines would indicate where the most effective improvements could be made in order to lower the total uncertainty in the core monitoring of a given reactor.

Concerning model uncertainty, improvements could be made by introducing modern core simulation in the monitoring of PWRs, using TIP data in an adaptive or non-adaptive way in the comparison with calculated local power. Fixed in-core detector data could also be used where applicable.

Further improvement and development

Improved 3-D simulator

There are now several advanced nodal core simulators commercially available. They do a better job than the old ones, especially for the more advanced fuel and core designs. Thus the new tools are available for the utility to use, and some are presently introducing the new simulator for on-line monitoring.

Using the same models/methods and data banks for both core monitoring and surveillance as for fuel and core design will of course give better overall agreement between core design and core follow, surveillance and monitoring results. This could allow the utility to use tighter design margins when doing the reload design.

Modern 3-D core simulators handle an order of magnitude more data for the calculation than older versions Therefore, the hardware also has to be upgraded when introducing the new simulators.

Improved thermohydraulic modules

Both PWR and BWR fuel designs have evolved steadily in the areas of neutronic and mechanical designs. The current hydraulic methods, while still adequate, have not fully kept pace with this development. Longer cycles, higher peaking and non-homogeneous bundle designs will continue to challenge the adequacy of the thermohydraulic models.

Thus the T&H models need to possess flexibility to handle the new fuel designs, including part length fuel rods and other hydraulic characteristics. As always, when modelling complicated physical phenomena, it is very useful to have access to experimental data for the validation of the improved models, e.g. void data for BWR fuel.

Improved core monitoring

Incorporation of on-line DNBR calculations for PWR plants can reduce uncertainty and conservatisms currently in use. This is, however, not a simple change to the present licensing strategy and will affect many parts of the safety analysis calculations as well as core monitoring.

Most modern core simulator codes have accurate xenon transient modelling methods incorporated into their basic design. Older core monitoring codes in use at some plant sites do not have adequate transient xenon models. With the current generation of mini-computer systems, all plant core physics personnel should have access to monitoring and prediction tools utilising time and power dependent transient methods.

As concerns incorporation of boron depletion routines in core monitoring routines for PWR plants, modern utilities have maintained a much tighter control of leaks in the plant and do not have to replenish the coolant as frequently as in the past. This results in the depletion of the ^{10}B abundance in the boron concentration in the core. Since the resultant depleted boron has less neutron poisoning effect, the boron concentration needs to be increased. The inclusion and tracking of the effects of boron depletion are not included in many PWR core monitoring systems. This can result in large discrepancies (~199 ppm boron) for plants that have long operating periods between shutdowns.

As already mentioned, the results of the on-line core simulator calculations can be treated in two principally different ways: the adaptive and the best estimate method, respectively.

In the adaptive method the calculated power distribution is adapted to the local power measurements. It is then re-expanded over the whole core in order to determine the limiting thermal margin in the core, which is then compared with a given limit value. Improved adaptive models are possible. One possibility is, e.g. to take into account more than the nearest detectors when adapting the power distribution.

In the best estimate method, we use the simulator directly to determine the thermal margins without adaptation of the power distribution. The measured detector readings are used to regularly evaluate the uncertainty of the difference between measured and calculated local power in the TIP measuring positions. Based on this uncertainty a statistical confidence interval is added to the calculated thermal margin and the result compared to the given limit value.

Conclusions and recommendations

Modern BWR fuel and core designs have become rather demanding from a neutronic and thermohydraulic modelling point of view. The BWR probably presents a greater challenge than the PWR in this respect. Introducing the new, advanced nodal core simulators in core design and in on-line core monitoring could lead to smaller design margins and better known operational margins. To reach this, however, it may in some cases also be necessary to improve the T&H modelling, which now tends to limit the accuracy in BWR core simulation.

Modern PWR core designs are becoming more non-homogeneous due to requirements of longer cycle times and reduced batch sizes. The current system of depending solely on once a month in-core flux mapping coupled with continuous monitoring from large volume ex-core detectors can contribute to unnecessary uncertainty in the monitoring of the core. A more aggressive licensing position can be taken by proving the accuracy of modern core simulators with confirmation from Aeroball or moveable in-core detection systems. These licensing strategies must be co-developed with the licensing authority, the utility and the vendor.

In this overview of core monitoring we have pointed out that although the present methodology is adequate, it has some shortcomings when applied to more advanced fuel and core designs. Several possible improvements were also mentioned that would lead to smaller uncertainties in design and to better known uncertainties and margins in operation. Many of these improvements are based on having better models for fuel and core calculations. Therefore it could be fruitful to have the issue analysed further within the frame of the NSC, perhaps in the form of a workshop.

Annex 2

Technical Programme

Monday, 4 October 1999

- *Tomas Lefvert, Enrico Sartori* – Opening address

Session I: Requirements on Core Monitoring Systems (*Chairman: Oddbjörn Sandervåg*)

- *Öivind Berg* – User Interface Design and System Integration Aspects of Core Monitoring Systems
- *Tell Andersson* – Functional Requirements for PWR Core Surveillance Systems
- *Juan Casal* – Uncertainty Assessment in BWR Core Monitoring

Session II: Sensors, Signal Processing and Evaluation (*Chairmen: Etsuro Saji, Öivind Berg*)

- *Jean Mourlevat*, Daniel Janvier, Holland Warren – Industrial Tests of Rh SPDs: The Golfech 2 Experiment
- *Ferenc Adorján*, I. Pos, S. Patai Szabó – Statistical Analysis of the Ratio of Measured and Predicted Rh SPND Signals for VVER-440/213 Reactor
- *Akihiro Fukao*, Etsuro Saji – The Study on the BWR In-Core Detector Response Calculation
- *Tsunemi Kakuta*, K. Suzuki, H. Yamagishi, H. Itoh, M. Urakami – Demonstration of Optical In-Core Monitoring System for Advanced Nuclear Power Reactors
- *Koki Inagaki*, Hironobu Shinohara, Satoru Yasue, Masaru Tamuro – Development of Advanced Digital Rod Position Indication System

Session III: Improved Core Models in Core Monitoring (*Chairmen: Allen Wells, Tomas Lefvert*)

- *Hoju Moon*, Allen Wells – Impact of Advanced BWR Core Physics Method on BWR Core Monitoring
- *Alejandro Noel*, Lorn Covington, Alf Nilsson, Daniel Greiner – Core Monitoring Based on Advanced Nodal Methods: Experience and Plans for Further Improvements and Development

- *Per Claesson* – JEF-2 Cross-Section Library for Casmo-4: Impact on Core Monitoring of OKG Reactors
- *Pär Lansåker* – BWR Core Stability Prediction On-Line with the Computer Code MATSTAB
- *Makoto Tsuiki* – VNEM: Variational Nodal Expansion Method for LWR Core Analysis

Tuesday, 5 October 1999

Session IV: Improved Core Monitoring Systems, Design and Operating Experience (*Chairmen: Yoichiro Shimazu, Stig Andersson*)

- *Marek Pecka*, Jiri Svarny, Jaroslav Kment – Some Aspects of the New Core Surveillance System at NPP Dukovany and First Experience
- *Martti Antila*, J. Kuusisto – Recent Improvements in On-Line Core Supervision at Loviisa NPP
- *Ivo Endrizzi*, Michael Beczkowiak, Guido Meier – Flexibility Enhancement of Siemens Core Monitoring Based on Aeroball and PDD In-Core Measuring Systems Using On-Line Core Monitoring Software
- *Yoichiro Shimazu* – Review of the Current Status of Core Monitoring System and Future Trend in PWRs in Japan
- *Sten Lundberg*, W. van Teeffelen, Jürgen Wenisch – Core Supervision Methods and Future Improvements of the Core Master PRESTO System at KKB
- *Henning Potstada*, Michael Beczkowiak, Martin Frank, Karl Linnenfelser – The Siemens Advanced Core Monitoring System FNR-K in KKI1, KKP1 and KKK
- *Per Kelfve*, Jesper Eriksson, Carl-Åke Jonsson, Stig Andersson – Design and Validation of the New ABB Core Monitoring System
- ***Discussion and conclusions. Closure of the workshop.***

Additional papers submitted (but not presented)

Session II

- J. Runkel, D. Stegemann, J. Fiedler, P. Heidemann, R. Blaser, F. Schmid, M. Trobitz, L. Hirsch, K. Thoma – New Technologies for Acceleration and Vibration Measurements Inside of Operating Nuclear Power Reactors
- Richard J. Cacciapouti, Joseph P. Gorski – Experience with Fixed Platinum In-Core Detectors

Session IV

- Moonghu Park – Introduction of Virtual Detectors for Core Monitoring System of Korean Standard Nuclear Power Plant

SESSION I

Requirements on Core Monitoring Systems

Chair: O. Sandervåg

USER INTERFACE DESIGN AND SYSTEM INTEGRATION ASPECTS OF CORE MONITORING SYSTEMS

Øivind Berg, Terje Bodal, Arne Hornæs, Jan Porsmyr
Institutt for Energiteknikk
OECD Halden Reactor Project
P.O. Box 173, N-1751 Halden, Norway

Abstract

The present paper describes our experience with the SCORPIO core monitoring system using generic building blocks for the MMI and system integration. In this context the different layers of the software system are discussed starting with the communication system, interfacing of various modules (e.g. physics codes), administration of several modules and generation of graphical user interfaces for different categories of end-users. A method by which re-use of software components can make the system development and maintenance more efficient is described. Examples are given from different system installation projects. The methodology adopted is considered particularly important in the future, as it is anticipated that core monitoring systems will be expanded with new functions (e.g. information from technical specifications, procedures, noise analysis, etc.). Further, efficient coupling of off-line tools for core physics calculations and on-line modules in core monitoring can pave the way for cost savings.

Introduction

Core monitoring and physics codes are becoming an integral part of the entire information system at the plants serving the reactor engineering group in core design, safety analyses and operation planning. During plant operation the control room operators rely on key information derived from on-line measurements and calculations for monitoring safety margins. Predictive simulations for planning operational manoeuvres are used for real-time core control optimisation. Core monitoring systems are gradually distributed to new user categories. The users of the information that is produced from these advanced systems are not only the specialists in core physics and simulation codes. Information distribution and easy access to core physics calculation tools have been one of the main motivation factors for the utilities installing the SCORPIO system, see Refs. [1-4].

Considering also the rapid development of information technology it is extremely important to design flexible core monitoring systems that can easily be adapted to the different user needs and integrated with other plant information functions. With the increasing number of different computerised support functions, it is important that the core monitoring system not be seen as a stand-alone system, but rather as a natural part of the entire information system (i.e. a unified MMI). This is of particular importance for the control room operators.

During recent years there has also been a shift from proprietary hardware and software towards more open off-the-shelf hardware and software solutions. As a consequence there is demand for a flexible framework for integrating the various modules into a core surveillance system. The present paper describes our experience in system design and development using generic building blocks for the MMI and system integration. In this context the different layers of the software system are discussed starting with the communication system, interfacing of various modules (e.g. physics codes), administration of several modules and generation of graphical user interfaces (GUIs) for different categories of end-users. The Picasso-3 User Interface Management System (UIMS) [5] supports object-oriented definition of GUIs in a distributed computer environment. It is described how re-use of interface components and other software modules can make the system development and maintenance more efficient. Examples are given from different system installation projects.

The SCORPIO system development history

The first version of the core surveillance system SCORPIO was installed at a Westinghouse plant, Ringhals Unit 2 in Sweden, in 1984. The main purpose was to provide a practical tool for reactor operators and reactor physicists for on-line monitoring and predictive analysis of core behaviour. It was implemented on Norsk Data mini-computers with a fully graphical user interface. Due to the technology offered at that time, both the hardware and software solutions were highly vendor specific and not portable to other computer systems without major modifications. However, the system concept and functions provided were appreciated by plant personnel. In particular, the MMI (implemented in Picasso-1) with the strategy generator provided a fast and accurate means for performing predictive core control analyses.

The second version of SCORPIO was developed in 1993-1995 and implemented on UNIX workstations. In addition to upgrading the system at Ringhals, the system was installed by Duke Power, USA, on four Westinghouse reactors and three B&W reactors. SCORPIO was also installed on the Sizewell-B reactor. The system could run on all major UNIX platforms in a distributed computer network using the TCP/IP communication standard. The Picasso-2 GUI system made it possible to develop the MMI independently from the physics codes and made the MMI portable by utilising the X-Windows standard.

Recently a new framework has been developed which further enhances the flexibility and capabilities for implementing core surveillance systems in different types of nuclear power plants. Modules can be added and replaced in an easy manner. It allows fast and reliable communication of data between modules based on the Software Bus tool [6] developed by IFE. Further, the Picasso-3 system supports efficient implementation of different user interfaces. Both UNIX and Windows NT platforms are supported.

The new framework has been applied in development and installation of a SCORPIO-VVER version [4] for the Dukovany NPP, Czech Republic. Here it was of particular importance to provide a flexible system for integration of modules originating from different companies. Development of a BWR version is now in progress. This means that SCORPIO will be available for all the major reactor types, and synergy is obtained by application of a common framework both with respect to system implementation and maintenance.

Integration framework

A framework for integrating the various modules into a core surveillance system should have the following features:

- The framework must be flexible, so that modules can be added or replaced in an easy manner, and so that the system can be developed fast and configured easily.
- Fast and reliable communication of data between all modules, including the graphical user interface (GUI), must be supported.
- The system should preferably allow for distributed computing and presentation. The different processes may thus be distributed over several computers in a network, with results available at more than one GUI.

In the following, the integration framework for SCORPIO is described. Figure 1 illustrates how the software system is organised in "layers", starting with the specific hardware platform, operating system and TCP/IP communication system. Above this level the system is implemented with higher level software tools which makes the core monitoring system less dependent on technology changes in lower level software components.

The information exchange is based on the Software Bus tool developed at IFE [6]. The Software Bus is an object-oriented communication system, based on the network protocol TCP/IP, and manages a dynamic set of distributed objects. In this context, an object will typically represent a variable, structure or function belonging to a certain module. Applications that use these objects are able to share data and functionality with other processes running on different systems across a network. A Software Bus C library enables communication to be performed through a set of high-level interface routines that are linked with every module. The Software Bus is presently implemented for UNIX workstations and Windows NT computers.

Figure 1. Software structure in "layers"

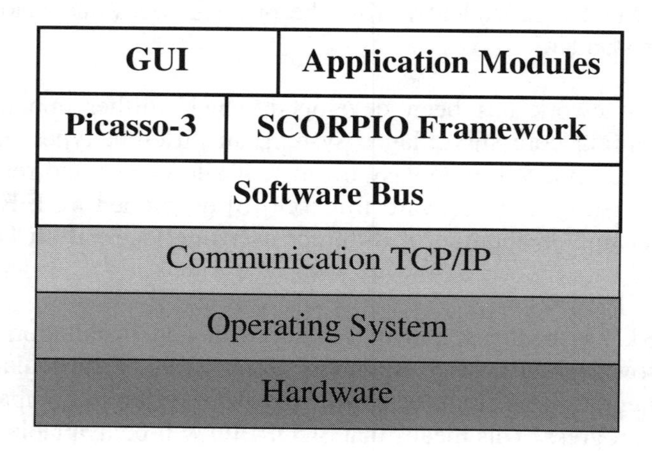

The GUI is built with Picasso-3, a user interface management system also developed at IFE [5]. Picasso-3 is closely related to Software Bus and is also designed to operate in a network environment. The SCORPIO framework contains the software supporting core surveillance applications. A program referred to as the Module Administrator connects the GUI to the other modules. The general structure of the integration framework is shown in Figure 2.

Figure 2. General view of the framework

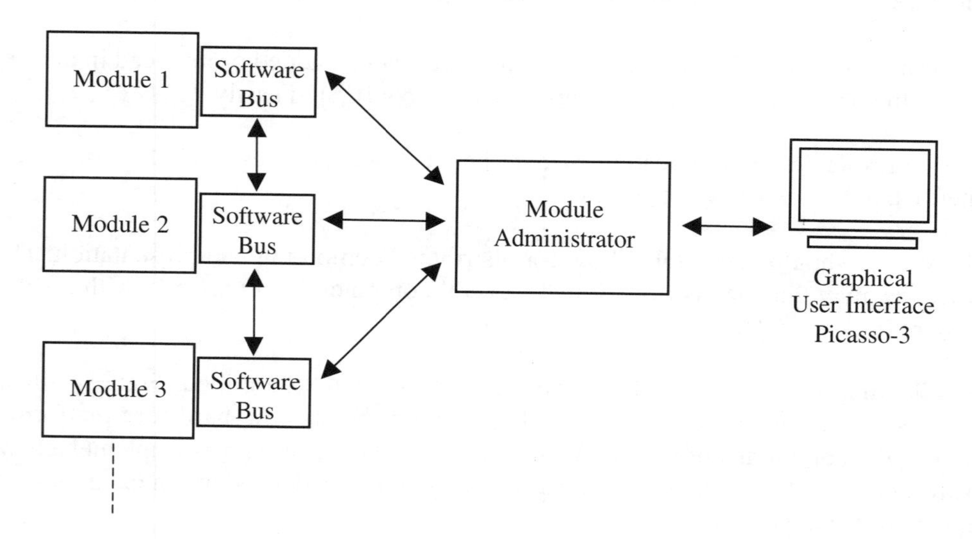

The module interface

Modules relevant for core surveillance are likely to be written in ANSI C/C++ or programming languages like FORTRAN, or any other having a C interface. To make the system as flexible as possible, the communication routines are preferably coded outside the module's source code. This is currently done in a number of C++ files collectively referred to as the Module Interface (MI). MI is linked with the various modules by use of global variables. This means that the module programmer does not need to know anything about the communication system, but merely has to declare the variables which are to be shared with other modules as global. For example, the requirement for FORTRAN modules would be to have all global variables declared in specified COMMON blocks.

For example, for the simulator module, variables that are shared with the rest of the system can be reactor power, core flow, inlet temperature, etc. All these variables need to be declared as Software Bus objects in the MI, so that they can be distributed to other parts of the system, as illustrated in Figure 3.

Figure 3. Data flow in the Software Bus system

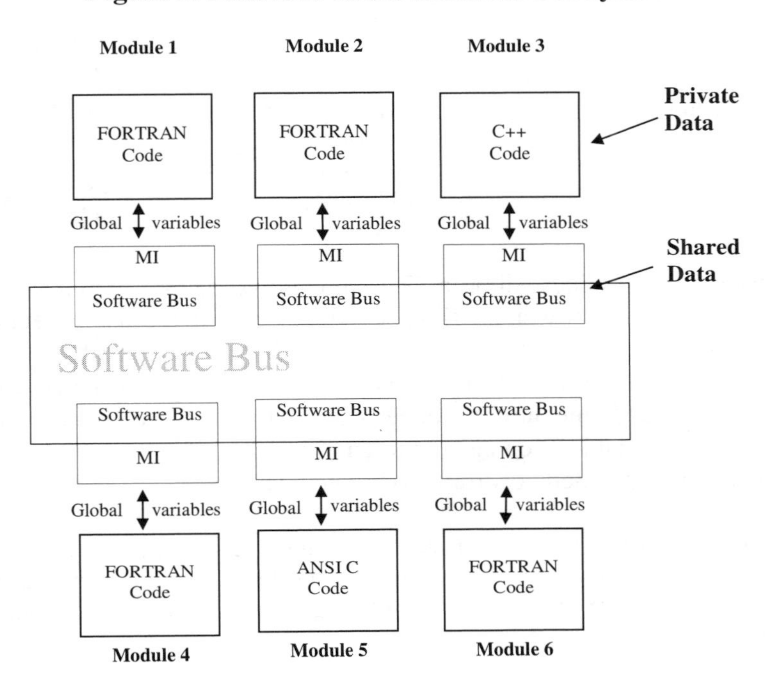

If a module contains many data items to be shared with other modules, manual declaration of Software Bus objects could be comprehensive. To relieve the system programmer of this tedious and possibly also error-prone task, a special program generates the part of the C++ code that defines and declares Software Bus objects and structures for all modules involved. It generates the C++ code by reading a configuration text file that contains information about the shared variables for all modules. This file also contains all structure declarations and all variables that are sent between the module administrator and GUI, see next section.

The Software Bus tool provides a small number of general functions that provides access to all Software Bus objects. With these functions it is possible for a module to get Software Bus objects that belong to other modules by simple function calls. This is how, for instance, an optimisation module would get information about parameters calculated in a simulator module. Sometimes it is also necessary to call subroutines in other modules. With the Software Bus, a module has the ability to call routines in other modules by declaring Software Bus remote procedures that are shared between the modules in the same way as variables.

The module administrator and graphical interface

The module administrator is a program that connects the modules to GUI. All data that are exchanged between the modules and the operator flow through this program. The Software Bus is used between the modules and the module administrator, in the same manner as it is used between the individual modules themselves. It is important to note that communication between the different modules is independent of the module administrator.

A central part of the Picasso-3 system is the Run Time Manager (RTM). RTM manages the actual creation and execution of user interfaces. A separate Trend Log RTM serves the different GUIs with trend data, see Figure 4.

Figure 4. The connection between the module administrator and GUI

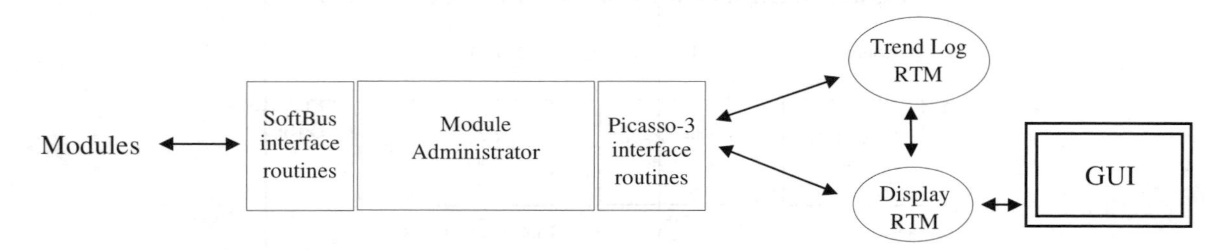

Among other things, the graphics editor in Picasso-3 allows the user to make dynamic presentations of results, such as trend curves, power distribution curves, etc. If the user changes a parameter, for example a measurement calibration factor, the new value will be sent through the module administrator to the modules that share this variable. In the same manner, when for instance a variable in the data acquisition module is updated, the new value is passed through the module administrator to the presentation screen. If this variable is connected to an object in the picture (e.g. a curve), the shape of this object is instantaneously updated according to the new value.

In addition to data sharing, GUI can issue commands to a program, e.g. to start calculations. This is possible by calling remote procedures in the modules. Dynamic pictures are required that connect for instance a button with the signal variable that is sent to the module administrator and then forwarded to the relevant module.

Framework applied for the SCORPIO-VVER version

The objective of the project was to develop and install a new reactor core monitoring system at the Dukovany NPP, Czech Republic [4]. Emphasis was put on creating a reliable, flexible, adaptable and user-friendly system which would be easy to maintain. The system comprises functions for on-line core monitoring and predictive analysis with interfaces to plant instrumentation and physics codes. Functions for system initialisation and maintenance are also included. In the development of the new core monitoring system for Dukovany NPP, we were able to utilise experience gained during the development of the SCORPIO core monitoring systems for western type PWRs.

Graphic user interface design

When constructing the graphic interface used in the SCORPIO-VVER project for Dukovany NPP emphasis was put on making the MMI as general as possible with re-usable components which could be adapted with little effort to meet the needs of the different user categories: operators, physicists and system supervisors. The MMI uses the whole screen for presentation of graphic information. The screen has been divided into four main areas:

- Header area.

- Drawing area.

- Message area.
- Menu area.

The layout is shown in Figure 5 and an example of a typical display is shown in Figure 7.

Figure 5. The screen layout

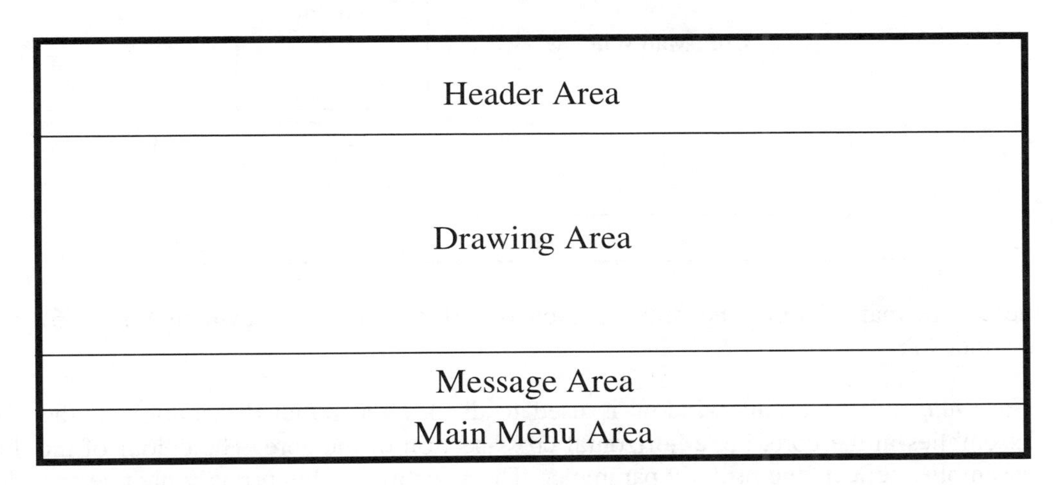

The purpose of the different areas is:

- *Header area.* This area is used to display important parameters. The layout of the information shown in this area is permanent, and used for all pictures.
- *Drawing area.* The layout of this area will depend on which sort of information is presented. Change of information is done by using the different buttons in the menu area. Basically the information which is presented in the area can be divided into two groups, i.e. core follow and predictive mode information. The background colour of this area is used to distinguish between these two modes of operation.
- *Message area.* This area is used to present an online message from the SCORPIO-VVER system to the user, for example an error message or a request.
- *Menu area.* This area is used to display "software buttons". These buttons are used to change between different layouts of information displayed in the drawing area. Special buttons are used to change between core follow and predictive mode operation. The set-up of the buttons will depend on the mode of operation. Thus either a core follow menu set or a predictive mode menu set are displayed. The background colour of this area indicates the user category (operator, physicist or system supervisor).

Core map pictures

Much of the information shown in the drawing area is presented as core map pictures. All the core map pictures in the SCORPIO-VVER system have a standard layout.

Figure 6. Layout of a core map picture

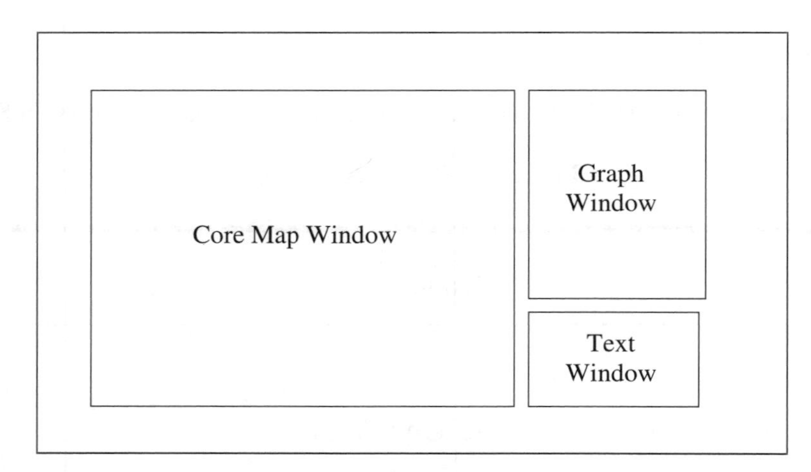

In these core map pictures the drawing area is further split (as shown in Figure 6), into the following windows:

- *Core map window.* This window is used to display the radial distribution of the different assemblies in the core, i.e. a horizontal cross-section of the core. The colour of the different assemblies reflects the primary parameter. The selection of this primary parameter is done by using the text window. By selecting one of the assembly cells, the axial distribution of relevant parameters will be shown in the graph window and actual values for the primary parameters will be shown in the text window.

- *Graph window.* This window is used to present axial distribution of relevant parameters for the selected assembly in the core map.

- *Text window.* This window is used for value presentation and command purpose. It is used to display the values of all the selectable parameters for the chosen core map picture. In addition it contains selection buttons used for selecting the primary parameter among all the selectable ones. The colour of the assembly cells in the core map will reflect the value of this parameter. An example of a core map picture is shown in Figure 7.

As mentioned earlier, much information is presented by using core map displays. To save construction time and maintenance time for such displays much effort has been put on making very general displays which can be used to display different kinds of information. Such a core map display is realised by displaying three different pictures in the drawing area. A picture containing the core map is displayed in the core map window, a second one containing the axial graph is presented in the graph window and a third picture presenting the selectable parameters is displayed in the text window.

Core map generating tool

The concept of presenting core map information is quite general and may be used in all core surveillance systems independent of the type of reactor. It does not even depend on if it is a PWR or a BWR reactor. But even if one only needs one or a few core map pictures showing the different assemblies in the core, it is time consuming to construct such pictures. To simplify this job a tool has been made to reduce the burden of constructing the core map.

Figure 7. An example of a core map picture in the SCORPIO-VVER system

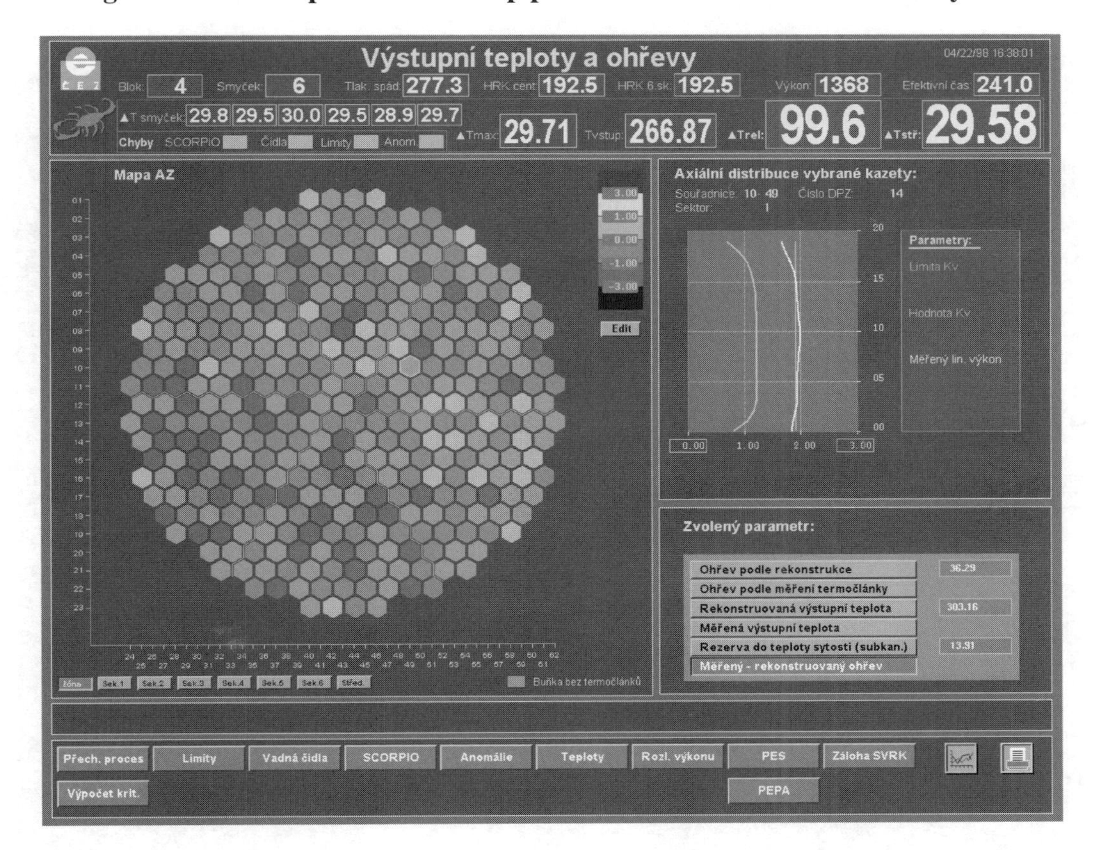

A special Picasso-3 library called "Tools" has been made containing a function called "CoreCreate". This function can be used to construct a new core map layout built up by a given assembly cell object. The picture shown in Figure 8 has been made using the "CoreCreate" tool with BWR assembly cells, saving time for the system developer.

Text labels

All text labels in the MMI are realised by using special text variables. Thus the labels themselves are not stored in the various MMI displays, but in special label definition files. The reason for this was that English labels were used under the development of the SCORPIO-VVER project for the Dukovany NPP, but for the end-user, Czech text was required. Instead of going through all the pictures to change the text labels it was more convenient and efficient to operate with two text label files, one with English text and the other one with Czech labels. To switch between the two languages is then only a matter of system configuration, and tools have been developed such that this can be performed on-line without having to restart the SCORPIO-VVER system.

Trend diagrams

An important class of pictures is trend diagrams, which are used by operators and physicists to correlate parameters for analysing various phenomena and possible problems in the core. Trend pictures present histories of measurements and results of predictive simulations. The Picasso-3 system

Figure 8. An example of a generated core map picture using the "CoreCreate" tool

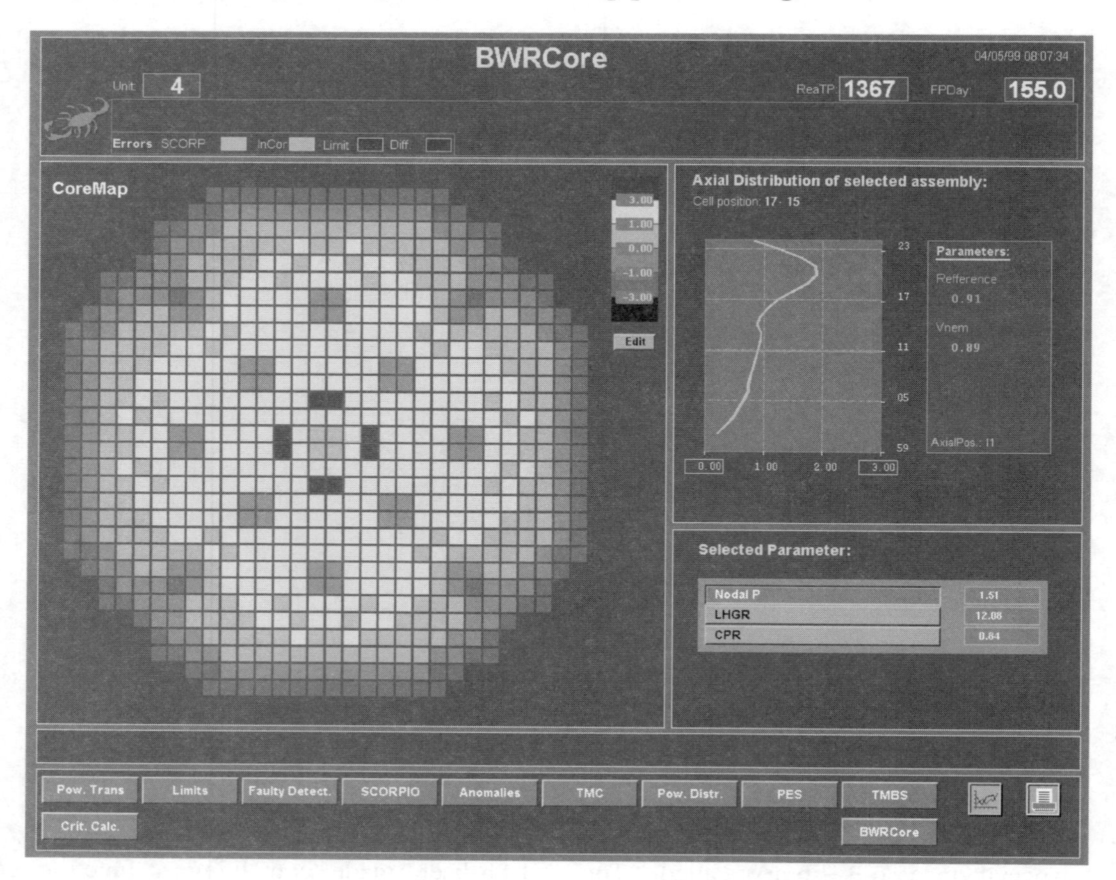

offers great flexibility for defining and configuring trend diagrams, e.g. multi-colour curves, a ruler for exact time/value read-outs from a curve, grid and time labels, scrolling back and forth in time, on-line change of variables, scaling, time span, colours, etc.

A separate part of the system, the trend log, takes care of logging the actual values of the variables specified for trending. A number of attributes can be specified for each variable, for example sampling time interval and maximum time span for storing data. An external trend log process (see Figure 4) can be set up to run independently of the display system to provide continuous data logging facilities even if the display system is not active.

Object libraries

The drawing of the different pictures has been systematised through the use of an object-oriented approach. Objects have been created for re-use in the pictures. Most of them are quite general and may be used in other projects. The objects have been organised in the following object libraries.

- *ButtonLib.* This library is used to store different buttons and associated dialogue actions. This library is quite general, and can be re-used in different applications.

- *LimitLib.* This library contains objects which are used to display values of process variables and their limit values. These objects may depend on the application.

- *UserLib.* This library is used to define specific user objects, e.g. different assembly cells used in the core map pictures.

- *InputFields.* This library contains different sorts of input fields with associated dialogue functions. This library is very generic.

- *MotifLib.* This library contains objects which are used to make the pictures "look and feel" motif-like. This makes the core surveillance system to appear as other motif-like applications in the control room.

Intranet coupling

Intranet solutions are introduced as one way to cope with the increasing demand for improved access and distribution of information in nuclear power plants. The information provided by advanced core monitoring systems is one type of information that may be requested. In the SCORPIO system installed in Ringhals, a module has been implemented which retrieves key data from the core surveillance system and generates HTML code for the WEB browser. In this way every person who has access to the intranet at Ringhals can obtain the status and main parameters from the core.

Conclusions

Efficient methods and software tools are essential for providing reliable and cost efficient implementation of core surveillance systems. This is particularly important in the future because rapid technology changes make it hard to keep the system updated over longer time horizons without major modification efforts. The system structure must be designed to be less vulnerable to technological changes.

The new SCORPIO framework for integrating core surveillance systems was developed with this in mind. The software system is organised in "layers" on top of the basic hardware/software platform. The framework consists of a communication system, software to facilitate interfacing of various modules (e.g. physics codes), administration of several modules, and a tool for generation of graphical user interfaces for different categories of end-users. Development of generic building blocks for the MMI and re-use of software components make the system development and maintenance more efficient. This was demonstrated during the implementation of the VVER version of SCORPIO for the Dukovany nuclear power plant in the Czech Republic. The framework is now utilised in development of a BWR version of SCORPIO. By using the SCORPIO framework, the development time is reduced and the maintenance work is carried out more efficiently, compared to developing systems with lower-level tools. For instance, the MMI can be developed and tested independently of the physics modules.

The methodology adopted is considered particularly important in the future, as it is anticipated that core monitoring systems will be expanded with new functions (e.g. technical specifications, procedures, noise analysis, etc.). Further, efficient coupling of off-line tools for core physics calculations and on-line modules in core monitoring can pave the way for cost savings. As an example, intranet solutions are introduced several places as one way to cope with the increasing demand for improved distribution of information in nuclear power plants. The information provided by advanced core monitoring systems is one type of information that is distributed in this way.

REFERENCES

[1] Ø. Berg, M. McEllin, M. Javadi, "Application of the Core Surveillance System SCORPIO at Sizewell-B", NEA/NSC International Specialists Meeting on In-Core Instrumentation and Reactor Core Assessment, Mito-shi, Japan, 14-17 October 1996.

[2] T. Andersson, Ø. Berg, T. Bodal, J. Porsmyr, K.A. Ådlandsvik, "SCORPIO – Core Monitoring System for PWRs, Operational Experience and New Developments", ANS Topical Meeting – Advances in Nuclear Fuel Management II, Myrtle Beach, South Carolina, USA, 23-26 March 1997.

[3] S.K. Gibby and S.C. Ballard, "Implementation of the Core Surveillance System SCORPIO at Duke Power Company", ANS Topical Meeting – Advances in Nuclear Fuel Management II, Myrtle Beach, South Carolina, USA, 23-26 March 1997.

[4] K. Zalesky, J. Svarny, L. Novak, J. Rosol, A. Hornæs, "SCORPIO-VVER Core Surveillance System", ENS International Topical Meeting on VVER Instrumentation and Control, Prague, Czech Republic, 21-24 April 1997.

[5] K.A. Barmsnes, T. Johnsen, C-V. Sundling, "Implementation of Graphical User Interfaces in Nuclear Applications", ENS International Topical Meeting on VVER Instrumentation and Control", Prague, Czech Republic, 21-24 April 1997.

[6] T. Akerbæk and M.N. Louka, "The Software Bus, An Object-Oriented Data Exchange System", Halden Work Report (HWR-446), May 1996.

FUNCTIONAL REQUIREMENTS FOR CORE SURVEILLANCE SYSTEMS

Tell Andersson
Ringhals AB, SE-430 22 Väröbacka, Sweden
E-mail: tean@ringhals.vattenfall.se

Abstract

Operating experience at Ringhals-2 has demonstrated the feasibility of a mixed core surveillance system comprised of fixed in-core detectors combined with the original movable detector system. A small number of fixed in-core detectors provide continuous measurement of the thermal margins while the movable detectors are used mainly at start-up to verify the expected power distribution. Reactor noise diagnostics and neural networks can further improve the monitoring system.

The reliability of the movable detector system can be improved by mechanical simplification. Wear and maintenance costs are lowered if the required flux-mapping frequency is reduced. Improved computer codes make the measurement uncertainties less dependent on the number of instrumented positions. A mixed system requires new types of technical specifications.

Ringhals 2 core surveillance system

Ringhals 2 is equipped with twelve strings of fixed in-core gamma thermometer strings. They replace the original movable detector guide thimbles. The remaining 38 positions are used by the original movable detector system for periodic flux-mapping.

A gamma thermometer string comprises nine sensors covering the active core height. An imbedded heater cable is used for electrical calibration.

A fixed in-core detector string is operable if five out of the nine sensors are operable including an operable heater cable for calibration. The fixed in-core detector system is operable if nine out of the twelve detector strings are operable. The fixed in-core detectors are electrically calibrated monthly.

The movable detector system is operable if 34 out of the 38 positions are operable.

A pre-calculated burn-up dependent detector-response function is used to convert specific heat in the gamma thermometer (W/g) to linear power in the surrounding fuel rods (W/cm). The burn-up of the fuel rods surrounding the detector is obtained from the CYGNUS code.

The pin power distribution is calculated every five minutes by the CECOR code with input from the gamma thermometers. The associated coupling coefficients are obtained from the SIMULATE code used by Vattenfall for in-core fuel management.

The TEL-HC program finally calculates the DNB margin. The input to TEL-HC is the hot rod axial power distribution, temperature, pressure and reactor coolant flow.

CECOR is measurement driven; it makes a straightforward extrapolation of measured local powers to a 3-D pin power map, as opposed to a core simulator which has to be "adapted" to fit measurement.

Comparisons with movable detector system

Comparisons made over a large number of cycles show that the power peaking factors F_Q and $F_{\Delta H}$ measured with the fixed in-core detector system generally agree better with expected values than the corresponding values obtained with the movable detector system. The result form the previous cycle is shown in Figure 1.

The conclusion is somewhat surprising considering the additional uncertainties associated with the fixed in-core system:

- *Fewer instrumented positions*: 12 instrumented positions versus 38 for the movable detectors.
- *Reduced axial resolution*: 9 detectors axially as opposed to 60 axial measurements.
- *Manufacturing variations*: each fixed in-core detector is an individual as a result of acceptable manufacturing tolerances.
- *Sensitivity drift*: The gamma thermometer's sensitivity varies with time, especially after a shutdown.

The difference between measured and expected integrated rod power for four of the detector strings are shown in Figure 2. Generally the deviation stays within ±3%. The absolute values from the gamma thermometers have been used in the comparison.

The DNB margin is measured continuously with TEL-HC. The appropriate DNB correlation is used by the code. Input to the DNB calculation is peak pin power distribution, reactor coolant flow and moderator temperature. The core flow is derived from the heat balance. The DNB margin after refuelling is typically 10%.

Anomaly detection

Detection of power distribution anomalies with the fixed in-core detectors

The fixed in-core detector system's ability to detect various types of anomalies in the core power distribution has been tested by simulation. The SIMULATE code was used to generate simulated detector signals proportional to power in the nodes surrounding the detectors for various types of core anomalies such as dropped and misaligned rods.

The simulated 108 detector signals (9 × 12) were then used as input to the CECOR code to calculate the power distribution and the associated power peaking factors. The result shows that CECOR will re-create the original SIMULATE power distribution. The difference between SIMULATE's original power peaking factor and the corresponding re-created CECOR $F_{\Delta H}$ peaking factor is less than 1% (see Table 1) for the anomalies tested [1].

A more brutal way to test the anomaly detection capability is to simply switch on the gamma thermometer heater cable at full power and record the impact on the measured peaking factors. When the heater cable is activated, the sensor's signal increases as if a power perturbation has taken place. The response from the system is immediate. As expected the effect on the peaking factors is higher for strings in the high power positions than for low power strings (Figure 3).

The conclusion is that a system with only 12 fixed in-core detector strings will detect significant core anomalies, with impact on F_Q or $F_{\Delta H}$, regardless of where in the core they appear. The reason for this is that a local anomaly will also give rise to a global power perturbation, which is picked up by the sensors even though they may be localised far away from the actual anomaly.

Control rod anomaly detection with movable detectors

Various applications of neural networks to detect anomalies have been tested in Ringhals in co-operation with Chalmers University of Technology (CTH – Department of Reactor Physics). According to the technical specifications, individual rods are not allowed to deviate more than 12 steps from the average bank position.

Indications of unacceptable rod misalignment occur approximately once per year at the Ringhals PWRs. A quick and accurate method to determine the position of a potentially misaligned rod is therefore essential.

It has been shown [2] that the position of the control rod can be determined with an accuracy of a few centimetres (Table 2). The sensitivity of the method with regard to reactor variables such as power level, burn-up and fuel management is presently under investigation.

The method has the following potential:

- Precise determination of a suspected misaligned rod position.
- Back-up function if the rod position indicators become inoperable.
- Back-up function if control rod computer system becomes inoperable.

We are also investigating the possibility of applying reactor noise diagnostics as an integral part of the core surveillance system with input from fixed in-core, movable or ex-core detectors.

The uncertainty in the MTC measurement at full power is presently a problem. We are therefore looking for alternative methods. An ongoing research project with CTH studies the possibility to improve the measurement by combining pre-calculated physics parameters with noise diagnostics. The difficulty so far has been to determine the absolute value of the MTC.

Advantages with the movable system

For certain types of applications, the movable detector system is more efficient than the fixed detector system. The superior axial resolution is useful for localisation of misaligned rods or axial off-set anomalies.

After refuelling good core coverage is necessary in order to detect possible anomalies in the core-loading pattern. For code verification good axial resolution, low measurement uncertainties as well as a large number of instrumented positions are required.

A conclusion of the above is that the fixed system and the movable system complement and functionally partially overlap each other. The fixed system permits continuous surveillance of the thermal margins, an obvious advantage. The movable system on the other hand is superior when it comes to precision measurements at steady state.

Mechanical reliability

Ringhals 2 was originally equipped with five movable detector drive units, each connected to ten core positions. After installation of the fixed in-core detectors one of the drive units became redundant and was subsequently removed. A further mechanical reduction to two drives is possible when the original system is eventually replaced.

The wear and maintenance cost for the movable detector system can be lowered if the required flux-mapping frequency is reduced. With an on-line system there is no need for periodic flux-mapping. The most important function for the movable detector is to perform a detailed flux-map at start-up in order to verify the core-loading pattern and expected power distribution. Once the start-up measurement is over, the on-line system is used to verify that the core is operated within the limitations given by the technical specifications. The movable detector system serves as a back-up system and a complementary system in case of suspected anomalies.

The mechanical reliability of the gamma thermometers has so far been excellent. Two types of failures have occurred: a heater cable failure in one of the first installed strings (1984) and cracks in the outer jacket tube in four of the strings as a result of the swaging process used in the manufacturing. If the swaging problem is omitted from the statistics (the recommended procedure is to draw the jacket tube onto the cable pack) only the heater cable problem remains.

The conclusion is that the lifetime of the gamma thermometer is limited by mechanical wear in the region around the lower core plate (investigated in the Studsvik hot cells). One difficulty is that EC wear measurements cannot be performed on the detector strings in the same way as for the movable detector guide thimbles. Heater cable calibration has been proved to be an efficient way to calibrate the instrument without reliance on movable mechanical components.

Potential economic benefits with a mixed system

A mixed system with continuous monitoring of the peaking factors can be used to improve the operating economy in a number of ways.

The uncertainty associated with load following and normal operating transients can be removed from the F_Q – leading to a gain of approximately 10%. A higher permitted F_Q in the core design facilitates the introduction of advanced loading concepts such as axial blankets.

If the $F_{\Delta H}$ limitation is replaced by a limitation on the DNB margin, an increase in $F_{\Delta H}$ of 5-10% is possible [3].

Continuous F_Q surveillance eliminated the need for a narrow delta-flux operating band other than as a guideline for the reactor operator. The band can be widened with little impact on the fuel management.

According to present technical specifications, power has to be maintained below 50% power for 24 hours if the delta-flux operating band is violated for more than one hour resulting in a production loss corresponding to half a full power day. The problem with the delta-flux operating band is the greatest at EOC when the boron concentration is low.

Mechanically the movable detector system can be simplified as previously mentioned by fewer drives in combination with a relaxation of the required measurement frequency. A further simplification is possible if the required minimum operable detector guide thimbles can be relaxed from the present criterion. This might be possible with the introduction of a new generation of computer codes with measurement uncertainties less dependent on the number of instrumented positions.

Technical specifications for a mixed system

The technical specifications for Ringhals 2 have only partially been adapted to the potential of a mixed core surveillance system. Three alternatives to combine the movable and fixed systems in the technical specifications have been discussed:

- Flux-map with the movable detectors at start-up only.
- Quarterly flux-maps with the movable detectors.
- Monthly flux-maps with the movable detectors.

A requirement to use the movable detector system at start-up assures that a thorough measurement is made to verify the expected power distribution. Once the start-up measurement is completed, the fixed in-core system takes over. In case of a suspected core anomaly, the movable detectors are activated to investigate and localise the anomaly.

The advantage with quarterly flux-maps is that the operability of the movable detector system, signal processing included, is checked more often than in the first alternative.

For the on-line system, the power peaking factor limitations can be replaced with a requirement to maintain a positive thermal margin related the most limiting of the DNB and linear heat rate constraints.

In case the fixed in-core system becomes inoperable, operation might continue if the delta-flux is maintained within the ±5% band. This band is an excellent general guideline for the reactor operator.

A risk with an on-line system is negative thermal margins caused by signal errors or software errors. The required action is to reduce power until thermal margin once again becomes positive. The experience with the Ringhals 2 system is that negative margins are the result of errors in the core coolant flow calculation or mechanical failures in the data logging system.

A computerised operability verification function is required in order to ensure that all operability requirements for the on-line system are met.

Conclusion

The fixed in-core detectors and the movable detectors are functionally complementary.

Noise diagnostics and neural networks are used to improve the core surveillance system's ability to detect anomalies in the core power distribution.

The fixed in-core detector system, based on only 12 strings, consistently performs better than the original movable detector system if one compares the deviation between the measured and expected power peaking factors. Consequently the reliability of the mixed system can be improved by the following:

- Mechanical simplification by a reduction of the number of drives.
- Reduced wear and maintenance costs by reduced flux-mapping frequency.
- Measurement uncertainties less dependent on the number of instrumented positions through better computer codes.

The unique features of the mixed system must be accounted for in the technical specifications of the plant.

REFERENCES

[1] T Andersson, "Simulering av härdanomalier", Ringhals Technical Report 898/91.

[2] Ninos S. Garis and Imre Pazsit, "Determination of PWR Control Rod Position by Core Physics and Neural Network Methods, Nuclear Technology, September 1998.

[3] Ringhals Unit 2, Required Thermal Margin Report, ABB-CE Report, 1994.

Figure 1. Ringhals 2, cycle 23 – comparison between IN-CORE and CECOR

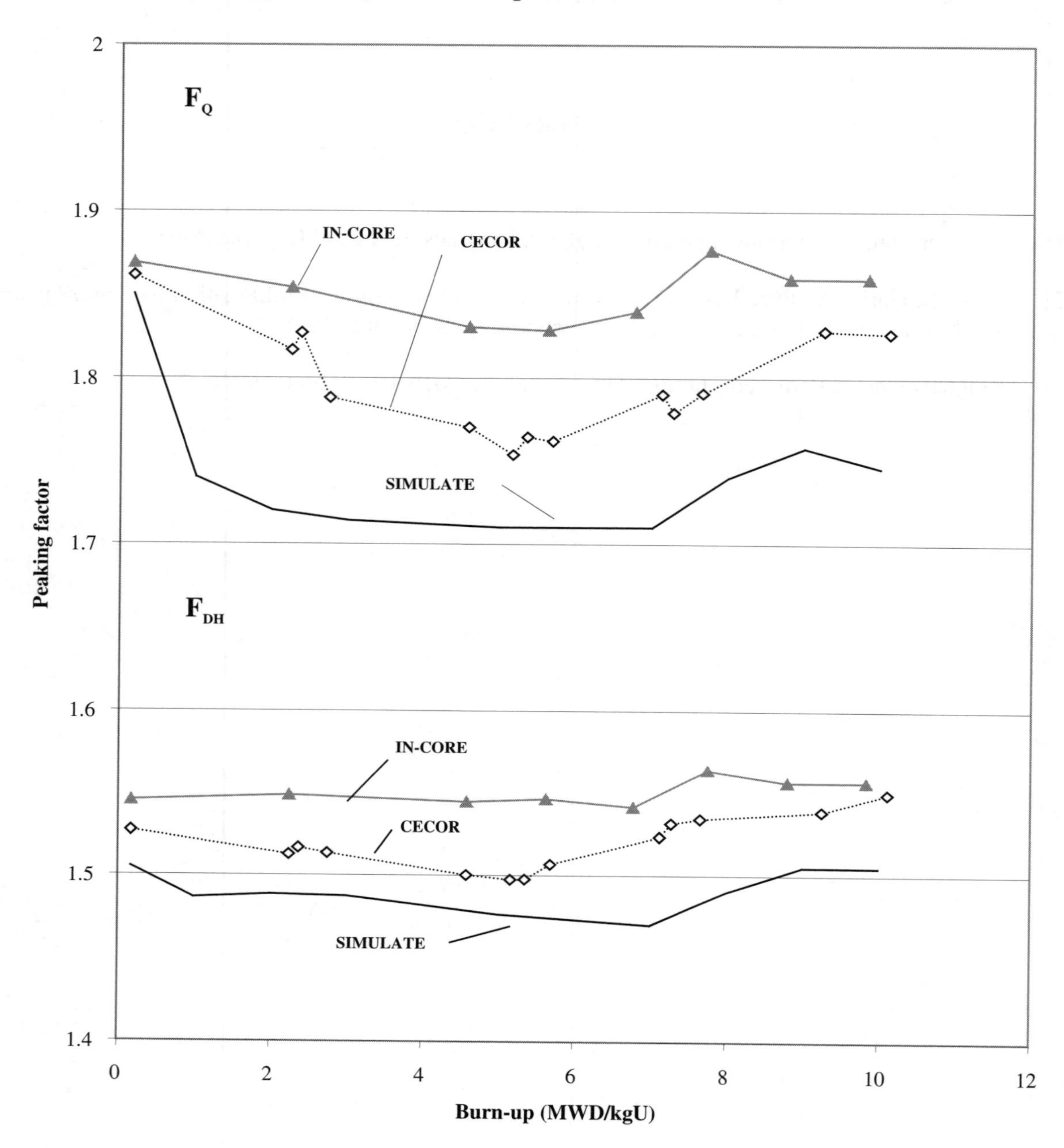

Figure 2. Ringhals 2, cycle 23, integrated gamma thermometer rod power

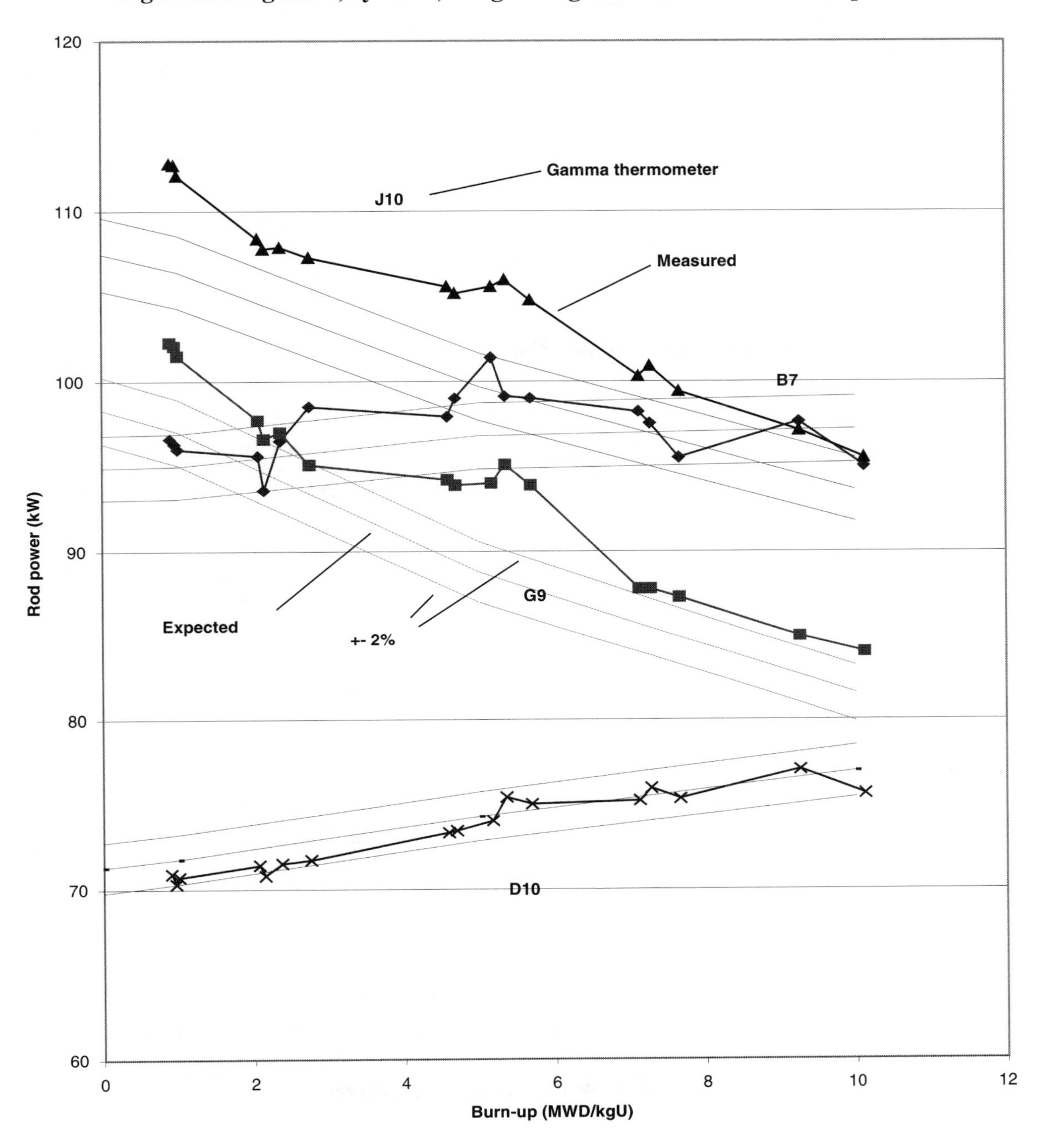

Figure 3. Anomaly sensitivity test

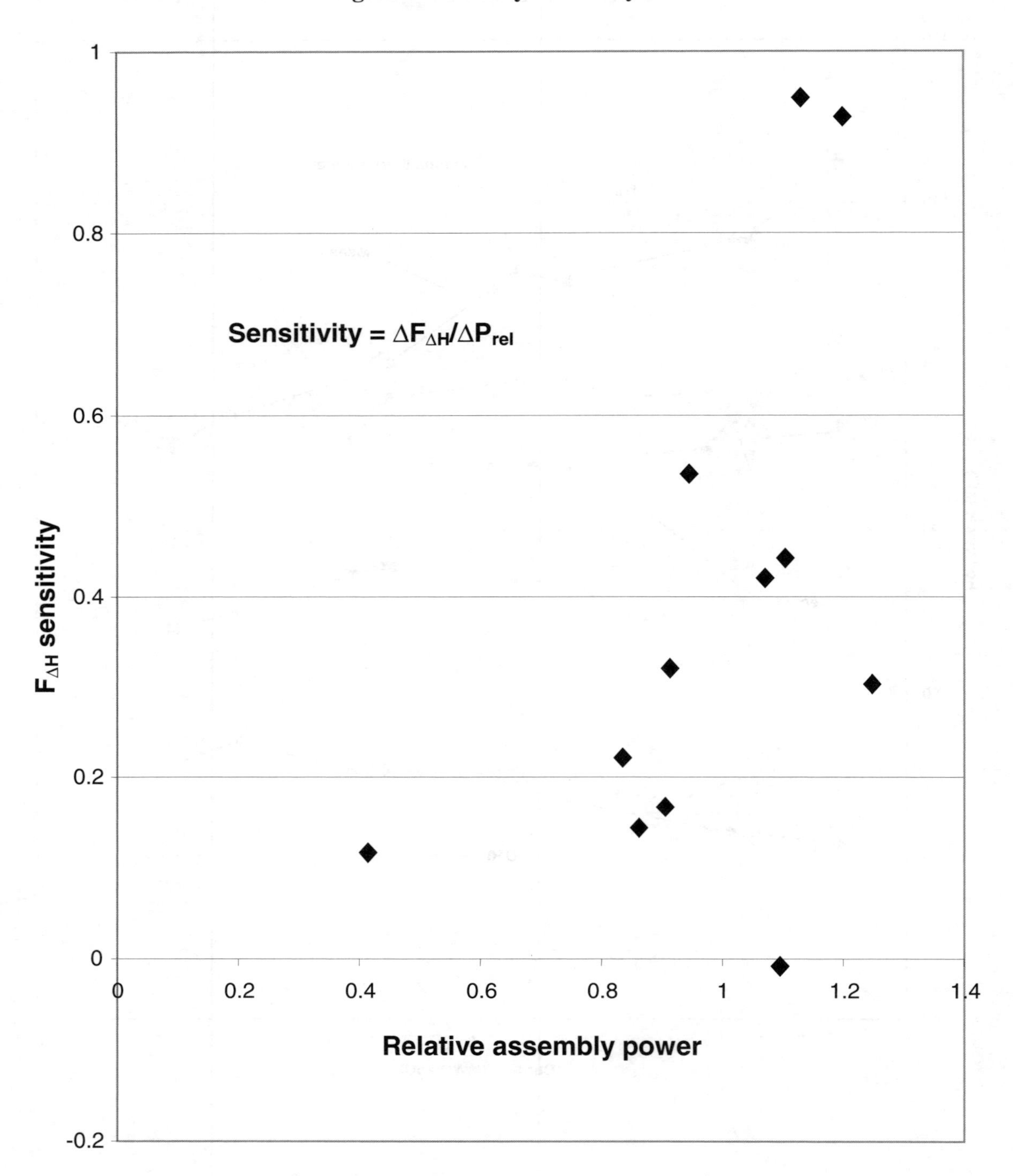

Table 1. Simulation of anomalies

Case	CBD position (steps)	Anomalous rod position (steps)	Original SIMULATE $F_{\Delta H}$	Re-created* CECOR $F_{\Delta H}$	Difference (%)
1	ARO	F14 = 150	1.536	1.527	-0.59
2	ARO	J7 = 0, J13 = 0	1.679	1.675	-0.24
3	170		1.541	1.530	-0.71
4	ARO	J7 = 100	1.568	1.578	0.64
5	ARO	J7 = 50	1.584	1.584	0.00
6	ARO	J7 = 0	1.597	1.596	0.06
7	ARO		1.523	1.519	-0.26
8	ARO	H2 = 180	1.536	1.536	0.00
9	190	J7 = 150	1.554	1.539	-0.97
10	100		1.577	1.567	-0.63
11	150		1.549	1.543	-0.39

* SIMULATE's nodal powers in the nodes surrounding the 108 gamma thermometers are used as input to CECOR to re-create SIMULATE's original power distribution.

Table 2. Efficiency of trained neural network

Control rod	Error (cm)
B08	2.0
F06	0.0
F11	-2.9
G14	-1.4
G09	-3.3

UNCERTAINTY ASSESMENT IN BWR CORE MONITORING

Juan José Casal
Research and Development
Nuclear Fuel Division
ABB Atom AB
Sweden

Abstract

Conventional boiling water reactor (BWR) core monitoring strategies, often based on adapted thermal margin estimations, are subjected to increasingly stringent requirements. The enhanced modelling capabilities of modern core simulators are improving the strength of these monitoring systems. However, core monitoring basically deals with risk of failure estimations; by operating closer to the limits, the "inaccuracies" of these estimations become more relevant. The present paper describes ABB's Core Watch concept, which relies on a comprehensive assessment of the uncertainties involved in thermal margin evaluations. These uncertainties originate both from the methods used and the modelling assumptions as well as from actual limitations in the available process signals. Increased reliability can be achieved through additional assessments in these areas and a number of tools to support these activities are herein described.

Background

Boiling water reactor core monitoring systems (CMS) utilising some kind of adaptive strategy to correct predicted values to match measured data have been successfully employed in many plants for more than two decades (ABB Atom delivered its first system in 1974). This positive experience is, partly, due to the fact that a relatively high degree of conservatism in the thermal margin evaluations has been required as a result of the complexity of the physical phenomena, which take place in the core. However, new and more stringent requirements have appeared in recent years as a result of power up-rates, the introduction of new fuel concepts and more aggressive operation strategies with smaller margins. An example of the latter is the trend towards longer cycles or higher discharge burn-ups which leads to increased levels of heterogeneity in the core conditions due to higher inter-assembly burn-up mismatch, higher enrichment and an extended use of burnable absorbers. The above mentioned conditions and requirements have the potential to erode the previous conservatism, forcing a reconsideration of the reliability of the current strategy.

Better use of the information and modelling capabilities available, together with a more comprehensive assessment of the current uncertainties in the thermal margins monitored should contribute to improve the reliability and robustness required for these core monitoring systems.

Thermal limits (revisited)

Although well known, it may be worthwhile to define the nature of the parameters monitored and the assumptions behind their definitions.

The purpose of core monitoring is to contribute to the safe operation of the reactor by avoiding fuel damage both under normal conditions and anticipated operational occurrences. The fuel integrity is guaranteed by the definition of thermal load limits that should not be overridden. These limits are established to cover different kinds of failures according to specific criteria.

For each thermal load (x) to be monitored a technical, or safety, limit (x_t) is defined prior to the start-up of the reactor. This is a design limit, normally defined by the fuel vendor, which can be expressed as a given value or as a more or less complex function of multiple variables. In addition, the reactor operator defines an operational limit (x_{op}) that will be applied during the cycle. This limit takes into account the above-mentioned technical limit together with a safety margin (s_{op}) based on the estimated uncertainty in the evaluated thermal load, i.e.

$$x_{op} = f\left(x_t, s_{op}\right)$$

The basic thermal loads monitored are:

- The critical power ratio (CPR) to avoid damage due to dry-out.
- The linear heat generation rate (LHGR) to avoid thermal mechanical damage and to protect against damage due to pellet-cladding interaction.
- The average planar linear heat generation rate (APLHGR) to protect the fuel rods in the event of a loss-of-coolant accident.

The thermal margin (M_x) is defined as the "distance" to the operating limit of each of these parameters, for a given reactor condition, with negative values indicating violations of this limit.

There is a clear trend, regarding the definition of technical and/or operational limits, towards a statistical evaluation of the risk of failure, and some limits are already expressed as "*a thermal load such that the probability of x% of the rods/bundles failing is less than y%*". In this case not only the uncertainty in the computed individual thermal loads but also the global core conditions would affect the predicted thermal margins. In this way the fact that several bundles come closer to the limit implies a higher failure risk, even when at the "hot spots" the thermal loads are not necessarily higher than in a case with very few limiting bundles.

Current core monitoring strategies

BWR core monitoring systems based on three-dimensional core simulators with relatively strong modelling capabilities have been in operation for almost two decades. As an example, ABB Atom has delivered its Core Master System including either the POLCA (Power On-Line CAlculations) or PRESTO core simulators to a number of different plants, as shown in Table 1. With the exception of the Leibstadt NPP, all the other plants operate with the POLCA (PRESTO)/UPDAT package using an adaptive strategy that corrects the calculated nodal power distribution according to calculated vs. measured TIP and LPRM signal differences prior to the calculation of the thermal margins. The successful experience, in the form of very limited fuel failures after about 225 reactor cycles, illustrates the soundness of this approach or, at least, the cleverness in the determination of the limits for safe operation.

Table 1. Core Master System deliveries and experience

Power plant	Country	In operation since	No. of cycles
Forsmark 1	Sweden	1980	18
Forsmark 2	Sweden	1981	17
Forsmark 3	Sweden	1985	14
Ringhals 1	Sweden	1976	23
Barsebäck 1	Sweden	1975	23
Barsebäck 2	Sweden	1977	20
Oskarshamn 1	Sweden	1976	23
Oskarshamn 2	Sweden	1974	24
Oskarshamn 3	Sweden	1985	14
Olkiluoto 1	Finland	1978	20
Olkiluoto 2	Finland	1980	18
Brunnsbüttel	Germany	1989	10
Leibstadt	Switzerland	1991	8

It is worth mentioning that the version of POLCA implemented in the CMS has always been functionally equivalent to the off-line version employed by ABB and some of its customers in core design and in-core fuel management applications. This has resulted in an advantageous coherence between off- and on-line calculations.

Present limitations

The present strategy suffers from a number of limitations that are briefly described below.

The adaptive core monitoring approach is based on a combination of calculations and measurement-based corrections to determine thermal loads. One of the drawbacks of this approach is the fact that it is almost impossible to estimate the uncertainty in these computed thermal loads. The reason for this is that, once modified via an approximate correction function, it is no longer possible to compare the adapted solution against any other reference data. When trying to estimate the conservatism (or lack of conservatism) of the assumptions involved in the computation of the thermal margins, this becomes a serious limitation.

In addition, the adaptive strategy relies more on the information provided by the detectors than on the predictions provided by the calculations, even when the available detector systems have inevitable design limitations, such as:

- The measured detector signal (gamma or neutron flux in the detector position, not power!) consists of contributions coming from the neighbouring assemblies, thus masking individual contributions. The adaptation of the power distribution in nodes within the core volume covered by the detectors is performed with some kind of "interpolation" among the corrections observed at the detector locations, with an (in-) accuracy that is difficult to assess.

- The correction performed is strongly dependent on the number of reliable detector signals available. Moreover, an important fraction of the core is not covered by any LPRM, leading to dubious extrapolations. The fraction of the reactor core not covered by LPRM detectors in the periphery (radial and at the top and bottom) could be more that 50% of the total volume.

- The measured signals themselves are affected by channel bow, detector displacement (both particularly sensitive in the case of neutron flux detectors), detector depletion, etc. This leads to an increased uncertainty in the interpretation of their relationship to nodal power.

This power adjustment could, in fact, be worsening the accuracy of the local thermal margin estimations, as in the case of unfortunate channel bow conditions that lead to lower flux in the detector position, which may then be wrongly interpreted as lower nodal power in the surrounding bundles. If sufficient conservatism has been built into the operational limits, such undesirable interpretations could go unnoticed (without fuel failure).

Components for improved monitoring

A better utilisation of the information provided by the calculations and the measurements is possible. With enhanced predictions (best estimates) from the core simulations and a more comprehensive uncertainty assessment frequently updated, a more robust system could be developed. This is the basic idea behind ABB's Core Watch concept, which will be described shortly.

Core simulation

The successful development of modern nodal methods paves the way for core simulators that can deliver both increased accuracy and more detailed information with near real-time performance, as required in a CMS. Even when the cores modelled are increasingly heterogeneous, the core simulator prediction capabilities become more reliable due to:

- More sophisticated nuclear data (cross-sections) homogenisation techniques.

- The ability to predict pin power and pin burn-up distributions.

- More detailed models to describe the (in-core) thermal-hydraulics of the system.

This is a continuous process and there are still challenges ahead, but the main core simulators commercially available clearly deliver useful estimations. At ABB, the core simulator having such capabilities is POLCA7, which is already included in the core monitoring systems delivered to the three Forsmark NPP Units in Sweden and to the Hope Creek NPP in the USA.

Uncertainty assessments

The technical and operating limits are defined based on assumptions that may influence the way the core monitoring is performed. As an example, the technical limit could be defined in such a conservative way that some of the parameters involved in its estimation do not need to be monitored; this could be the case of conservative assumptions about existing channel bow.

Best-estimate strategies, by relying more heavily on the calculations, impose additional requirements on the evaluation of the accuracy and reliability of both the core simulations and the process signals used.

Within the activities performed to define the operating limits for a given cycle, the uncertainties in the computed thermal load must be estimated or assumed prior to the start-up. These kind of assessments could also be repeated during the cycle, in order to track any potential degradation in the accuracy of the parameters. In both cases, the estimated uncertainty in the computed thermal load will be the result of several contributions:

- Uncertainties in the engineering data:
  - Fabrication tolerances.
  - Material composition.
  - Thermal expansion.
- Uncertainties in the process data affecting the calculation of the thermal load:
  - Reactor power.
  - Core flow measurements.
  - Temperature measurements.
  - Local flux measurements (TIP/LPRM), if they are utilised during the evaluation.
- Uncertainties in the models for the calculation of the thermal load:
  - Nodal or fuel rod power calculation.
  - Bundle power calculation.
- Uncertainties due to non-modelled phenomena:
  - Channel bow.
  - Void in inter-assembly bypasses.

- Uncertainties in the determination of the technical limit (if not already included in its own determination):
  - Uncertainties or limitations in empirical correlations.

Other applications

The field of core monitoring in power reactors is not static and new applications are becoming or may become available in the near future. Examples of this are stability monitors, more advanced data signal treatment with the use of noise analysis, the possibility of on-line re-evaluation of the operating limits and even transient calculations. Applications of this kind, when subjected to the same conditions as the traditional CMS, will extend the aid provided to the reactor operator.

ABB's Core Watch

Core Watch is an advanced core monitoring strategy developed by ABB in co-operation with the Swedish utilities. A prototype has been in implemented at the Ringhals NPP Unit 1 in Sweden in order to evaluate the robustness of its principles. The concept has also benefited from a positive experience at the Leibstadt NPP, where a monitoring strategy without any adaptation to detector signals had been in operation for several years. In this case, a parallel evaluation of calculation/measurement deviations is performed periodically in order to detect sudden losses of accuracy in the detector signal predictions (≈ power) performed by the CMS.

Basic principles

In a highly simplified way, one could say that Core Watch relies on three basic principles or assumptions:

- Safety limits are a measure of acceptable risk of failure.
- Thermal margin predictions are a measure of the risk level at given conditions.
- Process signals (i.e. on-line measurements) can be used as an indication of the reliability of these predictions.

Based on these principles, the key components of the Core Watch concept are:

- Evaluation of thermal margins with a 3-D core simulator (best-estimate).
- The uncertainties affecting the current thermal margins evaluation are taken into account.
- The detector signals are used for a continuous estimation of the calculation uncertainty.
- Punitive actions are introduced by the CMS if pre-assumed ("licensed") uncertainty levels are exceeded.
- Alternative operation modes are available in Core Watch in the event significant systematic deviations, which make the calculations unreliable, are observed. This could also happen in the case of faulty or insufficient process data.

Figure 1 shows the main structure of Core Watch, where a separate module performs the on-line uncertainty evaluation used to complement the best-estimate thermal margin calculations.

Figure 1. Core Watch structure

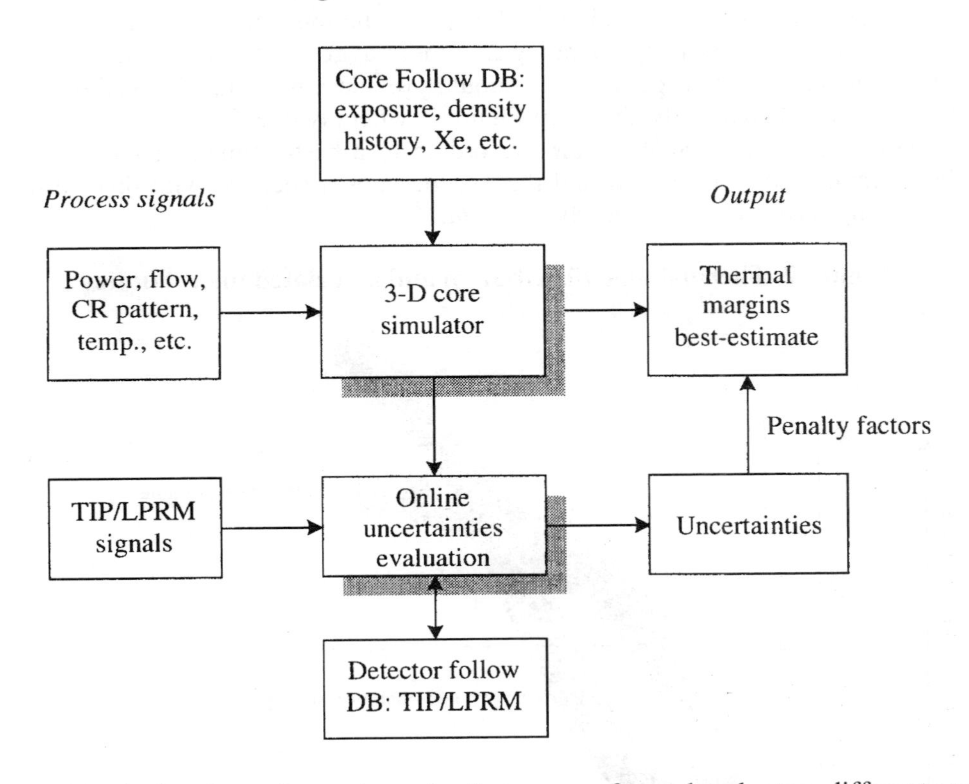

The basic idea is that thermal margin evaluations are performed under two different conditions:

- During core and fuel design and licensing of an upcoming cycle, where the different uncertainty contributions are estimated in a pre-established way.
- During operation, where on-line (updated) estimations of these uncertainty contributions are performed.

As long as the on-line estimated uncertainties are within the range of the uncertainties assumed during the licensing activities, the best estimates of the thermal margins are used as final results. If, for any reason, the on-line estimated uncertainties exceed the licensed values, penalty factors are imposed on the calculated thermal load (x_c, the best estimate). These penalty factors are calculated as the ratio between the licensed operating limit and an equivalent operating limit calculated with the on-line estimated, higher, uncertainty. Mathematically, this is expressed in the following way (for LHGR):

$x'_{op} = f\left(x_t, s'_{op}\right)$ *A new operating limit based on current uncertainties*

$f = x_{op} / x'_{op}$ *The penalty factor*

$M_x = x_{op} / (f \cdot x_c) - 1$ *The margin calculated with a penalised best estimate*

In this way, smaller margins are obtained, without changing the operating limit itself, which could be disturbing for the reactor operators. Applying the penalty factor in this way is equivalent to an even penalisation over all bundles or nodes, because no rearrangement of their relative margins is performed. This kind of penalisation on a global level could be somewhat over-conservative, because the highest deviations are generally found in the low power regions, but it is a reasonable trade-off to keep the system simple. Moreover, by operating in a more aggressive manner, several bundles could be close to the limits and a global penalisation should provide a more reliable estimation of the true safety margins. This is schematically shown in Figure 2, where two different operation conditions are illustrated. In both cases the thermal margin, according to a best-estimate calculation, is the same. However, the thermal load distributions and the uncertainties associated with them differ in such a way that the resulting risk of failure is clearly different.

Figure 2. Thermal load distribution and associated uncertainties

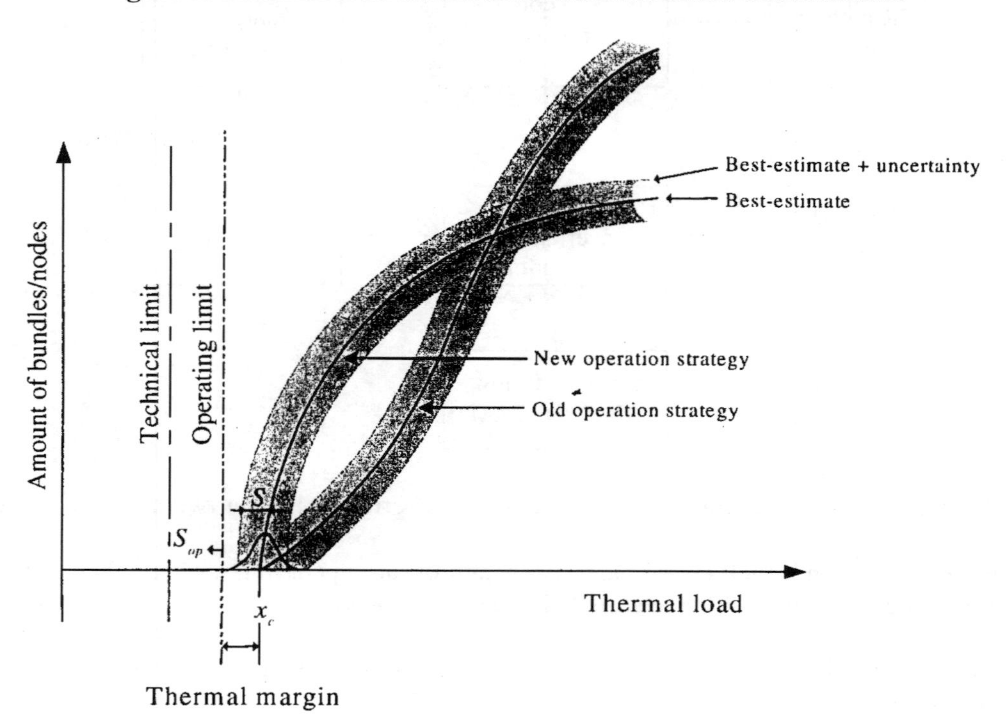

As a consequence of this observation, the calculated and measured detector signals are utilised in a different way in Core Watch than they are in the adaptive technique. Instead of being used to estimate local power corrections, as in the case of the adaptive strategy, they are used to produce global estimations of the calculation accuracy. In this way, the thermal margins and their estimated uncertainties are less dependent on individual measured-to-calculated detector discrepancies. Additionally, Core Watch can deal in a very natural way with cases where the operation is performed with a temporarily reduced number of LPRM, such as results from multiple detector failure. This is due to the fact that the confidence level of the uncertainty estimation is a function of the number of detectors involved in the comparisons. Even the frequency of LPRM calibrations could be adjusted with the support of the estimations performed by Core Watch in order to retain a given accuracy level.

Alternative operation modes are available to handle exceptional conditions that could lead to unreliable results if the normal mode, based primarily on core simulations, were used instead. These exceptional conditions could be related to either faulty or insufficient process data or to faulty or unavailable calculations. A restricted confidence in the core simulation is compensated with a stronger dependence on the TIP/LPRM signals. Two alternative operation modes have been proposed:

- CORR1D: it can be used in case significant axial systematic deviations are observed over a long period. In this case, an axial correction (1-D), based on average detector signal comparisons at each nodal level, is applied to the axial power distribution of each bundle, before the thermal margins are computed. This is a restricted adaptive strategy aimed to retain, as much as possible, the on-line uncertainty evaluation capabilities.

- CORR3D: it can be used in case the core simulation results are unavailable or clearly unreliable (e.g. if transient xenon calculation fails). In this case, a solution similar to the adaptive strategy (UPDAT) is applied, with a 3-D correction factor distribution based on individual detector comparisons applied on the latest available calculated nodal power distribution. In this operation mode, only fixed (licensed) uncertainty estimations can be used.

Both modes, but specially CORR3D, will result in higher uncertainties, compared to those obtained under normal mode, due to the more precarious understanding of the true core conditions. Consequently, these modes are restricted to be used in abnormal situations.

On-line uncertainty estimations

Core Watch depends on a comprehensive assessment, to be done during licensing (off-line) as well as during operation (on-line), of all the possible uncertainty contributions mentioned before (see the section entitled *Uncertainty assessments*). These contributions may be reactor-, cycle- and/or fuel-dependent and they could be difficult to estimate. However, it is this detailed and updated assessment that enables a better understanding of the status of the core, thus enhancing the reliability of the CMS predictions.

From the comprehensive list of factors contributing to the uncertainties in computed thermal margins described above, only two need to be monitored "on-line" during the cycle, namely:

- Calculated bundle power.

- Calculated nodal power.

For the remaining contributions an assessment performed during the licensing stage, prior to the start-up, suffices. Nevertheless, it should be kept in mind that this assessment could be restricted to certain reactor conditions, typically for reactor power/core coolant flow in a given range (e.g. at rated conditions). In this case it is necessary to define proper actions to be taken when the reactor conditions fall outside this range. This could be the case of thermal margins that are evaluated only when reactor power exceeds a certain level.

In order to assess the uncertainty in the computed bundle and nodal power, only TIP and LPRM signals are available for calculation-to-measurement comparisons. It is possible to use either TIP or LPRM signals, or a combination of both. The latter represents the recommended approach. In this case, uncertainties are calculated in connection with each LPRM calibration utilising the TIP signals. Between calibrations the LPRM signals are used instead. By selecting the maximum uncertainty predicted by any of them one has a conservative methodology that combines the more detailed information provided by the TIP strings with the more up-to-date, but scarcer, information provided by the LPRM detectors. This is schematically shown in Figure 3.

Figure 3. Combining TIP and LPRM based uncertainty estimations

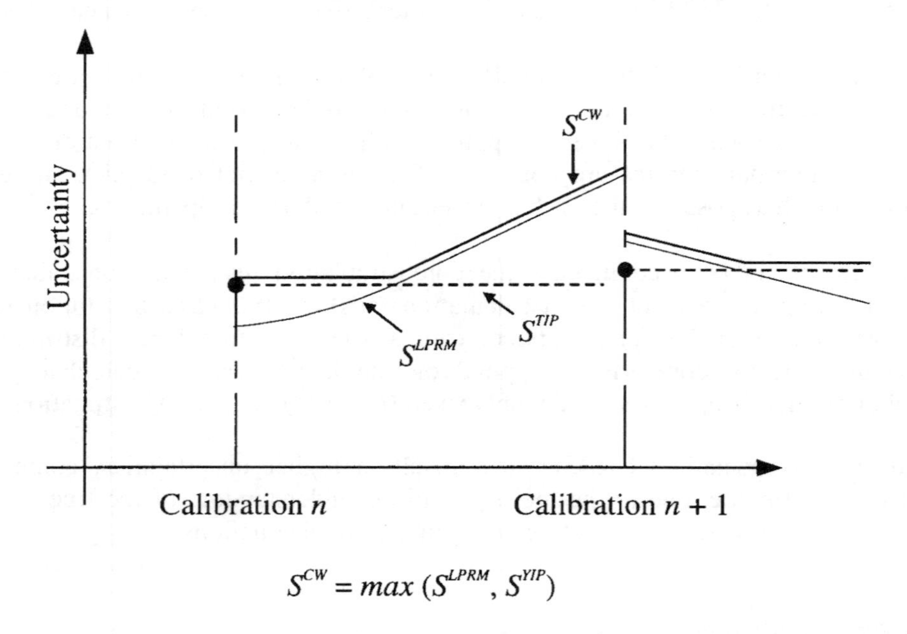

$$S^{CW} = max\,(S^{LPRM}, S^{YIP})$$

The calculated vs. measured comparison that can be performed during the on-line monitoring concerns detector signals (related to neutron or gamma flux). The observed differences will, therefore, be the combined contribution of three different uncertainties (or inaccuracies):

- Uncertainties in the power calculation of the bundles surrounding the detector string.
- Uncertainties in the detector model, which converts power (or flux) into computed detector signal.
- Uncertainties in the measured detector signals.

A proper model for the estimation of the last two contributions, together with the observed differences between calculated and measured signals, enables the estimation of the uncertainty in computed power. Core Watch will combine the estimated uncertainty in calculated power with the other pre-assumed (licensed) uncertainty contributions to quantify the uncertainty in the computed thermal load.

Tools for estimation of different uncertainties

Specific procedures are required to estimate the aforementioned contributions to the total uncertainty in computed thermal load. These assessments have to be carried out by different entities, e.g. fuel vendors will need to perform and provide the fuel specific analysis, leaving to the utility the reactor specific assessment. Proper statistical tools are required for these evaluations in order to take into account their quite different nature. Aiming to have a formally rigorous treatment of the uncertainties, the following factors must be considered:

- The potential consequences of normalisation and re-normalisation.
- Assumptions on variable independence/dependence (correlated vs. non-correlated variables).

- Systematic deviations.
- Relative vs. absolute uncertainties.
- Assumptions about errors following a normal distribution.
- Conversion of (other) uncertainty distributions to equivalent normal distributions.
- Combination of uncertainties in functions of multiple variables.

Realising that the plant measurement devices are limited in their functionality, an effort must be made to estimate the impact of different phenomena on the different uncertainties even beyond the capabilities of these instruments. In addition, it is also necessary to estimate the uncertainties introduced by the instruments themselves. As a result of this, alternative sources of information should be sought. Three of them, which have received special attention at ABB Atom, are described below. They include gamma scanning, experimental evaluation of neutron physics parameters and void measurements on modern (10×10) fuel assemblies.

Gamma scanning

There is a definite limitation in the TIP/LPRM capabilities: the lack of information about power distributions at the individual bundle and fuel rod level. From TIP and LPRM signal comparisons, it is necessary to infer the calculated power uncertainty at those levels without any support from the instrumentation. This lack of information can be partly resolved by means of gamma scanning of bundle and/or individual fuel rods during outages at the plant. By tracking, most typically, ^{140}Ba concentrations, a good estimation of the power predictions can be obtained. Two other "by-products" of this kind of exercise is the possibility of discovering exposure-related trends or deviations and the possibility of using these measurements to estimate the uncertainties of the TIP/LPRM system itself (in case the gamma scans are performed on TIP string neighbouring assemblies).

Therefore gamma-scanning measurements should be performed when significant changes in the core conditions appear. This could be the case with the introduction of radically new fuel concepts, the operation with increased discharge exposure or the existence of strong deviations from the assumed conditions (significant channel bow, CRUD, etc.).

ABB has the capability to perform ^{140}Ba gamma scanning both at the bundle and individual fuel rod level, and a number of measurement campaigns have been pursued over the last few years. The information provided by them has been used in the validation of ABB's simulation package. However, the same kind of information could also be used to estimate some of the uncertainties in the TIP/LPRM comparisons, as mentioned before.

Validation against critical experiments

Improved modelling capabilities, even when physically sound, should be validated against higher order methods and/or experimental evidence as far as possible. The increasing complexity of fuel concepts, including higher enrichment and extensive use of different burnable absorbers, should be correlated with validation activities. ABB Atom, EGL (owner of the Leibstadt NPP) and the Paul Scherrer Institute of Switzerland are undertaking, for that purpose, an ambitious experimental program named LWR-PROTEUS Phase I, utilising the PROTEUS facility at PSI.

The program includes extensive and accurate measurements in different critical configurations where the test zone consists of nine (3 × 3) full-scale SVEA-96 bundles. This is schematically shown in Figure 4. These experiments, which are performed under well characterised conditions, will provide detailed information about internal power distributions, reactivity worth of different components (fuel rods, control rods, etc.). In addition, it will be possible to study the influence of such parameters or perturbations as partial length rods, instrumentation, low coolant density and channel displacements.

It is expected that these results will help to quantify the modelling uncertainties and, eventually, to identify areas where further methods development could be required.

Figure 4. LWR-PROTEUS core layout

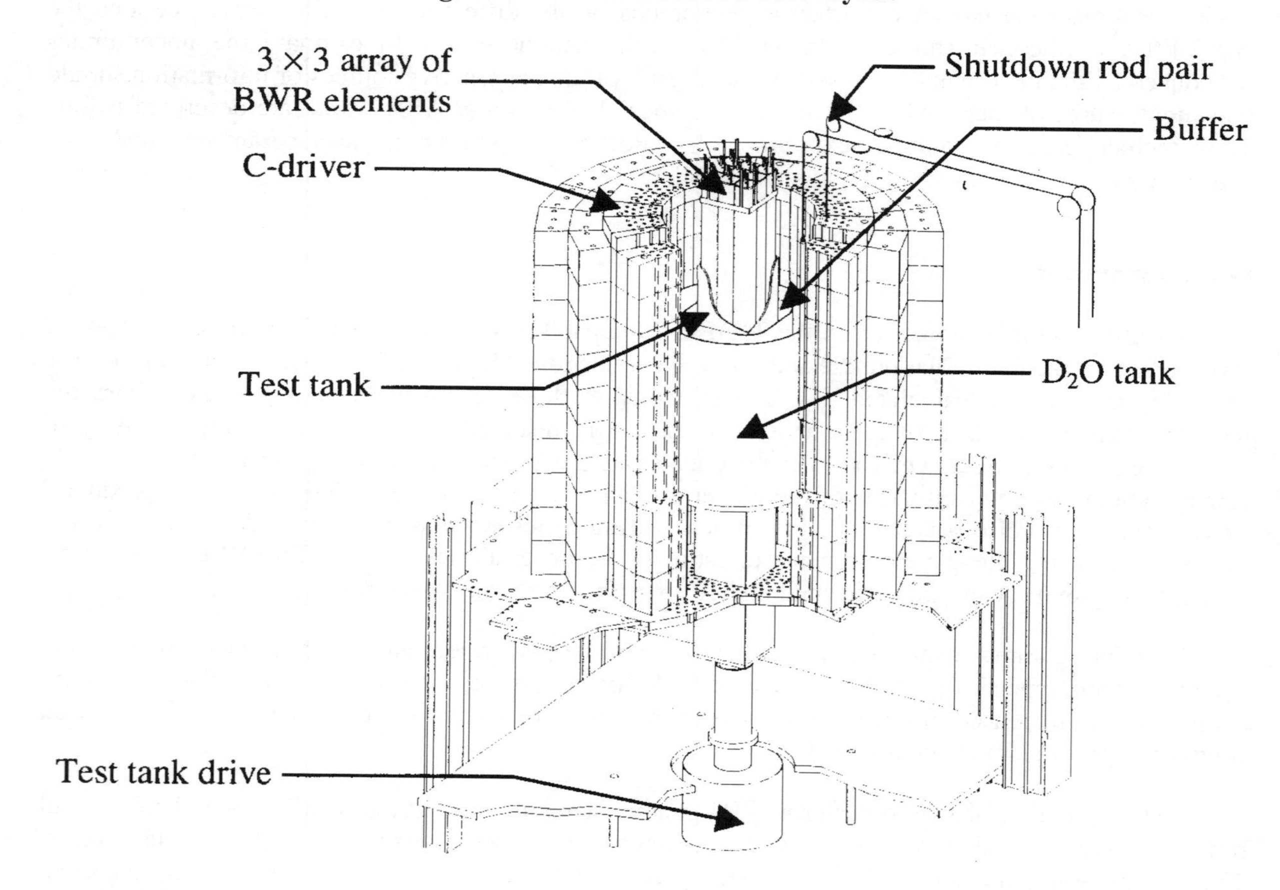

Void measurements

Void measurements in realistic geometries are required in order to develop void correlations and two-phase pressure drop correlations. The introduction of fuel designs with 10 × 10 rods together with water channels of different geometry, partial length rods, etc., leads to increasingly different conditions from previous void measurements performed in the late 60s with simple 8 × 8 bundles. In recognition of this fact, ABB has performed new measurements in its retrofitted FRIGG loop (see Figure 5) in Västerås, Sweden. These measurements will help not only to improve the modelling accuracy of the core simulator by means of a more reliable void correlation, but also to quantify the impact of potentially limited modelling capabilities in transition cores, which can be difficult to discover from TIP/LPRM comparisons.

Figure 5. Void measurements at FRIGG-loop (xy-plan)

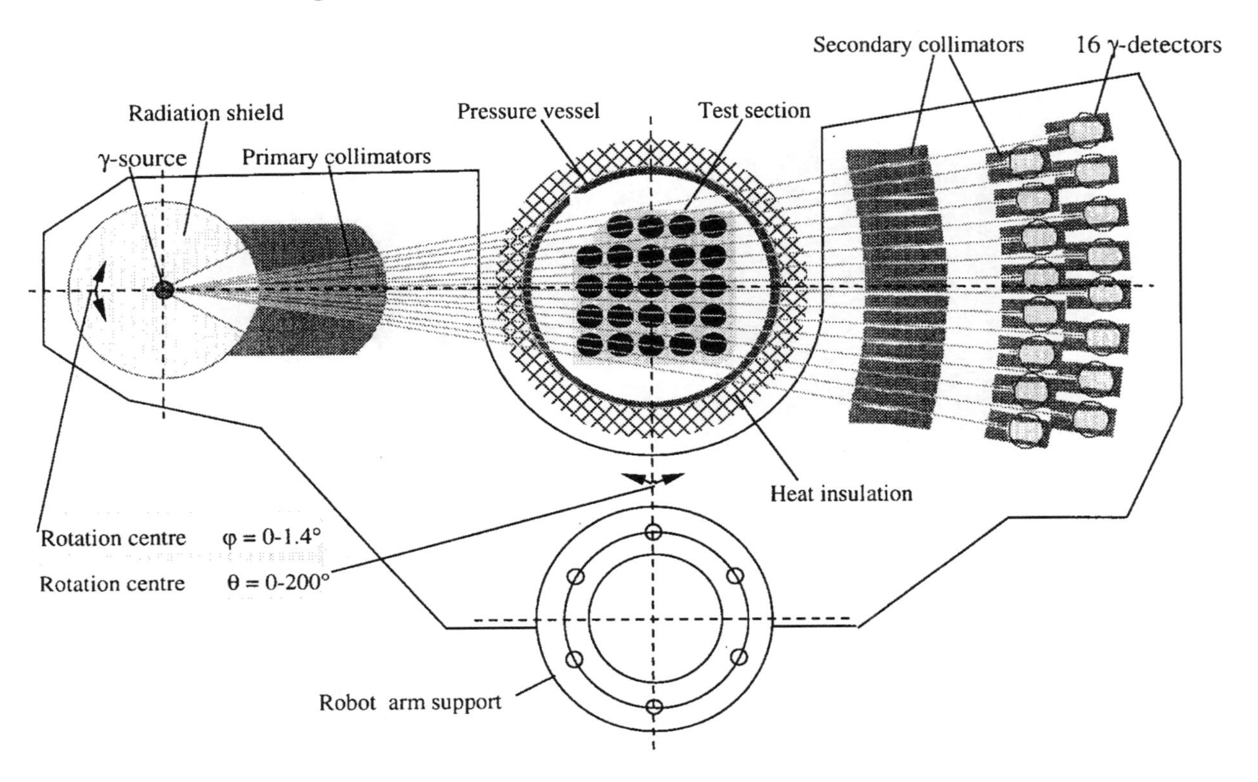

Conclusions

A comprehensive assessment of the uncertainties affecting ongoing thermal margin evaluations would enhance the reliability of a core monitoring system by allowing a better understanding of both the actual conditions of the reactor and the current modelling accuracy. This is the approach chosen in ABB's Core Watch, where a different strategy enables the accommodation to new requirements, without loosing flexibility in the day-to-day operations. This strategy, however, requires uncertainty assessments to be performed more frequently and/or in a more structured way. These assessments will eventually require the support of additional activities such as comparisons against experimentally generated data; some of these activities have been exemplified in the present paper.

SESSION II

Sensors, Signal Processing and Evaluation

Chairs: E. Saji and Ö. Berg

INDUSTRIAL TESTS OF RHODIUM SELF-POWERED DETECTORS: THE GOLFECH 2 EXPERIMENTATION

Jean L. Mourlevat
FRAMATOME
Tour FRAMATOME
92084 Paris La Défense Cedex France
Tel: 33 1 47 96 31 34 • Fax: 33 1 47 96 50 48
E-mail: jlmourlevat@framatome.fr

Daniel Janvier
EDF-Industry
Basic Design Department
12-14 Avenue Dutrievoz
69628 Villeurbanne Cedex France
Tel: 33 4 72 82 73 53 • Fax: 33 4 72 82 77 09
E-mail: daniel.janvier@edf.fr

Dr. Holland D. Warren
Senior Technical Consultant
FRAMATOME COGEMA FUEL
207 Nottingham Circle
Lynchburg, VA 24502-2748,USA

Abstract

In co-operation with Électricité de France (EDF), FRAMATOME has been testing two in-core strings which are equipped with rhodium self-powered detectors (SPDs) in the Golfech Unit 2 reactor (1 300 MW, 4 L plant) since August 1997. The rhodium SPDs and the strings which support them were designed and built by the US FRAMATOME subsidiary FRAMATOME-COGEMA-FUEL (FCF). The rhodium signals and some other plant parameters are acquired through the use of a specific device designed by the CEA (Commissariat à l'Énergie Atomique) and are processed off-line by FRAMATOME. This demonstration test is planned to last until mid-2000.

The following presentation is focused on the results obtained during the first demonstration cycle (from 08/97 to 12/98). The tests that have been conducted consist of checking the rhodium depletion and of comparing the rhodium signals to the movable probes. In order to compensate for the delay in the rhodium signals, a deconvolution algorithm has also been tested.

Up to now, the results are very satisfactory and a future large scale industrial application is being discussed with the EDF. The main objective of the next experimentation phase is to test – under industrial conditions – a prototype of an on-line monitoring unit known as the Partial In-Core Monitoring System (PIMS). This system will include 16 rhodium in-core strings and will use an on-line 3-D core model.

Introduction

In the present design of FRAMATOME plants, flux mapping is performed once a month as required in the technical specifications by the installed in-core instrumentation system using movable fission chambers. A full core measurement campaign takes about one hour and the signals are processed off-line a few hours later.

Core operating margins monitoring, often referred to as limiting conditions of operations (LCO) monitoring, is performed by the ex-core instrumentation. For the 1 300 MW 4 L plants, FRAMATOME has developed specific six-section chambers equipped with a fast neutron-filtering device. Such detectors are able to provide an accurate measurement of the axial power profile at the periphery of the core with a very short response time. The signals are processed by microprocessors in the alarm unit of the Système de Protection Intégré Numérique (SPIN – Digital Integrated Protection System) in order to provide the LOCA and DNBR margin values to the operators.

Quite obviously self-powered neutron detectors could improve the power distribution measurement accuracy and hence improve core operating margins. The implementation of a new nuclear instrumentation, however, is a costly operation, especially on the existing plants. Thus it is necessary to develop an attractive cost/benefit analysis to convince utilities.

The strategy chosen by FRAMATOME was to use the fixed in-core system for LCO monitoring and not as the reference flux mapping system, in such a way that the gain in accuracy could be converted directly into operating margins. Start-up tests phase reduction, operating flexibility increase or fuel cycle length increase by implementing a full low leakage fuel loading pattern may be searched for. Moreover, a cost reduction may be obtained because LCO monitoring equipment is not 1E classified.

A LCO monitoring system must be able to provide core operating margins even during power transients. This is the reason why the global response time of the system, including the sensor response time and the calculation time, must be rather low, i.e. less than 30 seconds.

The development of such a monitoring system can be reduced to three steps. The first step is to choose a sensor and to test it under real industrial conditions. The second step is to design a complete system that is able to provide an accurate measurement of the 3-D power distribution in the core and to install it in a power plant for testing. Finally, the third step consists of implementing on-line software and the associated calculation capabilities in order to process the sensor signals and to calculate the core operating margins in real time conditions. The second and the third steps may be combined.

The Golfech 2 experimentation is related to the first step.

The choice of the sensor and the mechanical design

At the end of the nineties, FRAMATOME decided – in collaboration with FCF – to use rhodium SPNDs as fixed in-core sensors. The advantages of rhodium are numerous:

- The physics is simple and well known.
- The response is due to neutrons.
- The compromise between sensitivity and lifetime is attractive.

- Rhodium depletion can be simulated easily by an empirical relationship eliminating the need to calibrate the sensors one against the others when several are used to measure the power distribution in the core.

- The response time is not large and can be accelerated by software if needed.

- Finally, the industrial fabrication is simple and is reproducible.

For all these reasons, FCF has strongly recommended to use the rhodium. FCF has a very large experience base with rhodium because the Babcock & Wilcox (B&W) plants that are presently operated in the USA are equipped with rhodium SPDs manufactured by FCF. The B&W plants have used a fixed in-core system as the reference instrumentation system since their original design. More than 200 reactor-years of feedback experience have been accumulated with this reference system and this experience shows that a rhodium based fixed in-core instrumentation is a very attractive solution.

The design of the rhodium SPD used at Golfech 2 is the most recent one set up by FCF. In addition to the cable which transports the useful signal coming from the rhodium, a second cable (compensation cable), is used to eliminate background which is generated equally by the gammas in the parts of both cables which are in the core. This second current is taken away from the first one by subtraction.

Eight rhodium SPDs (for a 14 ft. high core) are mounted in a mechanical assembly which is called the in-core detector assembly, or in-core string. Each rhodium detector is located at the middle of two adjacent fuel assembly grids. FCF succeeded in reducing the outside dimensions in such a way that it is possible to introduce the in-core strings inside the normal flux thimble used by the movable instrumentation (see Figure 1). The flux thimble is normally inserted in the instrumentation tube of the

Figure 1. In-core string view and cross-sections

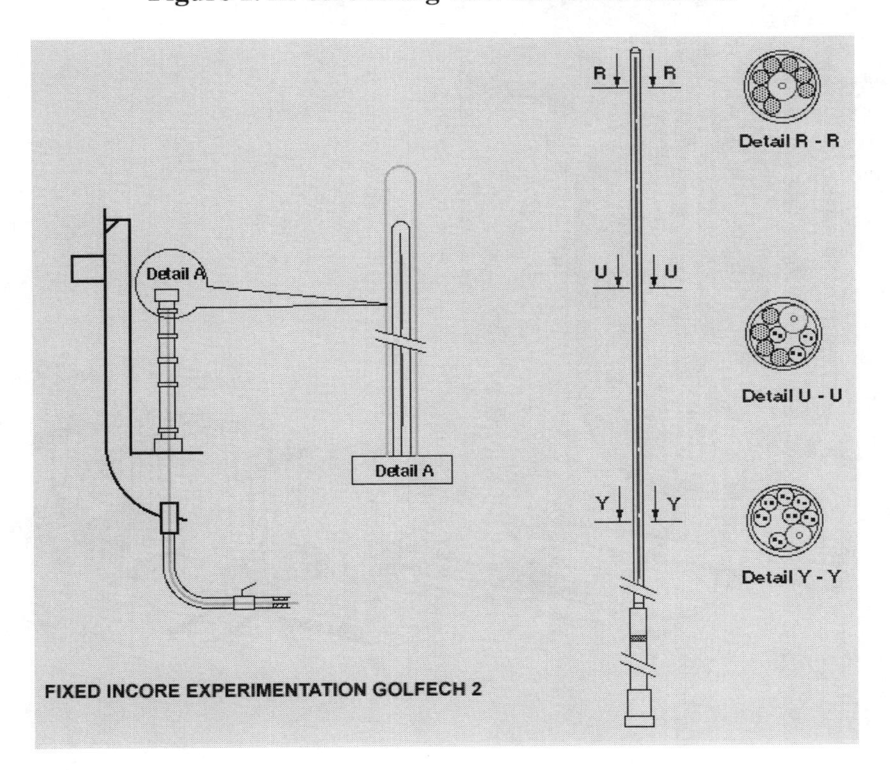

fuel assembly. Thus the mechanical interface with the primary circuit is not impacted because the flux thimble is not replaced and not modified. The mechanical seal between the flux thimble and the bottom vessel penetration tube is kept without any modification. Such a "dry" concept is a very important point to facilitate licensing.

A specific mechanical device, known as an "adapter", has been developed to make the mechanical interface between the bottom base of the string and the flux thimble. This adapter has been designed to be resistant to the primary pressure in case of a failure of the flux thimble.

All the bottom bases of the strings and the adapters are supported by a specific support frame which is earthquake-qualified. Figure 2 shows a general view of the in-core instrumentation room.

Figure 2. General view of the in-core room and vessel penetrations

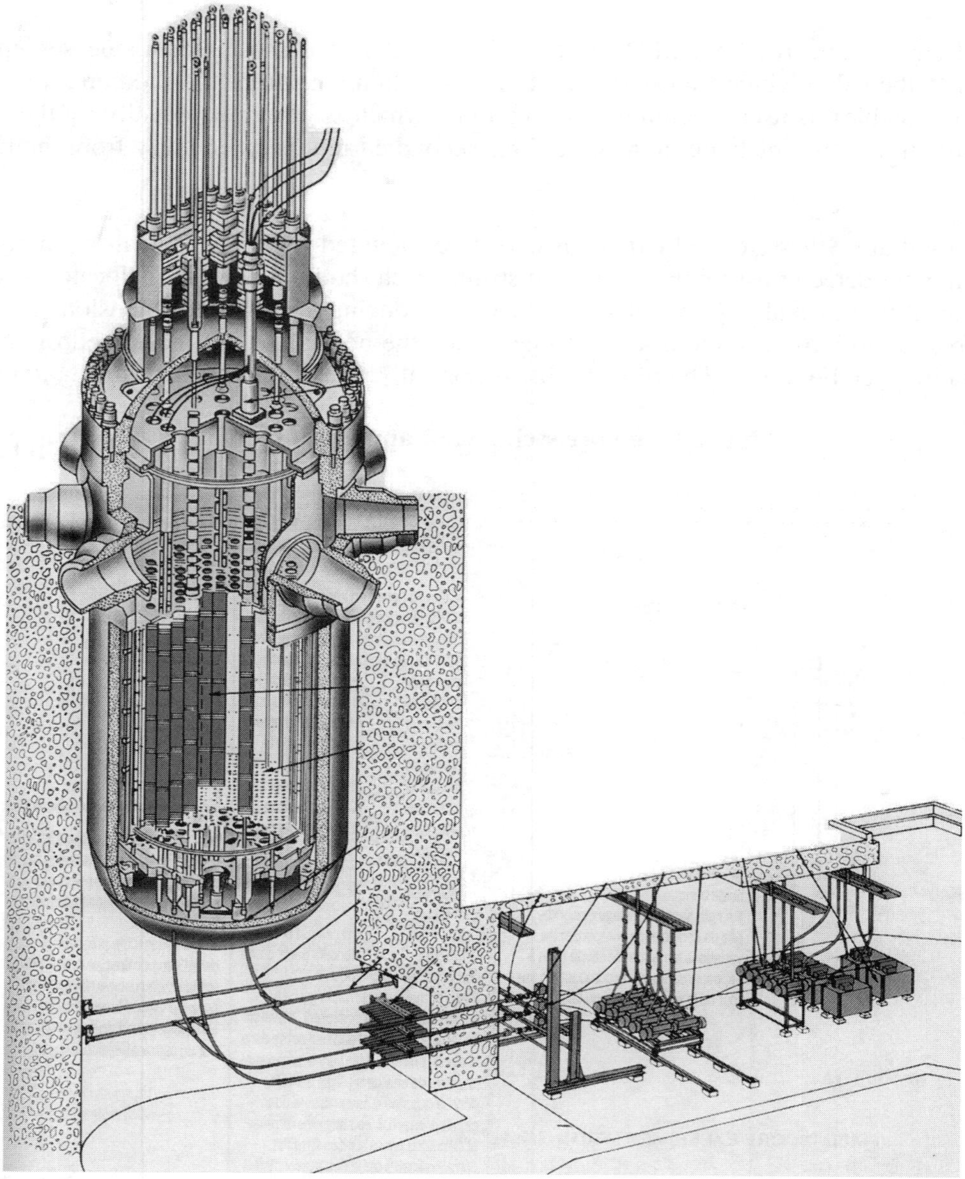

The Golfech 2 experimentation

Generalities

Golfech Unit 2 is the last unit of the 1 300 MW four-loop serie which is composed of 20 units. The first criticality occurred in May 1993 (the first criticality of the first unit of the 1 300 MW 4 L series (Paluel 1) took place in May 1984). The present fuel cycle is the sixth. All the French 1 300 MW units are equipped with a digital protection system known as SPIN. In case of low LOCA or DNBR margins, alarms are actuated by the alarm unit of SPIN. LOCA and DNBR margin values come from a calculation taking into account the axial power profile measured by the six-section ex-core chambers and measured and/or pre-tabulated hot spot radial factors.

The reference in-core instrumentation system consists of six movable fission chambers which are able to monitor 58 of the 193 fuel assemblies. As indicated before, a full core measurement campaign requires about one hour because ten passes are needed. Thus, core steady state conditions are required to perform an accurate flux map. The signals coming from the movable probes are acquired and digitised by the in-core instrumentation system computer, transmitted to the plant computer and then stored on a disk which is processed off-line by plant specialist personnel. The reference instrumentation system is used to check the core power distribution as required in the technical specifications and to calibrate the ex-cores.

General presentation of the Golfech 2 experimentation

Two in-core strings were introduced during the summer of 1997 for two cycles, i.e. three years. The experimentation will be completed by mid-2000. At this time, the strings and the corresponding flux thimbles will be discharged and stored in the fuel pool. The introduction of any experimental measurement device in the core uses vessel bottom instrumentation penetrations. This arrangement necessarily decreases the number of the fuel assemblies which can be explored by the movable probes. The two in-core strings were introduced in the Golfech 2 core at the H2 and J10 locations. This means that 56 fuel assemblies can be measured by the remaining movable fission chambers. Figure 3 shows the corresponding core cross-section. The number of two fixed in-core instrumented assemblies is a compromise between the wish to not induce a heavy perturbation in the plant operating and to have redundancy of the experimental strings.

The electronic acquisition facility has been designed and built by the French Commissariat à l'Énergie Atomique (CEA). The 32 analogue currents coming from the two in-core strings and some other variables such as the ex-core power level are acquired and digitised; then the signal coming from the compensation cable is subtracted from the signal coming from the Rhodium to give the useful signal to be processed. The useful signals are stored on a hard disk to be processed off-line.

The storage capacity is equivalent to three months approximately depending on the number of power transients. The values of the analogue currents can be checked before the digitalisation by a pico-ammeter. The display of each numerical value is possible also. This acquisition cabinet is located in the computer room near the main control-room. As indicated above, no processing is performed on-line: the data stored on the hard-disk are periodically moved on floppy disks in order to be processed at the FRAMATOME headquarters offices.

Figure 3. Core cross-section

Rhodium SPD locations
R, G1, … = Control banks
1, 2, … = Movable instrumentation channel number

	R	P	N	M	L	K	J	H	G	F	E	D	C	B	A
1							52			51					
2			53	SA		SB 43		R SPD		SB		SA			
3					G2		N2	56	N2	44	G2	50		41	
4		34	45	N1		SC		G1 32		SC		N1		SA	
5			G2		R 54				57		R 49		G2 31		
6	46	SB	55	SC		SD 48		N1 36		SD		SC		SB 37	
7			N2	47			33			39			N2 58		
8	27	R1	26	G1	35	N1	28	R		N1 38		G1 24	22	R1 40	
9		23	N2						13		2		N2		3
10		SB		SC	14	SD	SPD	N1		SD		SC 21		SB	
11	6		G2		R 30			4			R 15		G2		29
12		SA		N1		SC 12		G1	18	SC		N1 1		SA	
13			16		G2 7		N2	5	N2		G2			11	
14			8	SA		SB	19	R		SB 20		SA 10			
15					17			9							

The goals of the experimentation

The Golfech 2 experimentation had two main goals. The first one is related to the demonstration of the mechanical consistency of an in-core string with the mechanical design of the present in-core instrumentation system. The second one involves the functional aspect: it was necessary to check the rhodium depletion law and to validate the rhodium signal acceleration algorithm under dynamic conditions.

Because of its small outside diameter, it was possible to introduce the string easily inside the normal flux thimble and the adapter setting was accomplished without any particular problem. We can now consider that the first objective has been fully reached. Thus, the following results are solely focused on the functional aspects.

Results

Generalities

All results presented hereafter have been obtained from measurements acquired during the first cycle of experimentation from August 1997 to December 1998. Three kinds of results are described:

- Rhodium depletion.
- The acceleration of the rhodium signal.
- The comparison of the rhodium signals with the movable in-core ones.

Rhodium depletion

Rhodium depletion could be predicted with an empirical relationship. This relationship gives the rhodium sensitivity value to be used in the power distribution reconstruction. This sensitivity law has been defined following many experiments performed by FCF on a US plant. FCF has shown that the rhodium sensitivity is a function of the total electrical charge released by the rhodium since its initial introduction in the core. A correction is added to take into account the initial number of the rhodium atoms which is depending on the geometrical characteristics of each individual rhodium emitter after manufacturing.

It is obvious that the electric charge supplied by each rhodium detector becomes an important parameter and must be documented in order to be known at any time in case of an acquisition facility failure.

Figure 4 shows a general view of the rhodium currents as a function of the power level during the first power escalation after fuel reloading. The comparison with the ex-core signal is good; the strong variations which are observed at the 80% plateau are due to the axial xenon oscillation performed to calibrate the ex-cores.

Figure 5 shows the variation of the sensitivity of each rhodium detector as a function of time. After a quick variation period, a typical variation value equal to roughly 1%/month can be observed. These values are in accordance with the expected ones.

Deconvolution algorithm

The rhodium response time depends on the β-decay half time of the rhodium isotopes which are created with the neutron flux. Although these constants are not very high (around one minute for the highest one), the time delay induced by the β-decay is not suitable to follow the flux variation which occurs during load-follow transients. This is the reason why it is desirable to accelerate the signal.

One of the advantages of rhodium is the simplicity of the associated physics. It is therefore very easy to model this type of detector and to describe its transfer function (see Figure 6). It can be seen that in addition of the two β-decay paths (a,b) which are responsible of the most part of the total signal (and of the delay), a third path (c), corresponds to a fast process related to the gamma's influence. This component can be used to accelerate the signal.

Figure 4. First power escalation

SPD's currents versus time

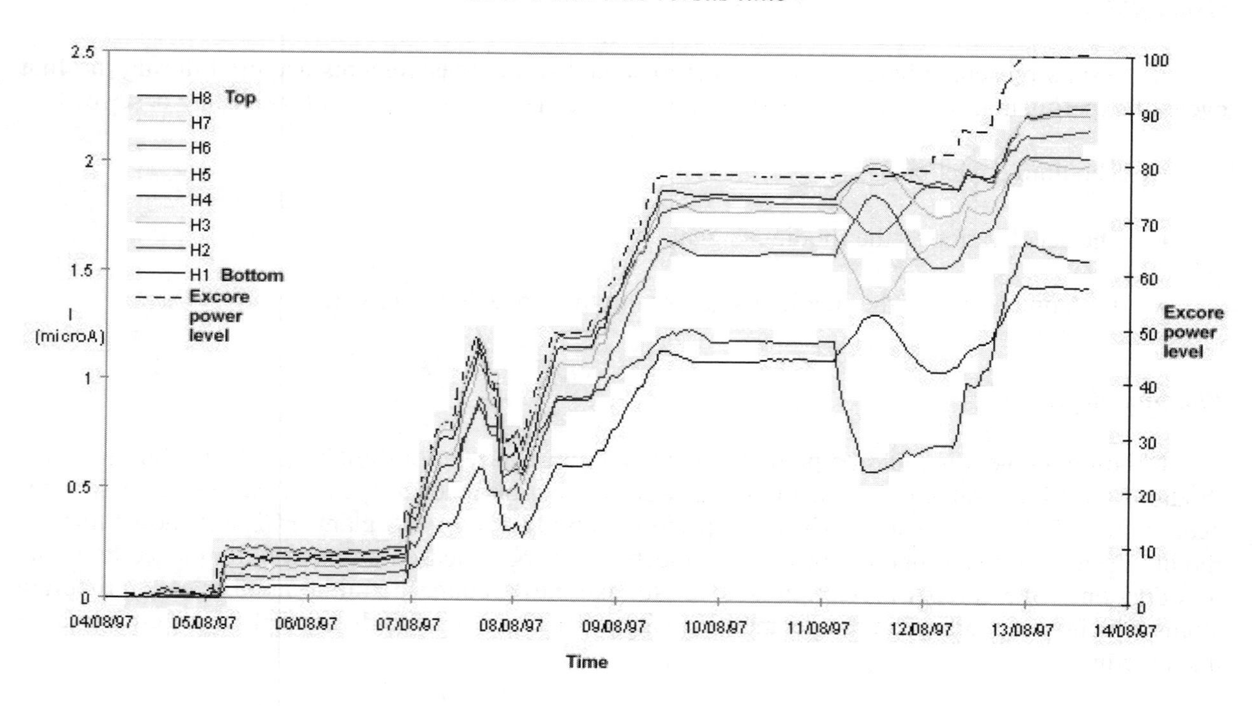

Figure 5. Variation of the rhodium's sensitivities versus time

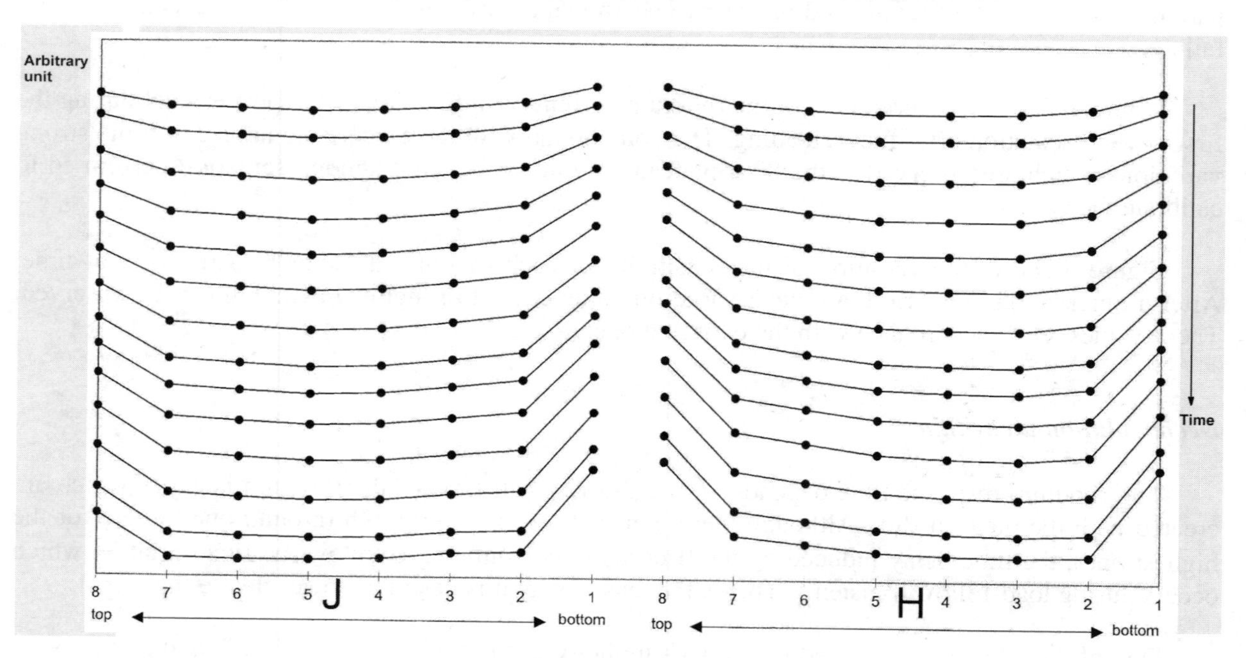

Figure 6. Rhodium transfer function

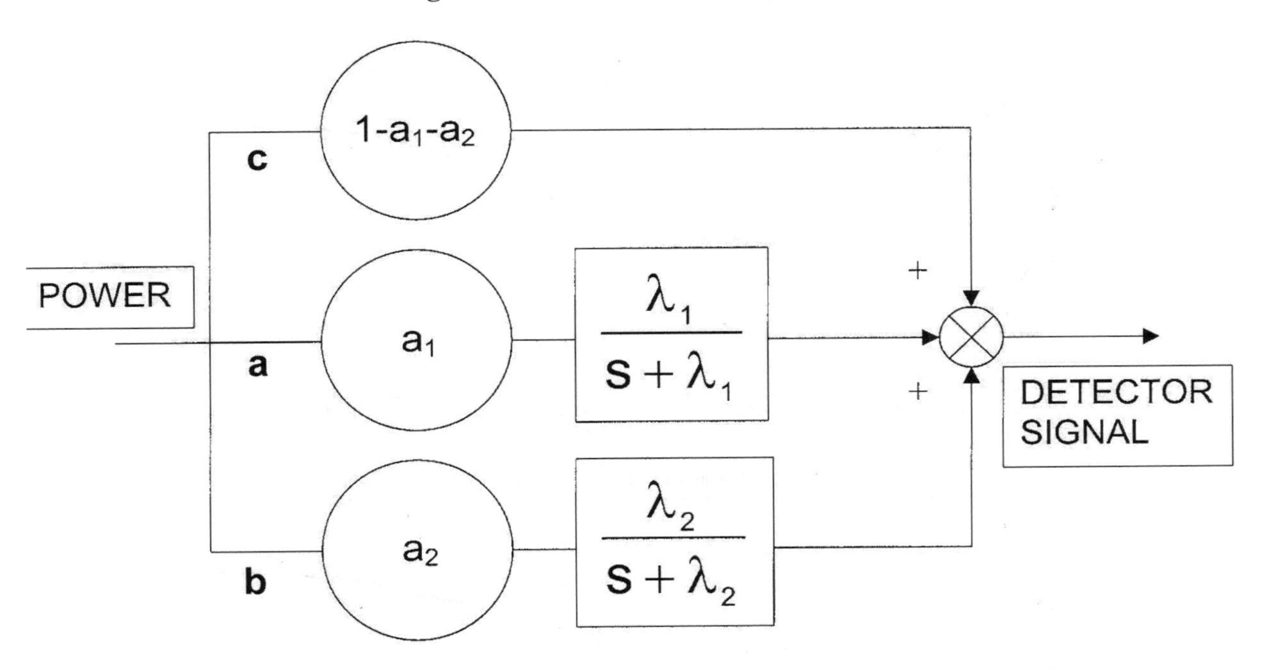

The use of a microprocessor allows on-line calculations. An on-line inversion of the transfer function could be considered. Such an algorithm is called "deconvolution algorithm".

In the frame of the Golfech 2 experimentation, the deconvolution algorithm is not implemented in the acquisition facility and the inversion of the transfer function is kept off-line. Obviously, the consequence of the deconvolution is to raise the noise and then to reduce the signal/noise ratio. It is therefore necessary to filter the signal and the time delay is once again increased. The final result is around 12 seconds with a noise/signal ratio close to 1%.

Compared to the rough value close to 80 seconds, the improvement is very significant. Many tests have been performed over several power transients. No numerical instabilities have been observed (see a typical example in Figure 7).

Comparison with the movable probes signals

The Golfech 2 configuration offers the opportunity to compare fixed and movable detector responses because both fixed and movable instrumentations are simultaneously present in the core. Obviously, it is not possible to insert a movable fission chamber in the fuel assembly where a fixed in-core string is already present.

It is possible, though, to make comparisons between the responses of the eight rhodium detectors installed in a fuel assembly and the signal of the fission chamber inserted in the symmetric fuel assembly when it exists or with the reconstructed axial power shape of this assembly. In this last case, the axial power shape of the fuel assembly which receives the string is reconstructed in the same way as for the assemblies which are not instrumented by a movable probe.

Figure 7. Rhodium signal deconvolution

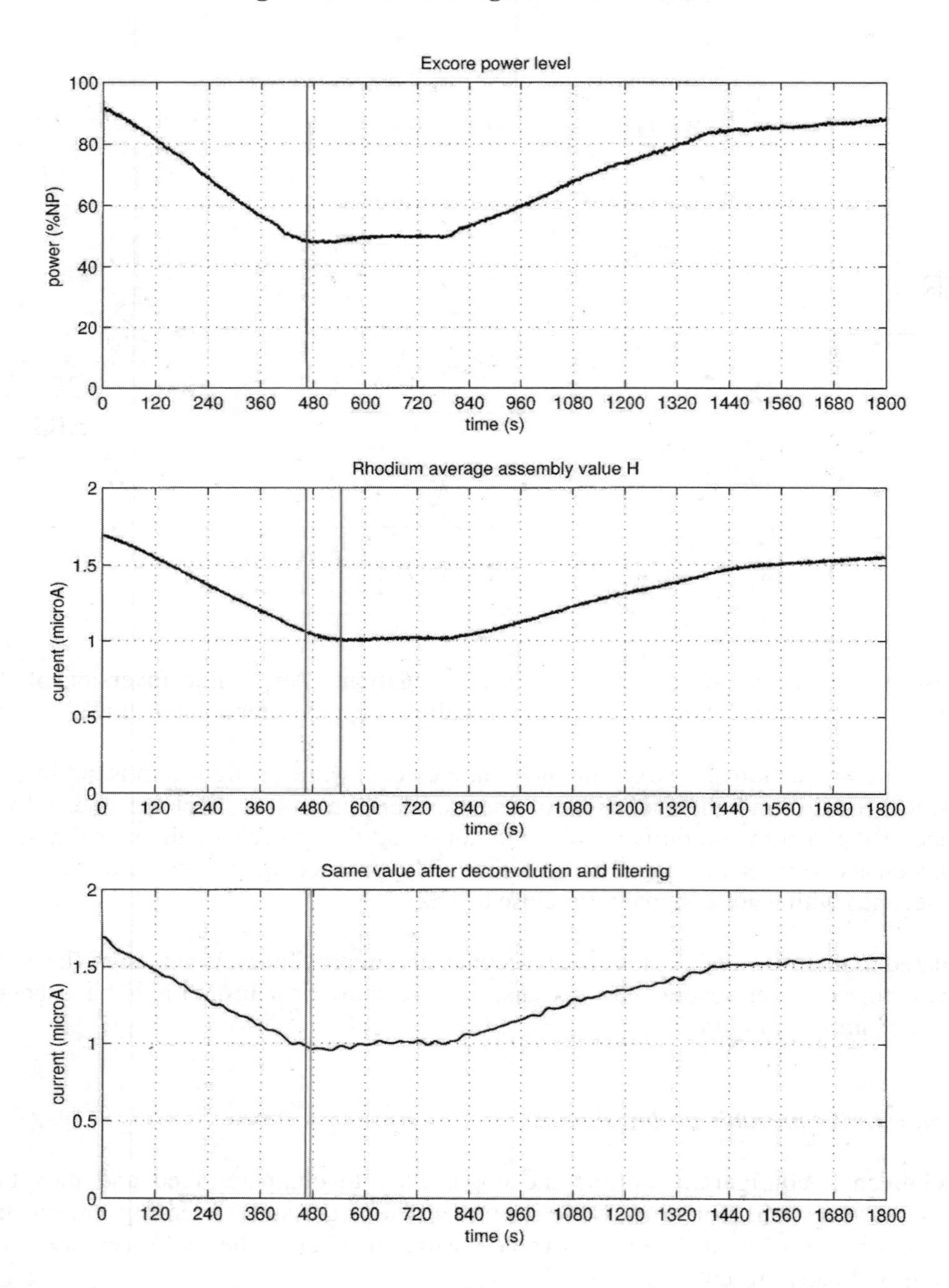

Two main steps are necessary in the rhodium signal processing before making such a comparison (we suppose that the processing of the movable fission chambers signals has been already done and is available).

The first step is the calculation of each rhodium detector sensitivity. This calculation needs to calculate the electrical charge they have released since the beginning of the experimentation (i.e. first neutron flux absorption) and to use the depletion relationship; the correction consists of taking into account the initial geometrical characteristics of each rhodium emitter, which must be done afterwards.

The second step is the calculation of the power to signal ratio. Nuclear design codes are necessary. By using a spectrum code, the absorption rate of the rhodium taking into account various effects like the self-shielding of the rhodium is calculated as a function of the spectrum of the assembly. FRAMATOME uses the APOLLO-2F code. A second calculation performed by means of a nodal code takes into account the local conditions. Thus, each rhodium detector is characterised by its own power/signal ratio. FRAMATOME uses the SMART nodal code; both codes make up the FRAMATOME nuclear design chain SCIENCE.

Figures 8 to 10 show a very good agreement between the rhodium responses and the reference axial power profile obtained during the flux mapping processing. This good comparison is a strong justification of the rhodium performances.

Figure 8. Comparison between rhodium signals and movable probes

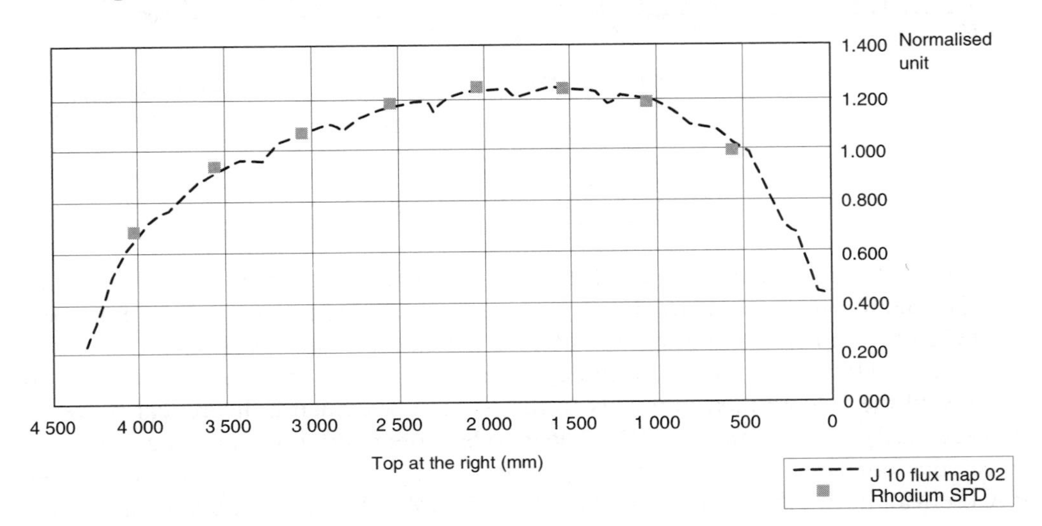

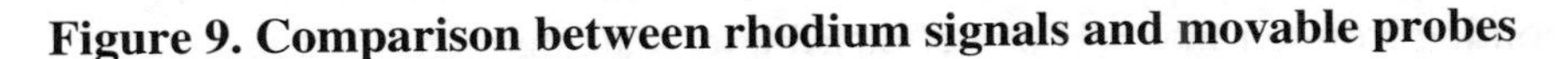
Figure 9. Comparison between rhodium signals and movable probes

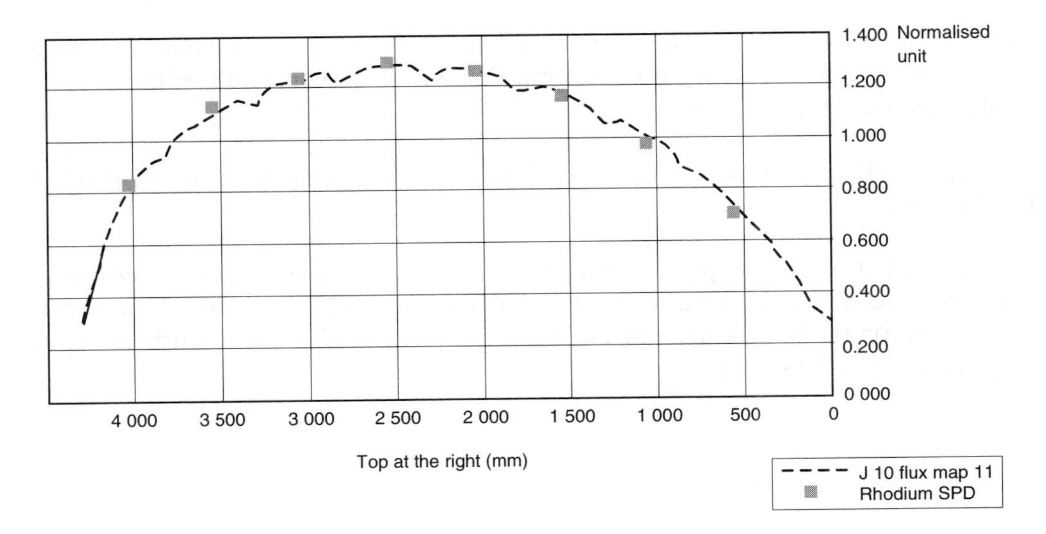

Figure 10. Comparison between rhodium signals and movable probes

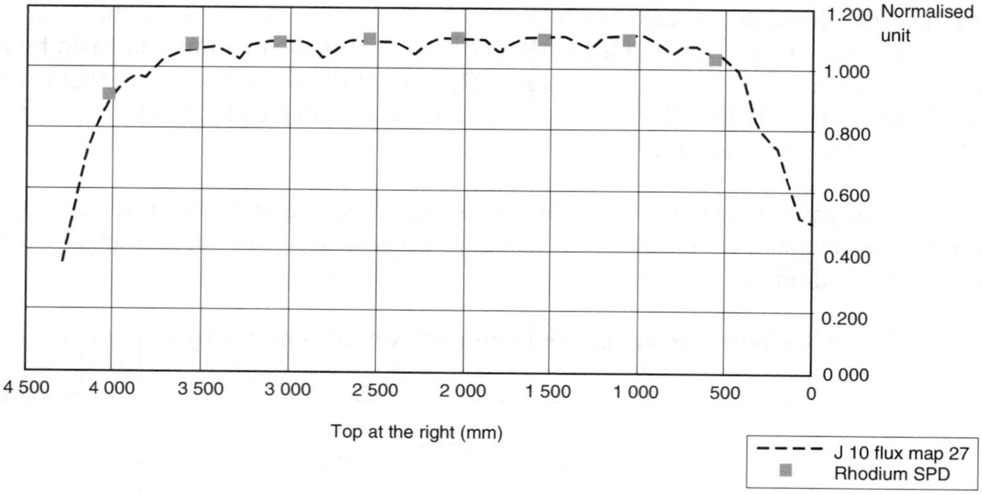

The future

The Golfech 2 experimentation was the first step of a more ambitious project: providing an advanced LCO monitoring system for Électricité de France. Preliminary calculations performed by FRAMATOME have shown that a minimum of 16 in-core strings are necessary to reconstruct the 3-D power distribution with an acceptable accuracy. In addition, a theoretical 3-D power distribution is needed to expend the in-core measurements to the whole core. Such a theoretical power distribution will come necessarily from an on-line 3-D nuclear code. Core operating margins are then calculated in a specific calculation. Thus, the monitoring equipment designed by FRAMATOME and known as the Partial In-Core Monitoring System will be based on the following two inputs: measurements coming from 16 in-core strings equipped with eight fixed sensors (rhodium SPDs) and a theoretical information coming from on-line 3-D power distribution calculations.

In the 1 300 MW, 4 L plant case, 42 fuel assemblies, which can be explored by the movable probes, remain. This number is enough to perform the normal flux mapping without any significant impact on the plant operation.

A prototype has been proposed to EDF (see Figure 11), which has recently decided to test it in a 1 300 MW 4 L plant.

The installation of the prototype is planned by mid 2002. A two fuel-cycle experiment is suitable to gain sufficient feedback experience before making the decision to install a Partial In-Core Monitoring System on the twenty units of the 1 300 MW series. The first industrial implementation could take place on 2006/2007.

Figure 11. In-core instrumentation (fixed and movable)

PIMS architecture

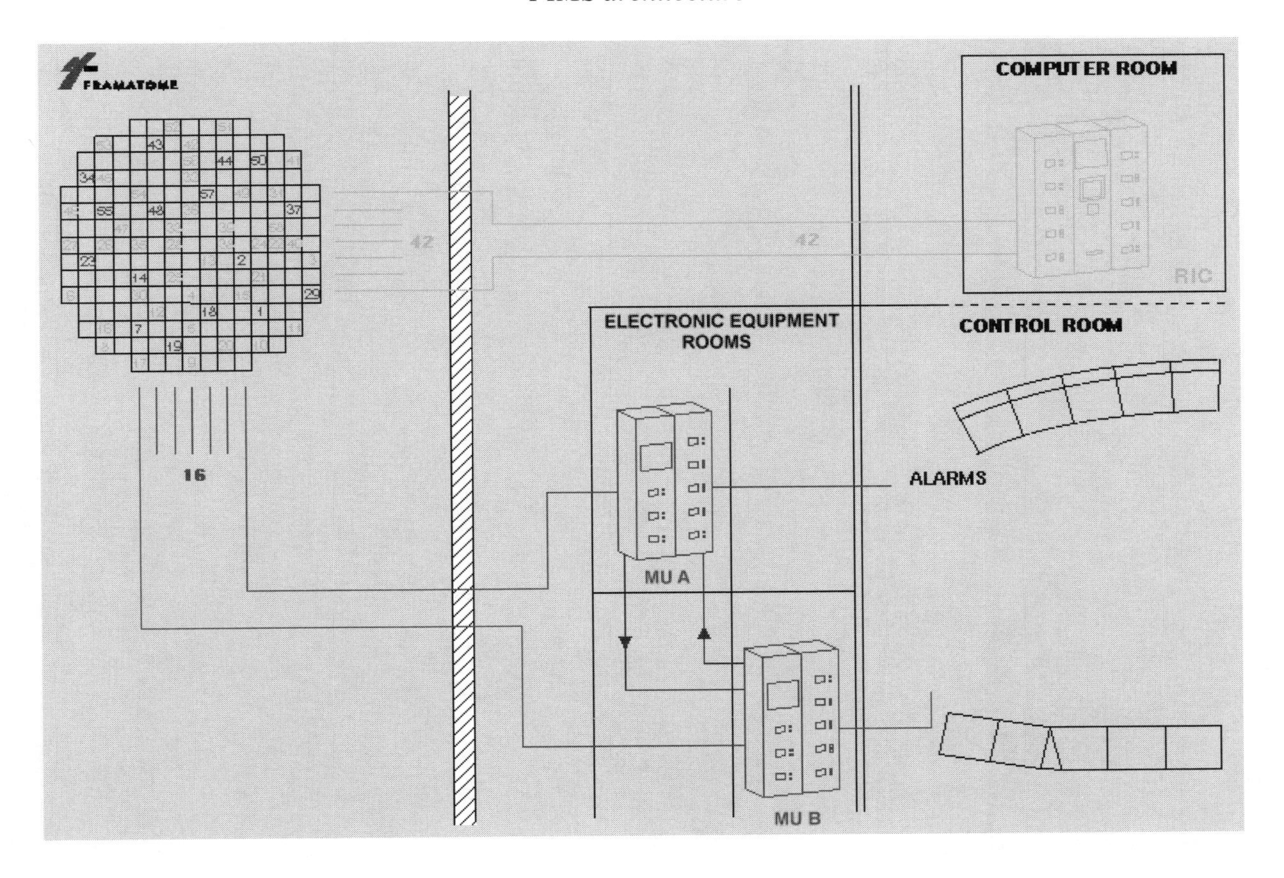

STATISTICAL ANALYSIS OF THE RATIO OF MEASURED AND PREDICTED Rh SPND SIGNALS IN A VVER-440/213 REACTOR

F. Adorján
KFKI Atomic Energy Research Institute
H-1525 Budapest P.O. Box 49
E-mail: adorjan@sunserv.kfki.hu

S. Patai Szabó, I. Pós
Paks Nuclear Power Plant
Hungary

Abstract

The goal of this work is to analyse the behaviour of the ratio of the measured and predicted Rh SPND in-core measurement signals in the environment of a VVER-440 reactor. The model we used for the detector signal prediction was based on the HELIOS transport code and the C-PORCA nodal core design code. A statistical sensitivity analysis versus some selected parameters (e.g. enrichment, burn-up) has been carried out by using a substantial amount of measured data. We also investigated the stability of the electron escape probability of the detectors versus their burn-up and other parameters with the aim of obtaining a tuned semi-empirical formula for the detector burn-up correction.

Introduction

One of the basic goals of the work has been to determine the fundamental characteristics of the signal of the Russian manufactured DPZ-1M type Rh SPND in-core detectors, used in the VVER-440/213 type reactors of Paks NPP, Hungary. The basic parameter to determine was the absolute sensitivity of the detectors and, naturally, their accuracy. To this end we used a great amount of measured SPND detector signals from four fuel cycles (Cycles 7-10) of Unit 4 of the Paks NPP. The reference data to compare with the measured detector signals were obtained by detailed reactor model calculations. It is well known that the calculation of the Rh SPND signals is rather complex, taking into account the adverse spectral shape of the absorption cross-section of the Rh. Nevertheless, the absolute sensitivity of the Rh detectors can be represented as the linear combination of the effective thermal and epithermal cross-sections of the detectors (taking into account its real geometry) and an average electron escape probability. There is also a known approach where two different electron escape probabilities are used for the electrons induced by thermal neutron absorption and for those induced by epithermal neutron absorption [5].

In our case, we applied the HELIOS/C-PORCA computational model [1,2] to predict the Rh reaction rates in the individual detectors, also taking into account their depletion in the computational model. Therefore the predicted detector reaction rates can be compared directly to the measured detector currents, i.e. the timely and the spatial behaviour of the ratio of the measured and predicted signals can be analysed.

Performing an appropriate statistical analysis on the detector signals, as well as on the ratios mentioned in the previous paragraph, we were able to evaluate both the quality of the measurements and the quality of the reaction rate prediction capability of the applied model. Unless the computational model contains some systematic error, the average of the ratios of the measured currents and the predicted reaction rates is nothing but the characteristic electron escape probability of the given detector type. This is true, however, only if this ratio has no systematic dependence on any of the basic parameters (e.g. the burn-up of the detectors, the neutron spectrum, etc.). Therefore, we had to investigate these dependencies, as well.

Method

Measurement handling

The complete set of measured data we used in this study consisted of snapshot files from approximately every 30 effective full-power days collected on Unit 4, during Cycles 7-10. We also used the cycle-long archive file (with 15 min. resolution) which was automatically collected by the VERONA core surveillance system [7] for Cycle 10. The individual axial position data of the detectors and the loading patterns of the cycles were also necessary auxiliary data for the evaluation.

The validity of the measured SPND data was re-evaluated. To this purpose we applied a special least squares fitting. Since the core contains 36 SPND strings, each consisting of seven detectors, we used the following approach. Let us denote the average measured axial distribution with a 7-element vector $\underline{\boldsymbol{v}}$ and define a 36-element vector $\underline{\boldsymbol{a}}$ as normalising factors for each of the SPND strings. We can obtain these two vectors by satisfying the following least squares condition:

$$Q=\sum_{j=1}^{7}\sum_{i=1}^{36} l_i d_{ij}\left(v_j a_i - I_{ij}\right)^2 = min \quad \text{and} \quad \frac{\sum_{i=1}^{36} l_i a_i}{\sum_{i=1}^{36} l_i} = 1$$

where $d_{ij} = 1$, whenever the measured current I_{ij} is valid, 0 otherwise; and $l_i = 1$, if the string i contains at least one valid detector, 0 otherwise. It can be shown that the above equation can be solved by fast converging iteration [3]. When the vectors $\underline{v}$ and $\underline{a}$ are determined, we can determine an average variance for each of the seven detector elevation layers as:

$$\sigma_j^2 = \frac{\sum_{i=1}^{36} l_i d_{ij}\left(v_j - I_{ij}/a_i\right)^2}{\sum_{i=1}^{36} l_i d_{ij} - 1}$$

and using that we flag a measurement I_{ij} as invalid if:

$$\max_{\{j\}}\left[\max_{\{i\}}\left(v_j - I_{ij}/a_i\right)/\sigma_j\right] > C$$

where C is a prescribed constant, corresponding to a sufficiently high confidence level according to the student distribution. Note that we applied such an algorithm that in one iteration step of this signal validation only a single detector having the highest student fraction – assuming it exceeded the above limit – was discarded, and this iteration was continued until no detector was found to discard. This iterative approach exhibits a much lower rate of judgement errors of both the first and second kind, compared to the more traditional method of marking all detectors as invalid, satisfying the above condition in a single step. In our case we used the value of C = 3.1, corresponding to a 0.99 confidence level.

Since this detector construction contains only a single lead wire compensation cable per string, the detector readings were corrected to get rid of their individual lead wire cable currents by using a spline technique. For the sake of signal validation all the detector readings were shifted to the nominal elevations of the detector layers, but the comparison of the real detector signals with the predicted detector reaction rates were carried out at the actual axial positions of the detectors.

To compare the real detector signals with predictions corresponding to fresh detectors, the integrated currents of each of the detectors according to the prediction model were used, since these were more accurate than the integrated measured currents. Note that this method needed the characteristic electron escape probability which is one of the results of this work.

Computational method

The calculation of the predicted reaction rate inside the detectors is based on a function describing the ratio of the local fast flux and the Rh reaction rate. To this end we needed two steps.

They are:

- To determine the local fast flux values at the assemblies and axial nodes where the Rh detectors reside using the C-PORCA code.

- To determine the reaction rate in the Rh wire at the centre of the assembly by using the HELIOS code; this should be accomplished according to the detailed 2-D geometry within the node.

On this basis we could obtain the ratio of the average fast flux in the node and the reaction rate of the detector at its centre. Based on these values this ratio can be parameterised according to some major specifications (e.g. burn-up, Xe, temperature, etc.) assuming linear and quadratic relation. The resulting constants were used later like the other group constants. The assembly model for the HELIOS code was analogous to that of the group constant calculations [6], except that the vicinity of the Rh detector was modelled in more detail and the Rh wire itself was divided into four concentric layers. Thus the Rh reaction rate can be obtained as:

$$r_C = \phi_1^{CP} \cdot S_f\left(E, B, C_B, T_m, T_f, w\right)$$

where r_C is the calculated Rh reaction rate (multiplied by the charge of a single electron), ϕ_1^{CP} is the nodal fast flux from the C-PORCA code, and $S_f(E,B,C_B,T_m,T_f,w)$ is the conversion function, being a function of the enrichment, burn-up, boric acid concentration, moderator temperature, fuel temperature and the local power rate. The detector current is related to the reaction rate as:

$$i_C = r_C \cdot \beta(G) \cdot \beta_x$$

where i_C is the calculated detector current, $\beta(G)$ is an empirical function describing the weak dependence of the electron escape probability as function of the integrated current or total delivered charge of the detector, and β_x is the characteristic electron escape probability of the detectors, which is assumed to be a constant.

To be more accurate, we need to take into account the detector depletion in the sensitivity function determination, as well. To this end, we assumed the generally applied empirical form of this dependence:

$$r_C = \phi_1^{CP} \cdot S_f\left(E, B, C_B, T_m, T_f, w\right) \cdot (1 - \eta \cdot G)^{\alpha}$$

where α and η are empirically fitted parameters which were determined by using the multi-layer HELIOS calculations. Thus, the non-linearity of this dependence comes purely from the self-shielding effects of the detector. By fitting the above form to our calculated results we obtained the:

$$\alpha = 0.55 \quad \text{and} \quad \eta = 0.00262[1/C]$$

values. Note that these parameters take into account only the change of the spectral self-shielding effect of detectors.

Having the necessary sensitivity functions, we performed a series of basically standard C-PORCA calculations from Cycle 4 of Unit 4 until Cycle 10, with 30 EFPDs burn-up steps. In these calculations in addition to the standard fuel burn-up evaluations we also evaluated the predicted detector currents and the detector depletions. At the cycle breaks we also took into account the appropriated SPND changes. As a result of these calculations we obtained the necessary predicted detector reaction rates for Cycles 7-10 and the results could be compared to the actually measured detector readings.

Statistical analysis

As stated above, the main parameter to investigate statistically was the ratio of the measured detector current and the predicted reaction rate of the detectors. We determined this parameter for each of the valid detector readings. The average of these ratios was also calculated over several specific subsets of detector, as well as the overall average. These average values were determined at every burn-up step, and they were also averaged over every refuelling cycle. The specific subsets of different detectors were defined as follows:

- The detector strings.
- The detector elevation levels.
- The axially central 3-5 detector layers.
- The detectors matching a spacer.
- The detectors between spacers.
- The detectors in assemblies with 3.6% and 2.4% enrichment, respectively.

Results

The investigation produced an extensive amount of results, but here we need to restrict ourselves to the most characteristic graphs and tables. The results in full detail are described in the final report of the investigation [6]. In Figure 1, one can see a typical distribution of the measured and predicted detector current ratios (M/C ratio). It is notable that the half width of the distribution for the central part of the core is not significantly smaller that the same parameter for the whole set of detectors.

One of the goals of the investigation was to seek dependencies of the measured/predicted ratio on some parameters. On the basis of the available data, we could only determine a systematic, though slight, dependence of this ratio on the burn-up of the surrounding fuel. No statistically significant dependence of the M/C ratio could be worked out versus the local power rate, the Xe concentration, the boric acid concentration and the detector burn-up. This suggests that the sensitivity function we used could not take into account this effect well enough. The typical dependence is illustrated in Figure 2. Note that we could not observe any systematic dependence of the M/C ratio on the detector burn-up, meaning that the major result of the non-linearity of this effect is due to the spectrally inhomogeneous self-shielding effect, which we accurately took into account through the HELIOS calculations.

Since the average M/C ratio is equal to the characteristic electron escape probability of the given detector type, it is of major importance. Table 1 summarises these results for each of the refuelling cycles and for the whole set of data. The detailed results of the dependency on the fuel burn-up are presented in Table 2.

Figure 1. The distribution of the measured/predicted detector currents (M/C ratios) for Unit 4, Cycle 10

The continuous line shows the distribution for all of the detectors, while the bars drawn with dashed lines for the detectors of layers 3-5

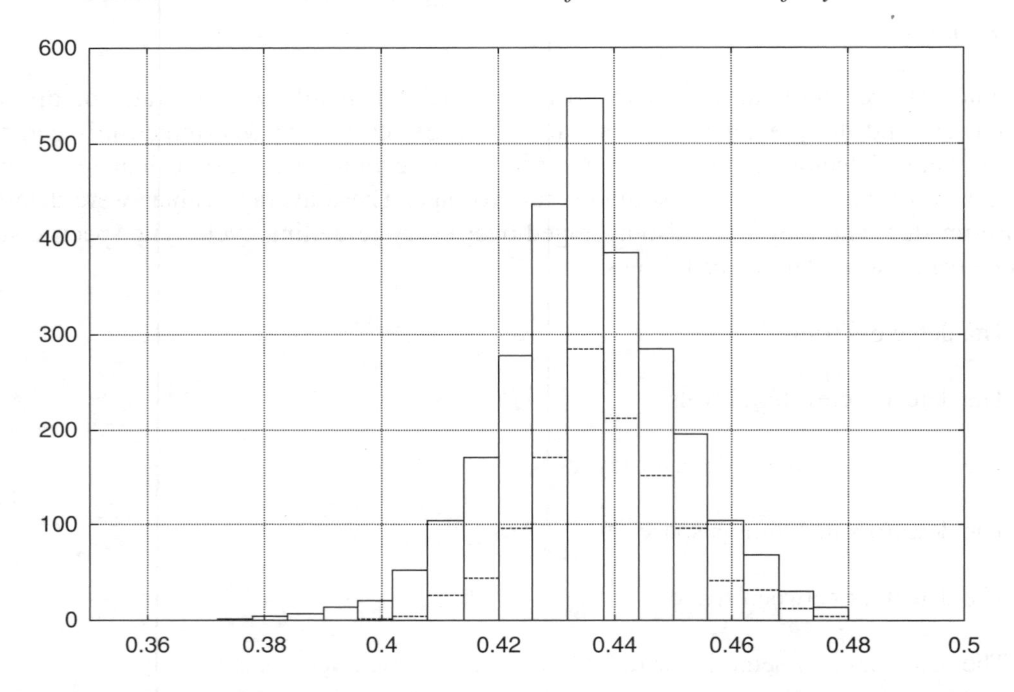

Figure 2. The M/C ratios as the function of the burn-up of the surrounding fuel

Unit 4, Cycle 9

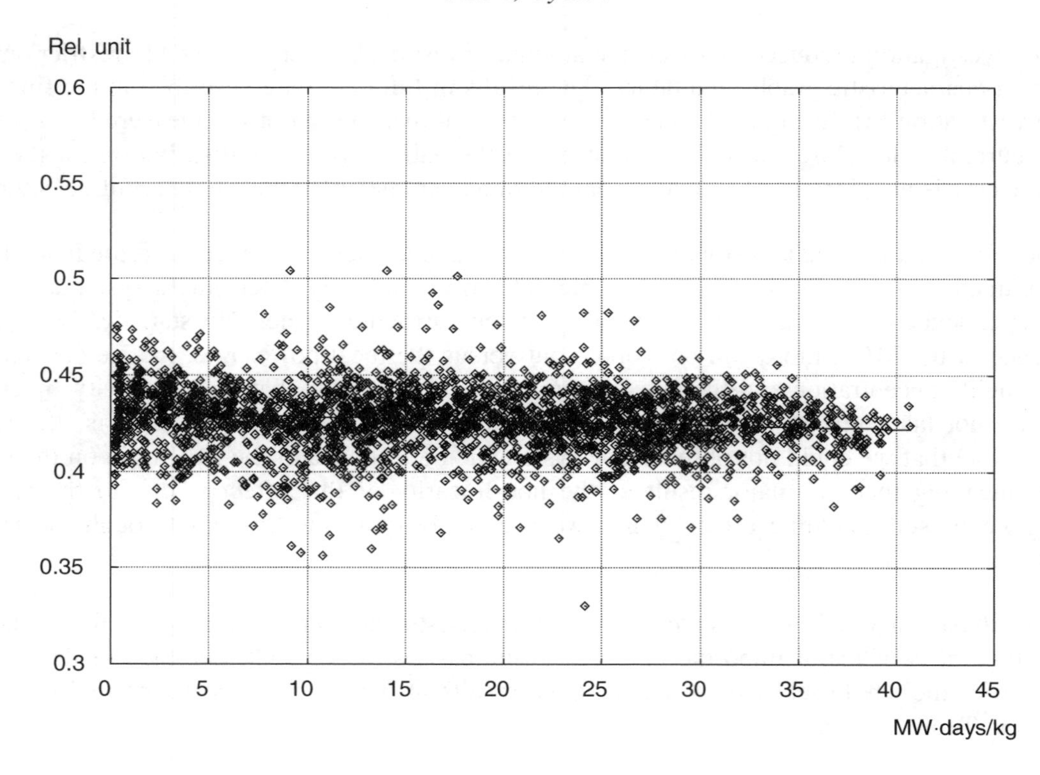

Table 1. Estimation of the electron escape probability

In the overall evaluation for Cycle 10 we used the VERONA archive data

Cycle	Average M/C = β	Error of β [%]	The variance of the M/C ratios [%]
7	0.4191	0.09	4.26
8	0.4258	0.10	4.83
9	0.4267	0.08	3.75
10	0.4319	0.08	4.10
10/archive	0.4354	0.07	3.44
Overall	**0.4273**	**0.045**	**4.45**

Table 2. Regression results of the fuel burn-up dependence of the M/C ratios

Cycle	Slope of burn-up dependence [kg/MW/days]	Zero burn-up value
7	-3.224e-4 ± 10.6%	0.4249 ± 0.17%
8	-3.334e-4 ± 12.0%	0.4318 ± 0.19%
9	-2.067e-4 ± 15.0%	0.4303 ± 0.15%
10	-2.620e-4 ± 13.0%	0.4366 ± 0.16%
10/archive	-2.036e-4 ± 14.0%	0.4391 ± 0.13%
Common regression	**-2.566E-4 ± 6.90%**	**0.4317 ± 0.08%**

Note that the zero burn-up value obtained from the common regression can be regarded as the most accurately evaluated characteristic electron escape probability of the given type of detectors. Thus, our evaluation has shown that the average electron escape probability value for the Russian manufactured DPZ-1M type Rh SPND detector is 0.4317 ± 0.0004, where the error means the single σ value. The sensitivity variance among the individual detectors is, however, 4.5%.

Summary

The most important result of the work is that the absolute sensitivity (the electron escape probability) of the Rh SPN detectors can be determined with reasonably high accuracy through this method. It is also important that this sensitivity does not depend significantly on the detector depletion in excess of the semi-empirical formula.

This analysis has also proven that by using the characteristic electron escape probability value and the predicted reaction rate, the delivered charge of detectors can be determined much better and easier from the calculations than from the measured currents.

REFERENCES

[1] J.J. Casal, R.J.J. Stamm'ler, E.A. Villarino, A.A. Ferri, "HELIOS: Geometric Capabilities of a New Fuel-Assembly Program", International Topical Meeting, Pittsburgh, 1991.

[2] I. Pós, "C-PORCA 4.0 Version Description and Validation Results", The Sixth Symposium of AER, Kirkkonummi, Finland, 23-26 September 1996.

[3] Z. Szatmáry, "Data Evaluation Problems in Reactor Physics, Theory of Program RFIT", KFKI-1977-43.

[4] F. Adorján, J. Végh, "Uncertainty Analysis of the Input Data of the VERONA-u System", a study ordered by Paks NPP (in Hungarian), 1996.

[5] W.A. Boyd, R.W. Miller, "The BEACON On-Line Core Monitoring System, Functional Upgrades and Applications", OECD/NEASC INCORE-96 Specialists Meeting on In-Core Instrumentation and Reactor Core Assessment, Mito, Japan, 1996.

[6] F. Adorján, "Analysis to Determine the Absolute Sensitivity of the Rh SPNDs of the Paks NPP", a study for Paks NPP (in Hungarian), 1998.

[7] F. Adorján, L. Bürger, I. Lux, J. Végh, Z. Kálya, I. Hamvas, "The Extended On-Line Core Monitoring Technology with the Latest VERONA-u Version", Proceedings of the IAEA OECD/NEANSC INCORE-96 Specialists Meeting on In-Core Instrumentation and Reactor Core Assessment, Mito, Japan, 1996.

THE STUDY ON THE BWR IN-CORE DETECTOR RESPONSE CALCULATION

Akihiro Fukao, Etsuro Saji
In-Core Fuel Management System Dept.
Toden Software, Inc. (TSI)
Japan

Abstract

The on-line core monitoring system for BWRs utilises the discrepancy between measured and calculated in-core detector responses for the purpose of predicting a "measured" power distribution. Hence, the calculated in-core detector response affects the on-line "measured" power distribution, and it is important for the on-line core monitoring system to produce calculated in-core detector response accurately. There are some BWRs which utilise thermal neutron detectors for the in-core detector system (i.e. TIP and LPRM). Thermal neutron flux values at the narrow-narrow corner are sensitive to various parameters, which are difficult to model correctly in the detector response calculation, for example channel box bowing or LPRM tube geometry. Considering these effects in the on-line core monitoring analysis, we were able to obtain sufficient level of accuracy in the monitored power distribution, which was verified by comparing with bundle-wise gamma scan measurement data.

Introduction

We have developed a BWR on-line core monitoring system based on CASMO/SIMULATE. This system has adopted the adaptive method [1], which corrects the calculated power distribution from an on-line core simulator so that the calculated in-core detector response matches the measured response. The validity of this method relies on the accuracy of the correlation between detector response and bundle power as well as reliable measurement.

There are some BWRs which use thermal neutron TIP and LPRM detectors for the in-core detector system. Thermal neutron flux values at the narrow-narrow corner are sensitive to various parameters which are difficult to model correctly in the detector response calculation, such as channel bowing, in-channel void distribution, water film on inner channel wall or LPRM tube geometry. Normally, these effects are not considered in the no-adaptive calculation. In the adaptive calculation, however, they may have to be considered.

The detector response calculation methodology of CASMO/SIMULATE is briefly described in the following section.

In our adaptive method, the ratio of calculated and measured detector response is directly applied to correct the calculated nodal power. In order to verify this method, bundle gamma scan measurement data was utilised to obtain the actual power distribution in an operating BWR plant loaded with modern fuel designs. By comparing the bundle-wise adapted axial power shapes adjacent to detector locations with the measured ones, we found some discrepancies between them. An example is shown in Figure 1.

Figure 1. Comparison of the ratio of measurement and calculation between power and TIP

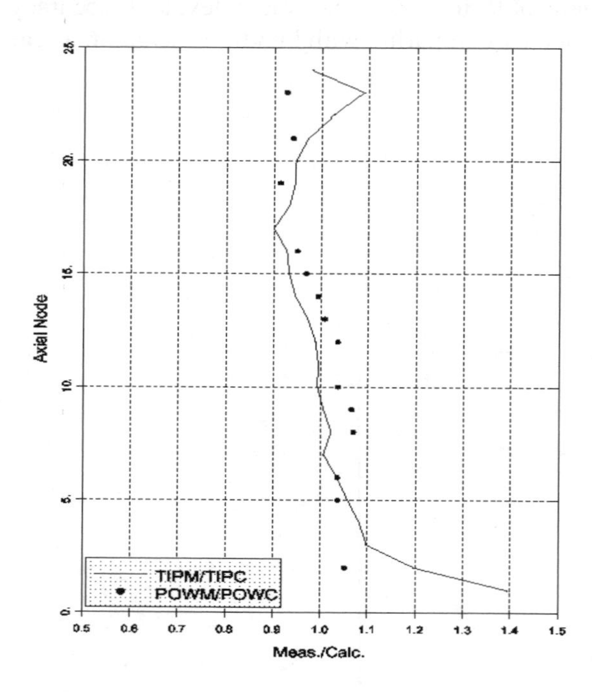

This figure shows the comparison between the ratio of measurement to calculation of the four-bundle power and the TIP surrounded by this four-bundle at the end of cycle just before the gamma scan measurement. The power distribution is the bundle-averaged power of the four-bundle surrounding detector. If these shapes were quite similar, the accuracy of power distribution would be improved with the adaptive calculation. Figure 1 is not exact, however, and there was not much improvement by using adaptive calculation.

As a result, we began to study the BWR in-core detector response calculation. We performed some sensitivity studies pertaining to the detector response calculation. For example, we modelled water film on the inner channel wall in the CASMO transport calculation. Some effect could be confirmed on the detector response when the thickness of water film is only 1 mm on the inner channel wall. However, it is difficult to estimate the thickness of water film in an actual core. The same thing can be said for in-channel void distribution.

In this paper, we introduce two sensitivity cases from our detector calculation study. One case examines the effect of modelling the LPRM tube geometry in CASMO. The other case looks at the channel box bowing effect. The adaptive power distributions considering these effects are then compared to gamma scan measurement data.

Methodology

Detector response calculation

The detector response calculations are carried out by CASMO transport calculation and SIMULATE advanced nodal diffusion calculation. Once SIMULATE calculates three-dimensional power distribution, detector responses on each detector location are calculated as a part of the pin power reconstruction model [2], which utilises the intra-nodal homogeneous flux shape by SIMULATE and the corner flux provided by the CASMO heterogeneous calculation. The geometry of four bundles surrounding a detector which is used for this method is shown in Figure 2.

Figure 2. Geometry of four bundles surrounding a detector

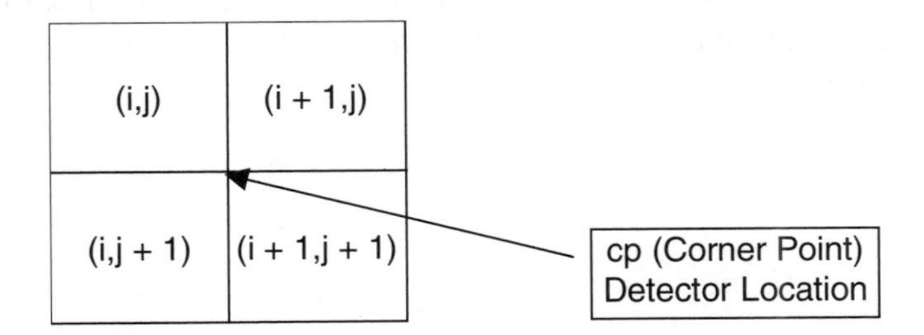

Heterogeneous corner point fluxes at the detector location are continuous in an actual core. SIMULATE also requires the same condition for pin power reconstruction. Heterogeneous corner point fluxes in SIMULATE are computed by homogeneous corner point fluxes and a correlation of homogeneous and heterogeneous fluxes at the corner point. This is called the "form function" and is determined by the CASMO heterogeneous calculation. The corner point fluxes are determined by averaging the four estimates of the heterogeneous corner point flux in each node as follows:

$$\phi_g^{het^{cp}} = \frac{1}{4}\left\{\phi_{g,i,j}^{hom\,cp} \cdot F_{g,i,j}^{cp} + \phi_{g,i+1,j}^{hom\,cp} \cdot F_{g,i+1,j}^{cp} + \phi_{g,i,j+1}^{hom\,cp} \cdot F_{g,i,j+1}^{cp} + \phi_{g,i+1,j+1}^{hom\,cp} \cdot F_{g,i+1,j+1}^{cp}\right\}$$

where $F = \dfrac{\phi_g^{het^{cp}}}{\phi_g^{hom^{cp}}}$ and F is form function (obtained by CASMO).

Thus, the SIMULATE corner point fluxes reflect not only the intra-nodal homogeneous flux but also the heterogeneous flux used for pin power reconstruction.

The reaction rates in the detectors are calculated from the reconstructed corner point fluxes in SIMULATE and the microscopic cross-sections for the detectors. In our case, we use thermal neutron detectors for the in-core detector system. The microscopic cross-sections of ^{235}U are applied in the detector response calculation.

$$R.R.^{\det} = \sum_{g} \sigma_g^{\det} \cdot \phi_g^{het^{cp}}$$

Adaptive calculation

Our adaptive method assumes that nodal power can be directly adapted by using the ratio of measured and calculated detector response, hereafter called the adaptive parameter. In other words, we assume that the adaptive parameter is equal to the ratio of actual nodal power and calculated nodal power.

$$\frac{R.R.^{\det}_{measured}}{R.R.^{\det}_{calculated}} = \frac{Power_{measured}}{Power_{calculated}}$$

Therefore:

$$Power_{measured} = \frac{R.R.^{\det}_{measured}}{R.R.^{\det}_{calculated}} \times Power_{calculated}$$

The adaptive parameters are normalised so that their average in each string is equal to unity. This is because the thermal neutron detector response provides local power information in the core and not global power information.

Result of study

Sensitivity study

Modelling of the LPRM tube

The LPRM string is composed of an inner tube which contains a TIP tube and is back-filled with a gas. This TIP tube and four LPRM detectors, which are located at axially fixed positions, are contained within a larger tube and filled with water. Both tubes are composed of stainless steel. The geometry of narrow-narrow water gap and LPRM string and LPRM tube modelling are shown in Figure 3.

For the actual LPRM tube geometry, a TIP tube is not concentric within an LPRM tube and there are cable tubes which exit from below the LPRM detectors. It is difficult to model this exact geometry in single bundle calculation. Thus we adopted a concentric annular model, which consists of air, stainless steel and water layers. The detector response calculation is performed based on this model using the characteristic transport method in CASMO-4.

Figure 3. Geometry of narrow-narrow corner and LPRM tube modelling

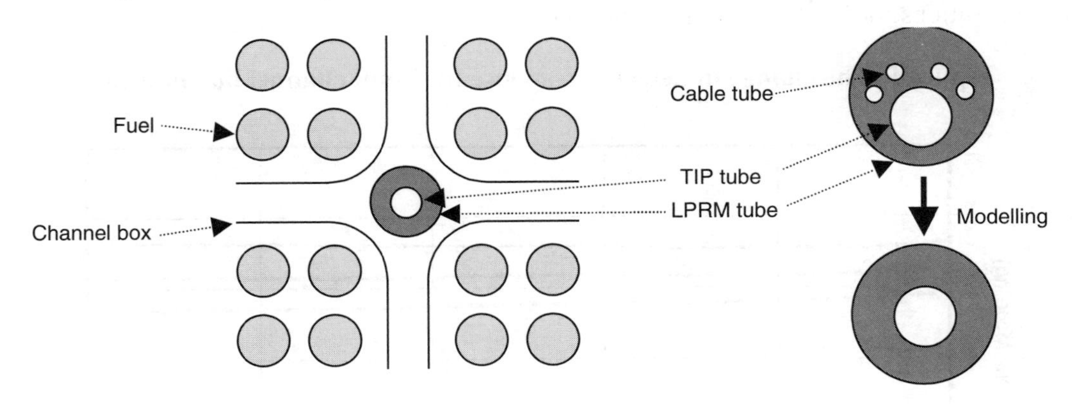

The effect of modelling the LPRM tube for each void fraction is shown in Figure 4.

Figure 4. The change in detector response by modelling the LPRM tube

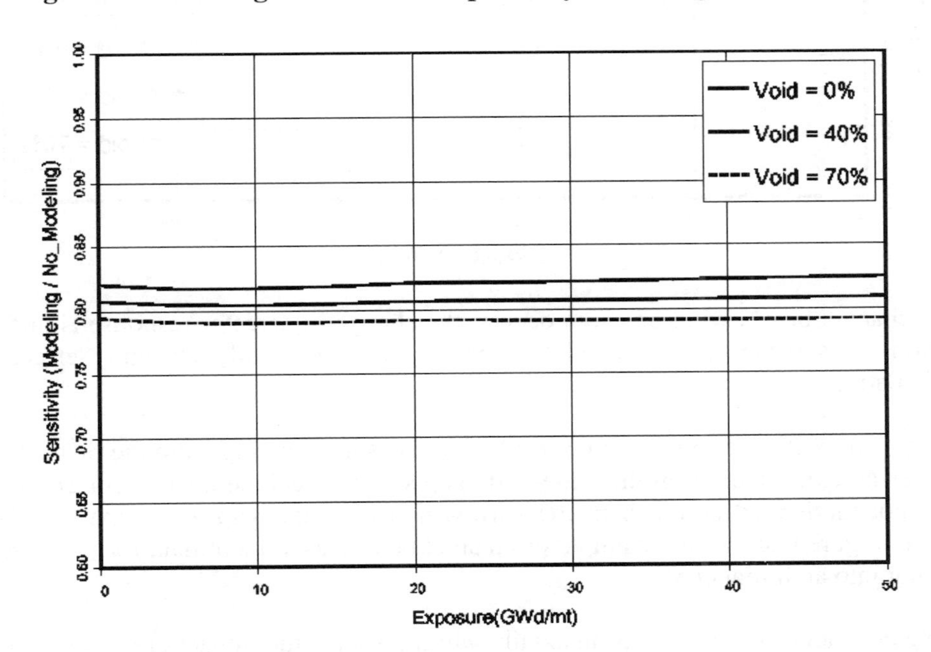

This figure shows the change in detector response by modelling the LPRM tube. When the LPRM tube is modelled, the change in detector response is different at each void condition. This means that LPRM tube modelling has axially changing effect on detector response calculation. It may have some effect on the adaptive calculation.

Channel box bowing effect

If a channel box is axially bent, an occurrence which is also known as channel box bowing, the distance between the detector and the corner pin changes from nominal value. In other words, the outer water gap increases or decreases when channel box bowing occurs. It may affect the neutron slowing down phenomenon in the narrow-narrow water gap and the thermal neutron detector response may be changed compared to nominal geometry.

The sensitivity of a 1 mm channel box bowing effect, which is calculated by CASMO assuming three void conditions for S-lattice, is shown in Figure 5.

Figure 5. The change in detector response by 1 mm channel box bowing

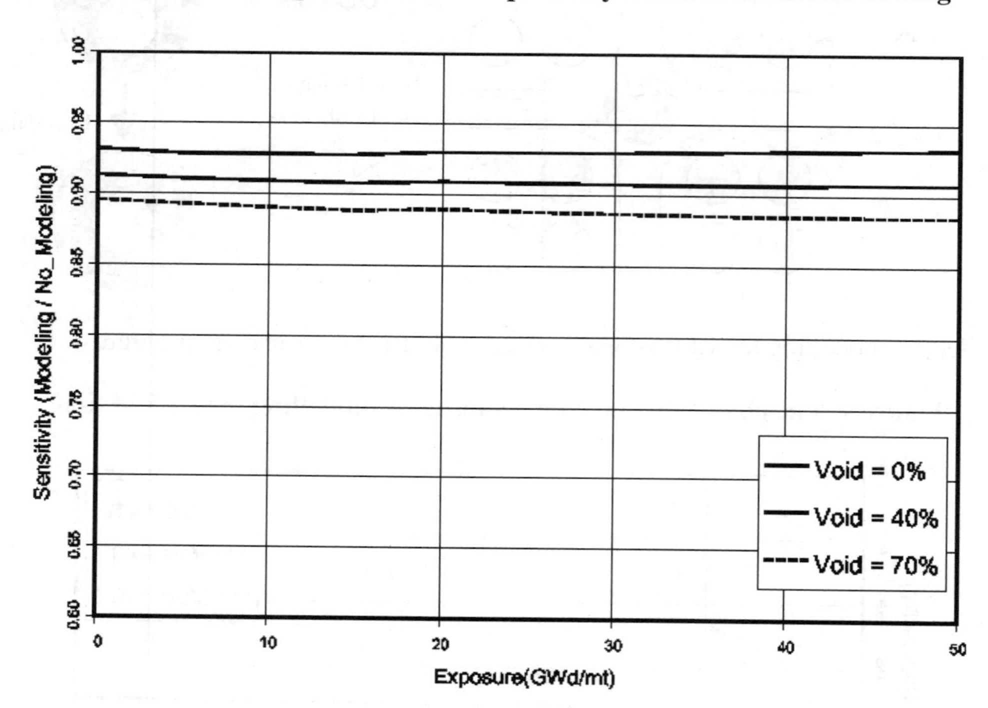

When a channel box bowing of 1 mm occurs, the change in detector response is different at each void condition. This means that channel box bowing has an axially changing effect on detector response calculation.

Normally, a new channel box with no irradiation has some bowing from the manufacture process which provides an extra margin on the wide-wide gap side to preclude a stuck control rod. Thus, fast neutrons are not moderated as much in the narrow-narrow water gap as compared to the nominal geometry case. Figure 6 shows an example of channel box measurement data for pre-bowed channels prior to loading into an initial core.

Since the channel box bowing has an axially varying shape, the impact of channel box bowing on detector response varies with axial elevation. Thus, it may also have some effect on the adaptive calculation.

In our system, it has recently become possible to model the channel box bowing effect using CASMO and SIMULATE. The sensitivity of the outer water gap thickness to the detector response is calculated for each bundle surface by CASMO. A sensitivity table of delta water gap is then generated for use in the SIMULATE channel box bowing calculation. SIMULATE can have three-dimensional channel box bowing distribution. The detector response is obtained using this sensitivity table of delta water gap and the three-dimensional channel box bowing distribution.

SIMULATE has a function to evaluate the growth of channel box bowing by irradiation. In this paper, however, only initial channel box bowing distribution without irradiation was considered.

Figure 6. Plant-A oriented channel bowing measurement data (normalised so that maximum bowing equal to unity)

negative side = control rod, positive side = detector

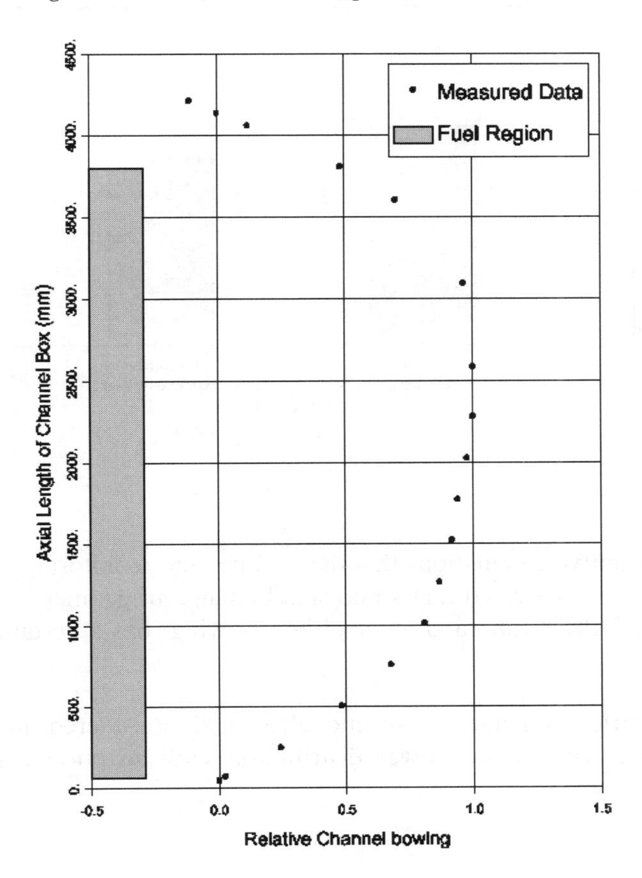

Gamma scan evaluation

From the sensitivity studies of the detector response calculation, we found that there might be some effect on the calculated axial detector response distribution by LPRM tube modelling and channel box bowing. As the adaptive parameter changes in proportion to the change in the calculated detector response, it affects "measured" power distribution. Thus, bundle gamma scan measurement data, which provide the actual power distribution, can be used to confirm these sensitivities in our adaptive calculation model.

We performed a comparison of power distribution accuracy for these four cases: without adaptation, adaptation without any effects, adaptation with LPRM tube modelling and adaptation with two effects. The root mean square (RMS) error of averaged axial power distribution comparison is shown in Figure 7.

This figure shows the RMS errors of averaged axial power distribution which are normalised so that "no adaptation" error is equal to unity. The RMS error of axial power distribution is certainly reduced by adopting LPRM tube modelling and channel box bowing. Consequently, on-line core monitoring accuracy of power distribution can be improved by adopting LPRM tube modelling and channel box bowing in the adaptive calculation.

Figure 7. RMS error of averaged axial power distribution (normalised so that "No Adaptation" error is equal to unity)

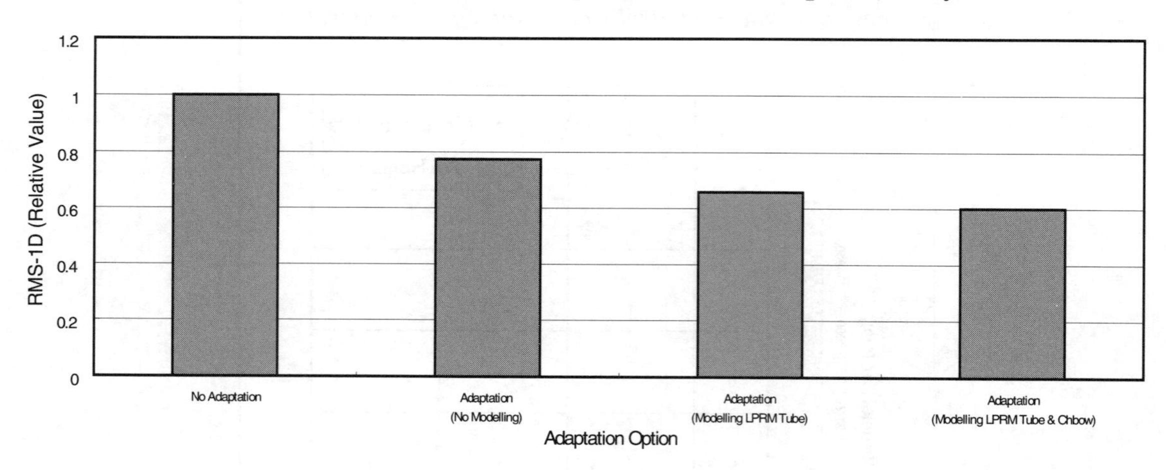

Conclusion

It is important for adaptive calculations that the on-line core monitoring system compute detector responses accurately. It was confirmed that some small change in geometrical configuration, such as explicit modelling of the LPRM tube and channel box bowing, has a certain impact on the neutron detector calculation.

When these geometrical changes were modelled and considered in the detector response calculation, sufficient accuracy in the power distribution with adaption could be attained for the on-line core monitoring system.

REFERENCES

[1] C.M. Kang and R.L. Crowther, "Adaptive Method for Three-Dimensional Boiling Water Reactor Simulation," *Trans. Am. Nucl. Soc.* 38, 340 (1981).

[2] K.R. Rempe, *et al.*, "SIMULATE-3 Pin Power Reconstruction: Methodology and Benchmarking," *Nucl. Sci. Eng.* 103, 334 (1989).

DEMONSTRATION OF OPTICAL IN-CORE MONITORING SYSTEM FOR ADVANCED NUCLEAR POWER REACTORS

T. Kakuta, K. Suzuki, H. Yamagishi, H. Itoh
Department of Nuclear Energy System, Japan Atomic Energy Research Institute
2-4 Shirakata-Shirane, Tokai-mura, Ibaraki-ken, 319-1195, Japan

M. Urakami
Research and Development Department, The Japan Atomic Power Company
Ohtemachi Bldg., 6-1, 1-chome, Ohtemachi, Chiyoda-ku, Tokyo, 100-0004, Japan

Abstract

A new kind of optical in-core monitoring system such as the power monitor and gamma ray thermometer using radio luminescence and thermal radiation due to gamma ray in optical fibres is described. Optical fibres retained their transmission characteristics, with strong radio luminescence wavelength ranging from 400 nm to 1 400 nm. Some peaks and thermal radiation in infrared were observed during heavy irradiation in a core region of the fission reactor. Intensities of radio luminescence and thermal radiation were found to be directly proportional to the reactor power. Also, temperatures could be estimated by using thermal radiation according to Planck's law. The optical measuring method using optical fibres could be composed simply and conveniently, serving as an in-core monitoring system of the reactor power and temperature.

Introduction

The application of optical fibres to instrumentation systems is a strong concern with regard to advanced nuclear power reactors which is currently under development at the Japan Atomic Energy Research Institute (JAERI). Up to now, the optical transmission characteristics of optical fibres were thought to be vulnerable to radiation. However, recently developed optical fibres were found to have good radiation resistance even in a reactor irradiated up to a neutron fluence of 10^{24}n/m^2 and gamma-ray doses of larger than 10^9 Gy [1].

This paper discusses the dynamic optical characteristics of optical fibres during heavy irradiation and their application to in-core monitoring systems as a power and temperature monitor. Moreover, future plans for core monitoring systems of advanced nuclear power reactors at JAERI are described.

Radiation resistant optical fibres

The most serious problem related to optical fibres in the radiation fields is the increase in signal transmission loss due to radiation-induced defects and formation of colour centres. It is known that radiation with high-energy particles causes ionisation and atomic displacements within the molecular bonding network of SiO_2 glass. The creation of defects and formation of colour centres on optical fibres, so called radiation-resistance, depends upon the structure of the glass, which is determined by changes in the refractive index of the core and cladding with dopants. In previous investigations for the development of radiation resistant optical fibres, it has been found that the pure silica (SiO_2) core optical fibre has better radiation resistance as compared with GeO_2 and/or P_2O_5 doped core fibres [2,3]. The structure of SiO_2 glass, which consists of common stable bonding, is much simpler than that of doped core glass. Further findings suggest that the content of oxyhydrate (OH) or fluorine to the SiO_2 core also improves the radiation content of OH or fluorine to the core, reducing the formation of E′-centre and non-bridging oxygen hole centres (NBOHC) in the SiO_2 core glass. Figure 1 shows some methods for improving the radiation resistance of SiO_2 core glass [1].

Figure 1. Some methods for improving the radiation resistance of SiO_2 glass

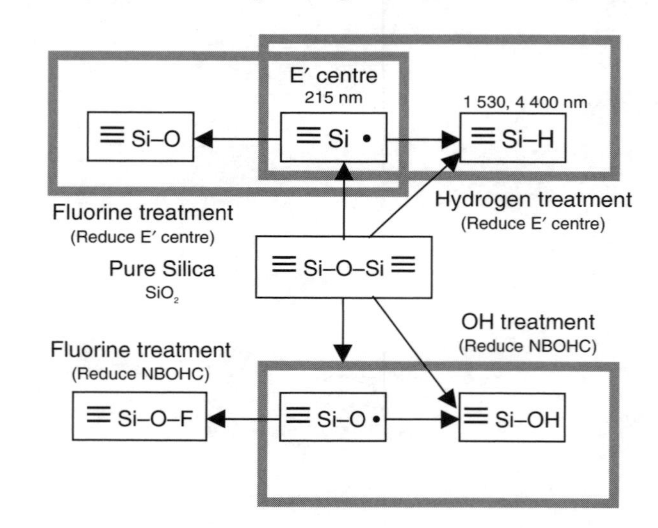

Of the two types of radiation resistant optical fibres developed, the SiO_2 with fluorine contained core fibre having a diameter of 0.2/0.25 mm, and the OH contained an SiO_2 core with carbon coated fibre having a diameter of 0.8/1.0 mm (see Table 1).

Table 1. Example of radiation resistant optical fibres

Type of Fibres	Core		Cladding	
	Composition	Diameter	Composition	Diameter
Step index OH contained	SiO_2 OH: 800 ppm	0.8 mm	SiO_2-F F: 4.0 wt%	1.0 mm*
Step index Fluorine contained	SiO_2-F F: 1.6 wt%	0.2 mm	SiO_2-F F: 5.6 wt%	0.25 mm

* With carbon coating

An example of radiation resistance, the radiation-induced optical absorption of fluorine contained core fibre, is shown in Figure 2. The core optical fibre containing fluorine kept good optical transmission characteristics during heavy irradiation up to neutron fluence of 10^{24} n/m^2 and gamma ray doses of larger than 10^9 Gy. Core optical fibre containing OH with carbon coating has good heat resistance up to a temperature of 1 100 K.

Figure 2. Example of radiation resistance

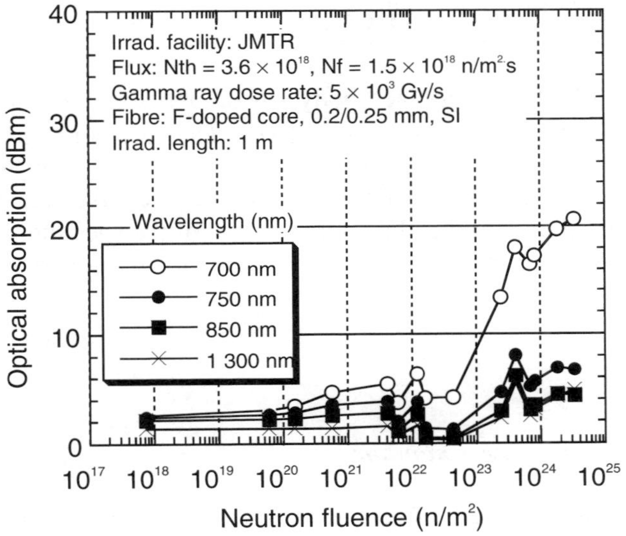

Experimental details

The measuring procedure for irradiation experiments is illustrated in Figure 3. Experiments were carried out in the Japan Material Testing Reactor (JMTR) at JAERI. An irradiation capsule in the core region of was used to measure the radio luminescence and thermal radiation from optical fibres. The fast ($E > 1$ MeV) and the thermal ($E < 0.678$ eV) neutron fluxes are 1.5×10^{18} n/m^2·s and 3.6×10^{18} n/m^2·s, respectively, and the gamma ray dose rate is 5.0 W/g (5.0×10^3 Gy/s) at the full reactor power of 50 MW. The radio luminescence and thermal radiation from optical fibres were measured with optical spectrum analyser AQ-6315. The optical spectrum analyser AQ-6315, whose wavelength range from 350 nm to 1 750 nm measured the optical power density and its sensitivity, which is about -90 dBm.

The SiO_2 with core optical fibre containing fluorine, whose diameter is 0.2/0.25 mm, was used to measure the radio luminescence from the optical fibre itself.

Figure 3. Experimental procedures for in-core monitoring

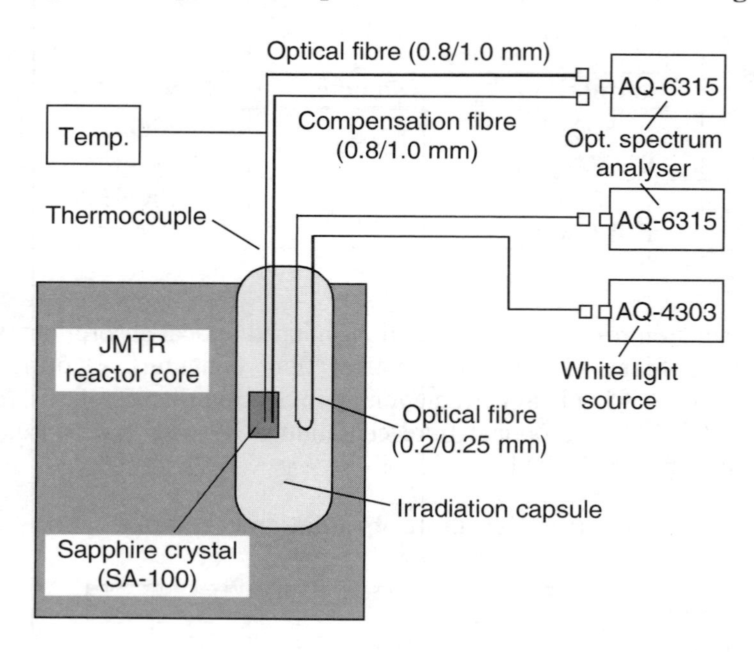

The sapphire crystal SA-100 has a black body whose purity is 99.99%, and detects the thermal radiation by gamma ray heating at high reactor power. The OH contained SiO_2 core with carbon coated fibre, whose diameter is 0.8/1.0 mm, was used for transmission of thermal radiation from the sapphire crystal. The OH contained optical fibre and the thermocouple was inserted into the sapphire crystal to measure the thermal radiation and the temperature of the specimen. To compensate the radio luminescence from the optical fibre itself, the same type of OH contained fibre without sapphire crystal was accommodated in the capsule.

Results and discussions

Radio luminescence for in-core power monitor

One of the more significant optical effects to be observed in heavy radiation fields, namely radio luminescence in an optical fibre through a fluorescence process, was observed [6,7,8]. Figure 4 shows an example of the spectral radio luminescence in the SiO_2 with core optical fibre containing fluorine itself. The radio luminescence, which ranged from 400 nm to 1 400 nm, with sharp peak intensity at about 1 270 nm [1] was observed. A broadband radio luminescence peak at 450 nm was also observed [7,8]. Distribution of radio luminescence is approximately inversely proportional to the cube of the wavelength in the range from 700 nm to 1 400 nm. The distribution of radio luminescence in the wavelength range from 700 nm to 1 400 nm is thought to be due to Cerenkov radiation [9]. The relation of reactor power and optical intensity of the 1 270 nm peak is shown in Figure 5. The optical intensity is directly proportional to the reactor power. Also, some of observed radio luminescence peaks such as that at 450 nm in the SiO_2 core of optical fibres, and Cerenkov radiation, had intensities proportional to the reactor power.

During heavy irradiation, the absorption of fluorine-containing radiation resistant optical fibre was quite low in the wavelength range from 400 nm to 1 400 nm, except the NBOHC absorption whose wavelength centred at 630 nm. The radio luminescence is not affected by the absorption of optical fibre itself.

Figure 4. Observed radio luminescence in fluorine contained optical fibre

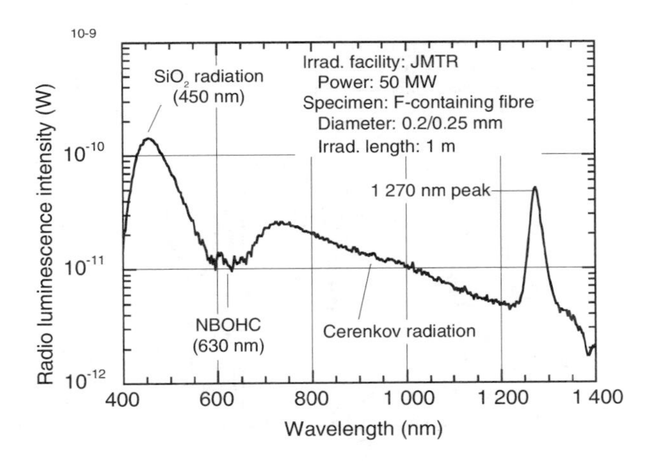

Figure 5. Relation of reactor power and optical intensity

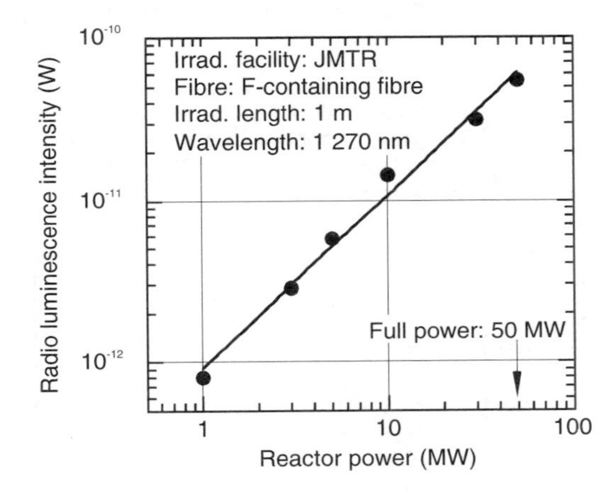

The result of radio luminescence measurement demonstrates the possibility of using the radiation resistant optical fibre as a new optical sensing for in-core instrumentation system such as in-core power monitoring and radiation monitoring.

Thermal radiation for temperature monitor

The temperature of sapphire crystal due to the gamma ray heating dependent on reactor power was measured by a thermal radiation. With the increase of the reactor power, the radio luminescence peak from SiO_2 at a wavelength of 450 nm and thermal radiation from sapphire crystal in the infrared region were observed. Observed luminescence from SiO_2 fibre and sapphire crystal at the high reactor power of 10 MW and 50 MW is shown in Figure 6.

From the intensity of the thermal radiation, the temperature of the sapphire crystal could be estimated according to Planck's law [10]. Figure 7 shows the estimated temperatures as a function of temperatures measured by a conventional thermocouple. They showed good agreement, which indicated that the radiation resistant optical fibre with sapphire sensor system could be used as a temperature monitor in a core region of the reactors. The present measuring system could be composed simply and conveniently in an in-core monitoring system as a gamma thermometer.

Figure 6. Observed thermal radiation from sapphire crystal

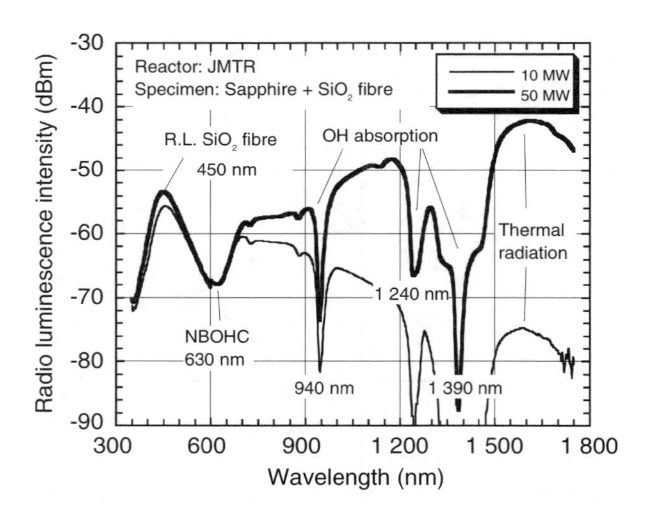

Figure 7. Estimated temperatures by thermal radiation

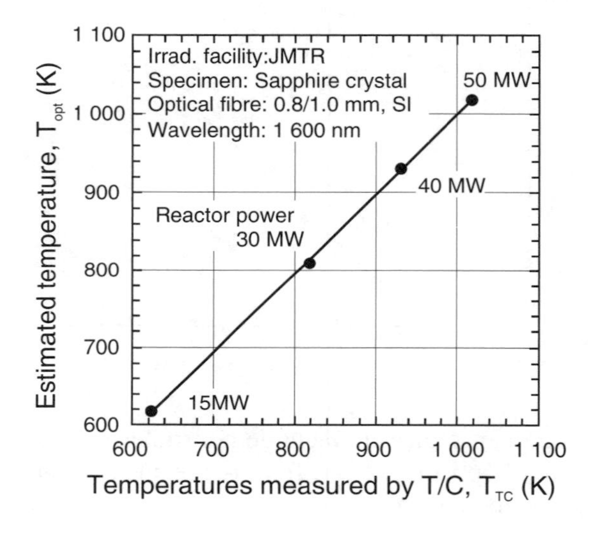

Conclusion

For the development of a new optical instrumentation system such as the in-core power monitor and gamma ray thermometer, heat and radiation resistant optical fibres were developed and their dynamic optical phenomena were observed in a core region of the JMTR fission reactor.

During heavy irradiation, strong radio luminescence wavelength ranged from 400 nm to 1 400 nm with some peaks, and thermal radiation in the infrared region was observed. Intensities of radio luminescence and thermal radiation were found to be directly proportional to the reactor power. Also, temperatures could be estimated by thermal radiation according to Planck's law. The results have demonstrated that the power and temperature of the in-core region can be monitored by measurements of radio luminescence and thermal radiation. The newly developed heat and radiation resistant optical fibres can be applied to a core region of a power reactor and the present optical measuring method could be composed simply and conveniently in an in-core monitoring system.

The new optical monitoring system using heat and radiation resistant optical fibres is planned to be applied in the core monitoring systems for advanced power reactors such as the high temperature engineering test reactor (HTTR) and the reduced-moderation water reactor (RMWR), both under development at JAERI.

Acknowledgements

The research was performed under contract with the Japan Atomic Energy Research Institute (JAERI) and Japan Atomic Power Company (JAPCO). The authors would like to thank Dr. M. Nakagawa of JAERI and Dr. M. Yoshimura of JAPCO for the their continuous support of this joint research, and also express appreciation to Dr. N. Shamoto of Fujikura, Ltd. for fabrication of the heat and radiation resistant optical fibres.

REFERENCES

[1] T. Kakuta, T. Shikama, M. Narui and T. Sagawa, "Behavior of Optical Fibers under Heavy Irradiation", *Fusion Engineering and Design,* 41, pp. 201-205 (1998).

[2] E.J. Friebele and M.E. Gingerlich, *J. Non-Crist. Solid*, 38 &39, 245 (1980).

[3] T. Kakuta, N. Wakayama, K. Sanada, O. Fukuda, K. Inada, T. Suematsu and M. Yatsuhashi, "Radiation Resistance Characteristics of Optical Fibers", *Journal of Lightwave Technology*, Vol. LT-4, No. 8, pp. 1139-1143, August (1986).

[4] N. Nakai, T. Tokunaga and K. Ishikawa, "Gamma Ray Irradiation Effects on SiO_2 Based Optical Fibers", IEEJ, EIM-82-28, pp. 45-54 (1982).

[5] T. Kakuta, K. Ara, N. Shamoto, T. Tsumanuma, K. Sanada and K. Inada, "Radiation Resistance Characteristics of Fluorine Doped Silica Core Fiber", *The Fujikura Giho*, No.86, pp. 50-54 (1994).

[6] P.D. Morgan, Proc. 17th Symp. Fusion Technol., Rome, 14-18 Sept. (1992).

[7] T. Tanabe, "Photon Emission Induced by Neutron and Ions", Dynamic Effects of Irradiation in Ceramics", A US/Japan Workshop, Santa Fe, New Mexico, pp. 231-256 (1992).

[8] T. Tanabe, S. Tanaka, K. Yamaguchi, N. Otsuki, T. Iida, M. Yamawaki, "Neutron Induced Luminescence of Ceramics", *Journal of Nuclear Materials*, 212-215, pp. 1050-1055 (1994).

[9] T. Shikama, M. Narui, T. Kakuta, H. Kayano, T. Sagawa and K. Sanada, "Study of Optical Radiation from SiO_2 During Reactor Irradiation", *Nuclear Instruments and Method in Physics Research*, B-91, pp. 342-345 (1994).

[10] M. Gottiebm and G.B. Brandt, "Fiber-Optic Temperature Sensor Based on Internally Generated Thermal Radiation", *Applied Optics*, Vol. 20, No. 19, pp. 3408-3414 (1981).

NEW TECHNOLOGIES FOR ACCELERATION AND VIBRATION MEASUREMENTS INSIDE OPERATING NUCLEAR POWER REACTORS*

J. Runkel, D. Stegemann, J. Fiedler, P. Heidemann
Institute of Nuclear Engineering and Non-Destructive Testing
University of Hannover (IKPH)
Elbestr. 38 A, D-30419 Hannover, Germany
E-mail: runkel@mbox.ikph.uni-hannover.de

R. Blaser, F. Schmid
Vibro-Meter SA
P.O. Box 1071, CH-1701 Fribourg, Switzerland
E-mail: rbla@vmfr.vibro-meter.ch

M. Trobitz, L. Hirsch, K. Thoma
Nuclear Power Plant Gundremmingen (KRB)
P.O. Box 300, D-89355 Gundremmingen, Germany

Abstract

A miniature bi-axial in-core accelerometer has been inserted temporarily inside the travelling in-core probe (TIP) systems of operating 1 300 MW_{el} boiling water reactors (BWR) during full power operation. In-core acceleration measurements can be performed in any position of the TIP system. This provides new features of control technologies to preserve the integrity of reactor internals.

The radial and axial position where fretting or impacting of instrumentation string tubes or other structures might occur can be localised inside the reactor pressure vessel. The efficiency and long-term performance of subsequent improvements of the mechanical or operating conditions can be controlled with high local resolution and sensitivity. Low frequency vibrations of the instrumentation tubes were measured inside the core. Neutron-mechanical scale factors were determined from neutron noise, measured by the standard in-core neutron instrumentation and from displacements of the TIP tubes, calculated by integration of the measured in-core acceleration signals. The scale factors contribute to qualitative and quantitative monitoring of BWR internals' vibrations only by the use of neutron signals.

---

* This paper was not presented orally at the workshop.

Introduction

A miniature bi-axial accelerometer was developed for acceleration measurements in radioactive and high temperature environments by Vibro-Meter. In co-operation with IKPH a special version of the sensor, which can be inserted inside the travelling in-core probe (TIP) tubes of nuclear power reactors instead of the TIP, was developed [1]. The geometrical dimensions of the accelerometer and the TIP are similar, such that the accelerometer can be moved in the same way as a TIP. The coupling between the transducer and the inner wall of the tube is performed by means of a specially designed adapter spring (Figure 1).

Figure 1. Bi-axial accelerometer with adapter spring and inserted inside a TIP tube

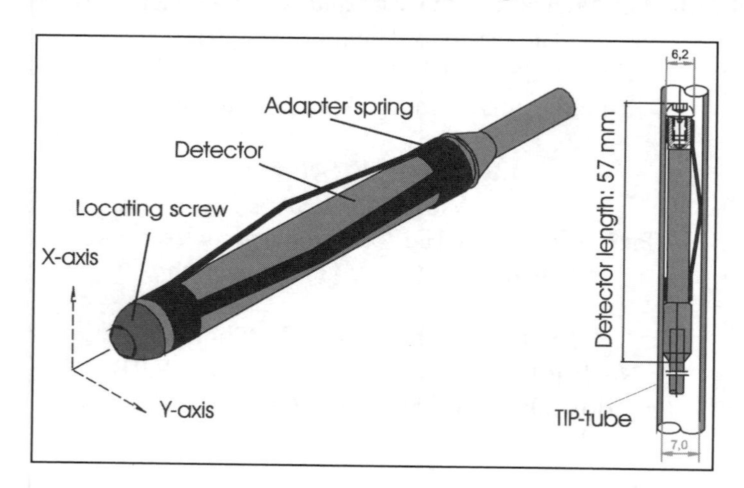

A sufficiently high coupling force for acceleration transmission and a sufficiently low frictional resistance for moving the sensor by means of pushing the cable has been verified in a full scale model of a TIP tube and instrument string. The new sensor has been inserted in all TIP tubes of both units of the KRB during full power operation. The low frequency vibrations of the reactor and core internals were investigated [2,3]. In higher frequencies flow induced impacts of some in-core instrument tube structures were localised. The German company Siemens KWU developed a technology to strengthen the impacting tubes with a special sleeve tube. The sleeves were mounted during regular outages of the plants. After restarting it could be verified by repetitive reactor external and internal acceleration measurements that the impact intensity had been drastically reduced due to the improvement of the construction.

Sensor design

The design of the bi-axial accelerometer is effectively imposed by the geometrical envelope permitted, i.e. that it should be attached with the longer axis parallel to the tube that needs vibration measuring. This means that the transducer is sensitive in x and y directions and without sensitivity in the z axis where the length is relatively unimportant. The bi-axial accelerometer works in the so-called "bender" mode patented in 1992 (European Patent No. 316498, US Patent No. 5.117.696).

The whole assembly consisting of transducer head and spring (Figure 1) does not have sharp edges and is designed to be easily moved forward and backward instead of the travelling neutron probe inside TIP tubes. The sensing elements (Figure 2) use the piezoelectric effect to measure

Figure 2. Inner construction

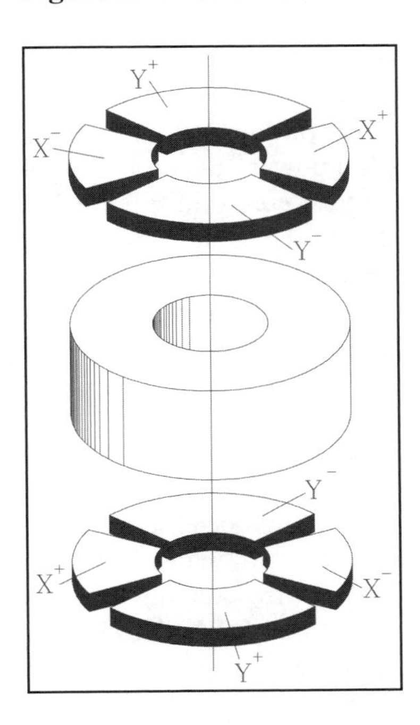

vibrations. The charge output of a piezoelectric ring is proportional to the dynamic compression applied to it. The piezoelectric stack is held together by means of a pre-stressed rod. On the top of it there is a seismic mass fixed to the piezoelectric stack.

In order to obtain a sufficient sensitivity the number of crystals together with the total seismic mass operating on them has to be optimised. The total mass working in the crystal stacks is composed of a high-density material (tungsten) in order to get the maximum mass within the minimum volume. The whole sensing element is bound together under a high pre-stress so that no parts may move with respect to each other, especially under the maximum shock that is encountered. It is essential that the pre-stress does not sensibly change with temperature, which would necessitate the incorporation of a thermal compensation material to match the rates of expansion between the diverse parts within the sensing element and the pre-stressed rod. Under the influence of inertial forces, which act perpendicular to the axis of the pre-stressed rod, seismic mass bending load occurs. This results in an increase of the compression at one side and in a decrease of the compression at the other side of the piezoelectric rings.

Each piezoelectric ring has a four-segment electrode mounted on its top face and a four-segment electrode mounted on its bottom face. The opposite pairs of segments are connected together so as to provide a bi-axial accelerometer. The integrally attached hard-line cable compromises a MgO insulation and four conductors which are connected to the electrically separated sensing elements. The choice of the piezoelectric material is driven by the operational temperature, the radiation and the design itself which requires very homogenous and stable materials. The natural hydrogen piezoelectric crystal defined as "Vibro-Meter VC2" was chosen so far. Although this material has been found non-linear in terms of sensitivity under high radiation at low frequencies (see next section), the previous applications show that the bi-axial accelerometer can be successfully used for acceleration and vibration measurements inside operating power reactors. Further developments will be in finding the best piezoelectric material for the application and its evaluation under high radiation.

Results of low frequency vibration measurements in operating 1 300 MW_{el} BWRs

The vibrations of BWR internals were analysed in the low frequency range (below 10 Hz) by use of the simultaneously measured signals of in-core neutron detectors and the in-core bi-axial accelerometer. The later one was temporarily positioned at the neutron detector height inside of several instrumentation tubes during full power operation. The bi-axial sensor could be moved inside the TIP tube to any position up to the upper core grid (Figure 3). It was inserted in the TIP tube at the position of the cable reel and pushed with the cable itself. Pre-operational tests in a full scale model of the TIP system have been performed for the successful optimisation of the adapter spring, which must be weak enough so that the detector can be moved and strong enough so that the coupling force is sufficient to measure the tube vibrations in the frequency rage of interest (up to 1 kHz in case of impact detection, see next section).

The normalised auto power spectral densities (NAPSD) of in-core neutron noise signals, measured in one of the 42 instrument strings of the BWR (Figure 4, top of right side) show a significant peak at 2.5 Hz. It is known that this peak is caused by a vibration of the instrumentation tubes between the upper and the lower core grid, without impacts or contacts of the guide tube to the surrounding fuel channel boxes [6,7]. The corresponding auto power spectral densities (APSD) of the accelerometer signals (Figure 4, left side) have a similar shape at the measuring positions inside the core, while they deviate as expected at measuring positions below the lower core grid and outside of the reactor pressure vessel. The amplitudes of the movements can be inferred from the acceleration signals by double integration in time domain as well as in frequency domain and from the neutron noise spectra through neutron mechanical scale factors relating the random displacements to the relative change of the neutron signal. From the APSD of displacement the root mean square (RMS) value in x and y direction and the resulting vector can be calculated [5]. The procedure of calculation of the RMS is based on integrating the shape of a peak given by a linear vibration equation, but the effect of non-linear sensitivity of the accelerometer in the low frequency range and under the influence of radiation has to be considered (see following paragraph). The resulting $RMS_{x,y}$ has the dimension of mm. The RMS_n value of neutron signals is calculated from the NAPSD in the same frequency range correspondingly. For correlated acceleration and neutron noise signals, neutron-mechanical scale factors for different vibration modes of BWR instrument tubes (at 1.6 Hz and at 2.5 Hz) and for a combined vibration of instrument tubes and fuel channel boxes at approximately 4 Hz were determined [2,3]. This enables to estimate the displacements of the corresponding vibrating components inside the core of the given BWR through the scale factors from the neutron noise spectra of signals measured by the standard in-core instrumentation. For a PWR application some corresponding results have been published [4,5].

Under the influence of the high radiation inside the core a remarkable increase of the measured amplitudes of the acceleration signals was observed in the low frequency range (Figure 5). This reversible, frequency and radiation dependent sensitivity change of the bi-axial in-core accelerometer has not been fully clarified until now. The effect must be taken into account, however, and corrected for the determination of the scale factors.

Further investigations will be take place to find new piezoelectric materials for the specific application in high radiation and high temperature environment in order to determine and minimise the radiation influence to the low frequency signal of the in-core accelerometer.

Figure 3. In-core instrumentation of a 1 300 MW_{el} BWR

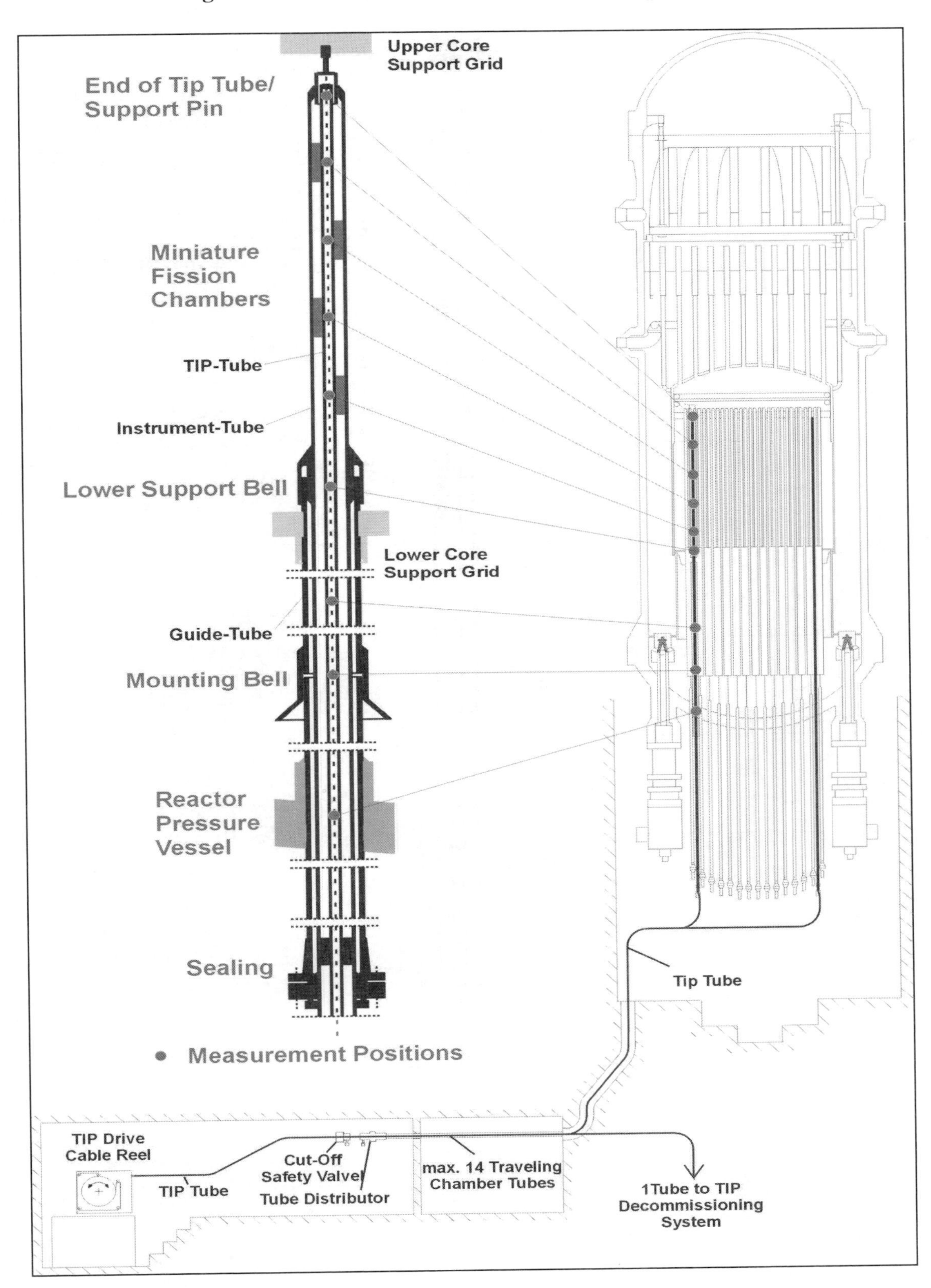

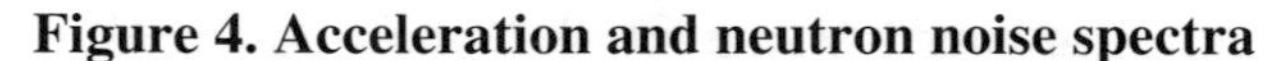

Figure 4. Acceleration and neutron noise spectra

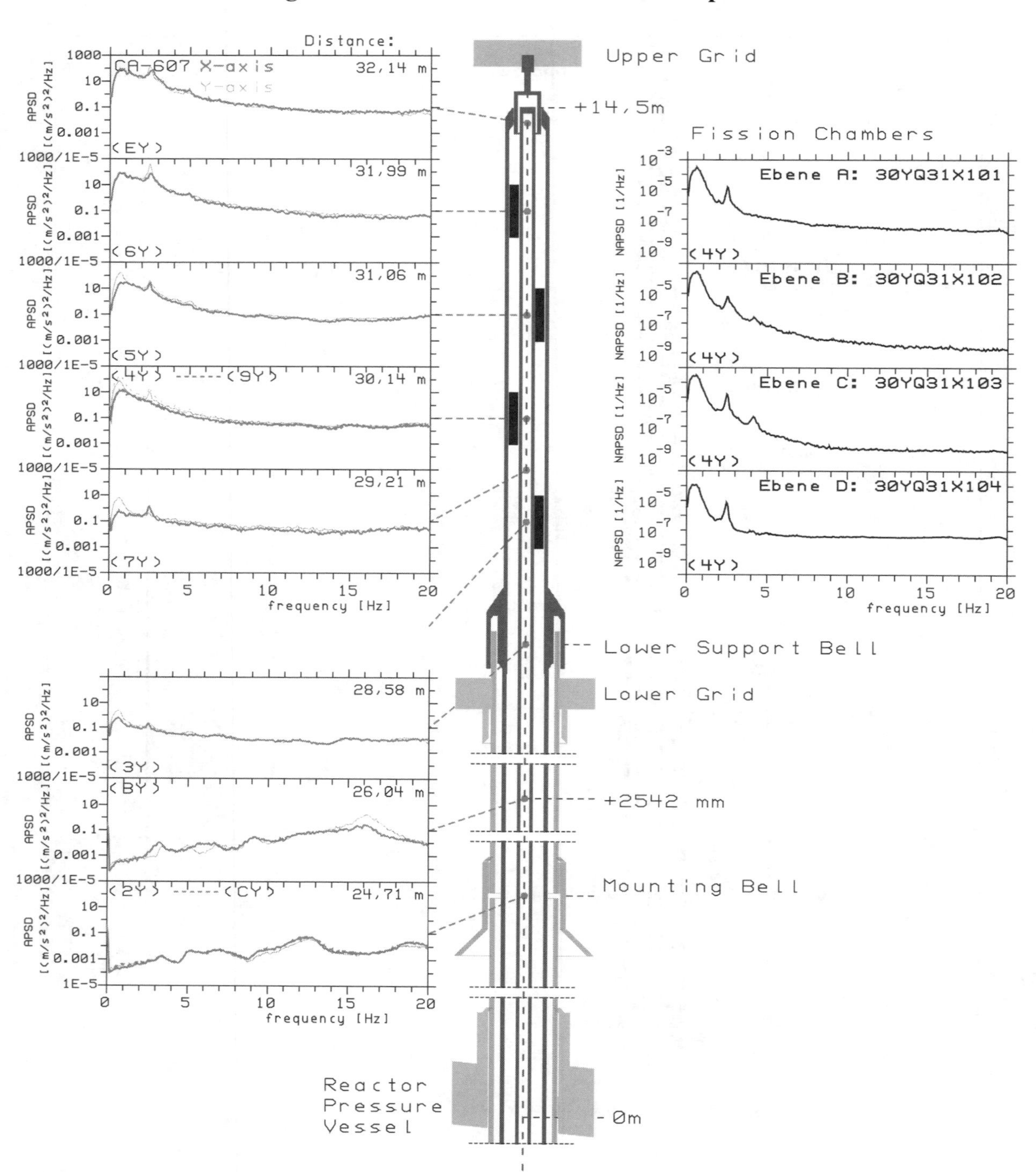

Figure 5. Reversible and frequency dependent sensitivity change of the accelerometer due to high radiation

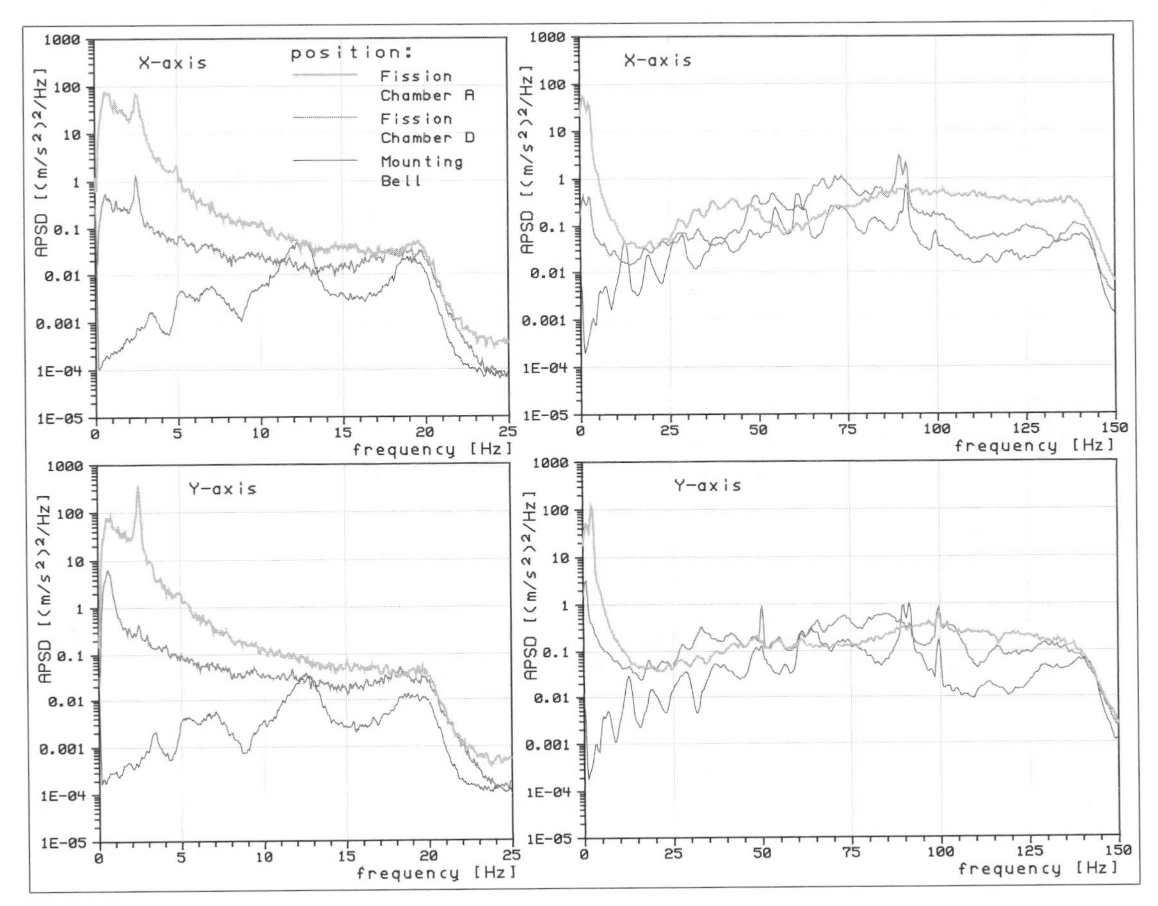

Detection of impacting in-core instrumentation tubes inside operating 1 300 MW_{el} BWRs

The vibrations of all in-core instrumentation tubes of both units of the KRB are frequently analysed twice in each fuel cycle by use of neutron noise analysis. Particular vibrations of single instrument strings were found and subsequently monitored [6]. The instrument tube vibrations and the effect of possible impacts to surrounding structures on the vibrations were analysed and analytically modelled [8,9]. The theoretical results were used to interpret the particular instrument tube vibrations. This interpretation gave some hints that impacts of at least one of the instrument tubes against the surrounding structures occurred. A detailed inspection of the relevant tube during a subsequent regular shut down was made, and it was found that the instrument guide tube was damaged at the position of the mounting bell (Figure 3), and that flow-induced impacts of the three telescoped tubes of that particular instrument string had caused a degradation of the structure in the area of the mounting bell.

In order to detect even weak impacts, which might not influence the vibrations of the instrument tubes inside the core region and which therefore cannot be detected by neutron noise analysis, the new bi-axial accelerometer has been inserted in all TIP tubes during full power operation.

In the frequency range up to approximately 1 kHz (due to the sensor construction and the adaptation by a spring (Figure 1) this is the maximum frequency of the measuring system) flow-induced impacts of some other in-core instrument tube structures were also localised.

The acceleration signals (*x* and *y* direction) measured inside of the TIP tubes at the mounting bell position of all instrument strings are shown in Figure 6 in time domain. In a few core position bursts, caused by impacting structures, they are clearly visible. At those core positions, the bi-axial in-core accelerometer has been moved to several axial positions up to the upper end of the TIP tube in order to localise were the impacts were transmitted to the TIP tube. In Figure 7 some examples of these measurements in one of the instrument strings are given.

Comparing the signals in Figure 7, it is not clearly seen in what axial position the strongest bursts are measured. In order to indicate the position were the impacts are created, statistical values of the acceleration signals were calculated and some of the results are in Figure 8. The standard deviation (Figure 8, upper part) characterises the energy content respectively the averaged amplitudes of the signals. The Kurtosis factor (Figure 8, middle part) is used to quantify the burst content of the signals. If the signal fluctuations are normally distributed and are not influenced by bursts, then the value of the Kurtosis factor is three, while bursts cause an increase of it. For weighing the energy and burst content of the measured acceleration signals we defined the product of the standard deviation and of the Kurtosis factor as an intensity value of possible impacts.

The comparison of the statistical values calculated from measured acceleration signals at several positions inside a single TIP tube before and after the guide tube has been sleeved (see below) shows relatively high standard deviations and consequently relatively high intensity values (product of standard deviation and Kurtosis factor) in the core region (Figure 8). This is caused by the obvious increase of the sensitivity of the bi-axial accelerometer when it was exposed to the high flux inside the core. But this has no influence to the Kurtosis factors, which show a relative maximum in the axial vicinity of the in-core instrumentation guide tube mounting bell where the impacts occurred. Obviously the bursts were transmitted from the guide tube to the inner TIP tube through the water inside the gaps between the telescoped tubes. This enables the localisation not only of the radial but also of the axial position where fretting or impacting of instrumentation tube components or other structures might occur inside the reactor pressure vessel and core.

Siemens KWU developed a technology to strengthen the impacting tubes by special sleeve tubes inside the instrument guide tube at the position of the mounting bell. The sleeves were mounted during regular outages of the plants. After restarting it was verified through repetitive reactor external and internal acceleration measurements that the impacts have been reduced drastically due to the improvement of the construction (Figure 9). After this improvement the Kurtosis factors calculated from the acceleration signals measured at the same positions as before (Figure 8) still show a maximum at the mounting bell position, because weak impacts were registered after strengthening the guide tube, but the intensity values demonstrate the remarkable reduction of the impact energies.

In Figure 10 the impact intensity values of all in-core instrument tubes measured at the axial position of the guide tube mounting bell are compared before and after strengthening two of the guide tubes (those which had the highest values before). It can be clearly seen that the sleeves reduced the impact intensity at these positions and that the intensity did not increase at other positions, where low intensity was detected previously and consequently no sleeves had been mounted. Repetitive measurements of the in-core acceleration inside all of the TIP tubes allows the early detection of degradation due to fretting and impacting. The efficiency and long-term performance of subsequent improvements (sleeves) can be controlled with high local resolution and sensitivity.

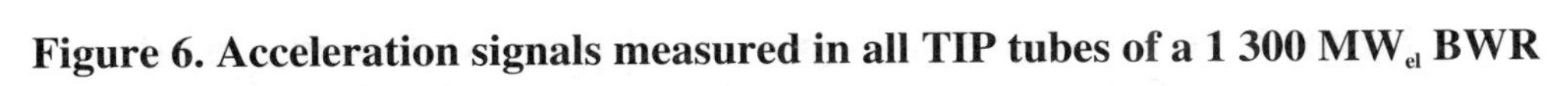

Figure 6. Acceleration signals measured in all TIP tubes of a 1 300 MW_{el} BWR

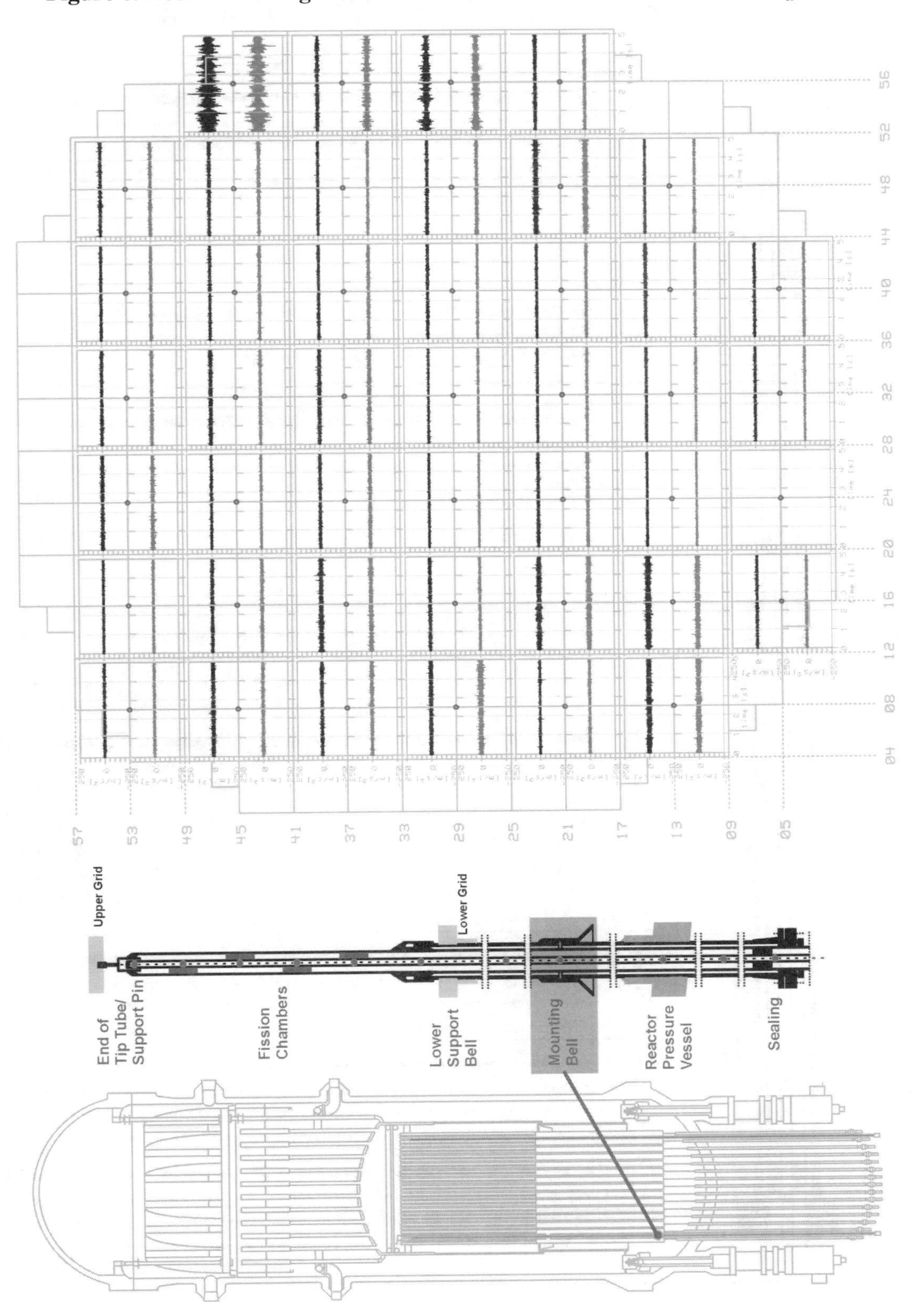

Figure 7. Acceleration signals measured in one TIP tube at several axial positions inside the reactor pressure vessel and the core

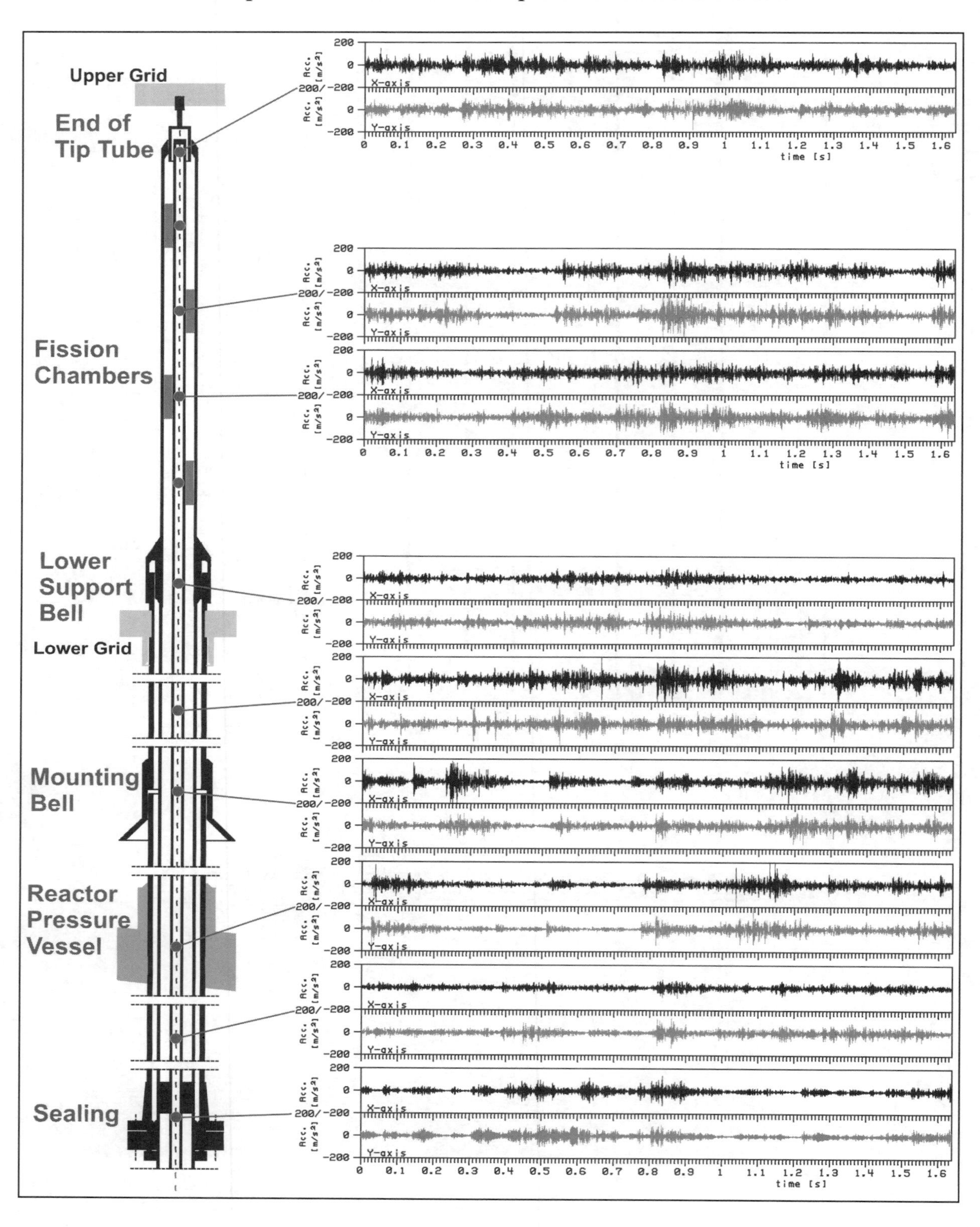

Figure 8. Statistical analysis of acceleration signals measured in one TIP tube at several axial positions

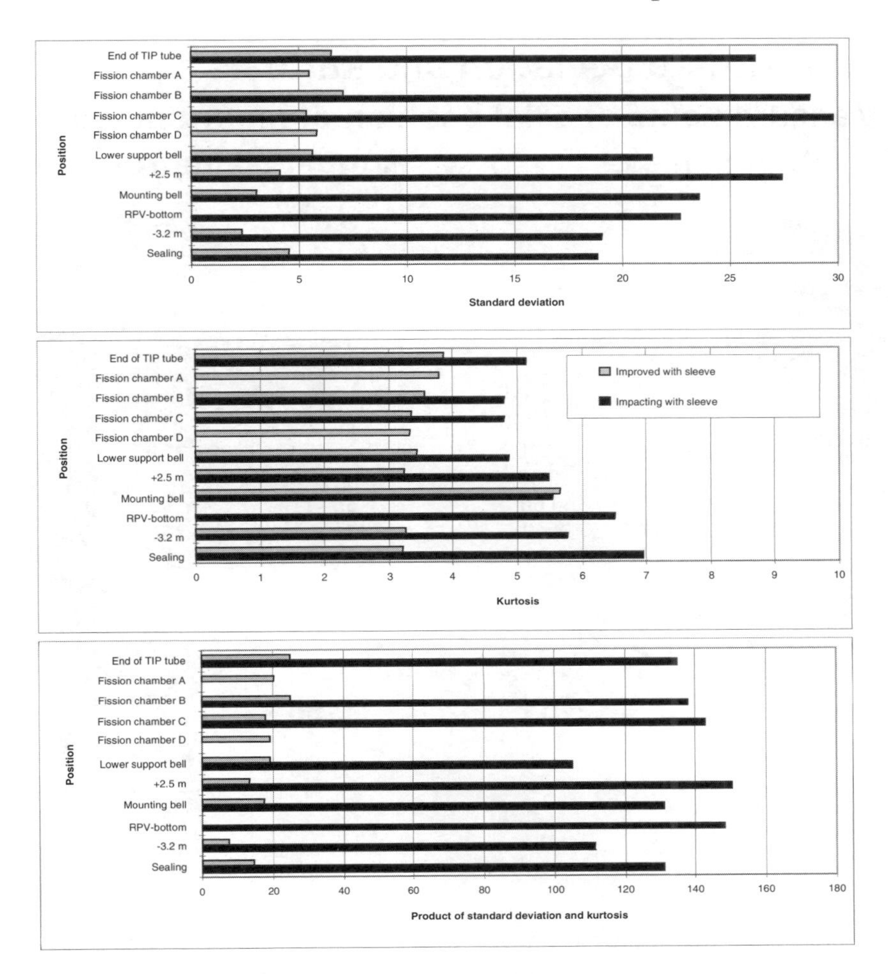

Figure 9. Acceleration signals measured in a TIP tube before and after the installation of a sleeve

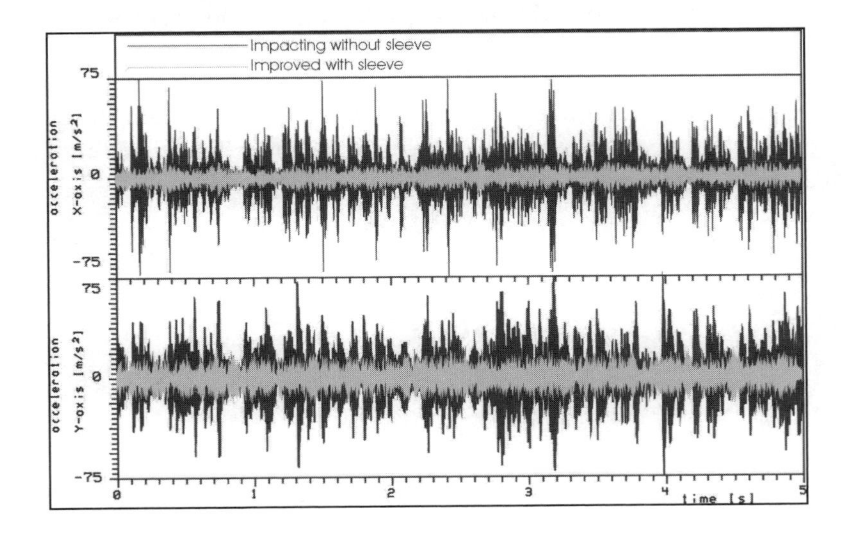

Figure 10. Reduction of impacts due to the improvement of the construction of guide tubes, comparison of statistical values

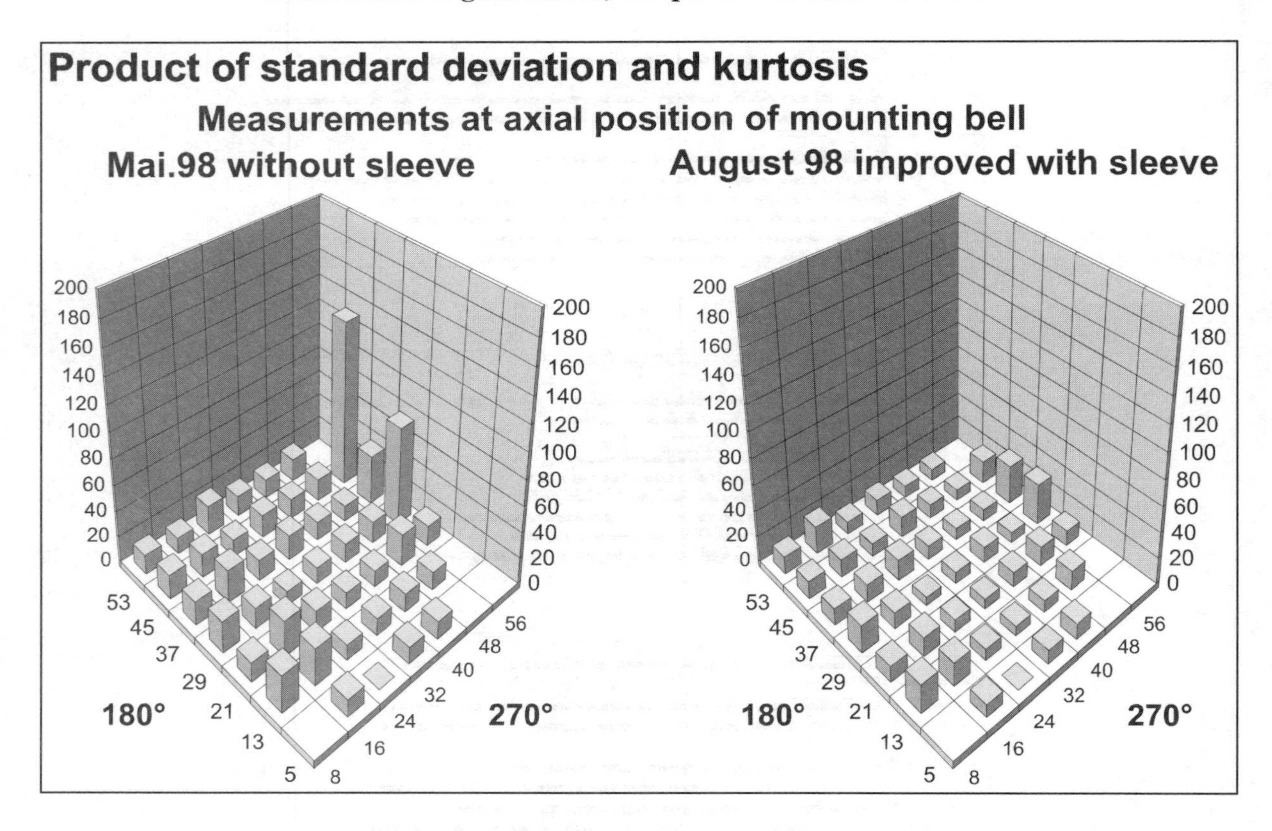

Conclusions

A miniature bi-axial in-core accelerometer was successfully used to detect even weak impacts and to localise the radial and axial position of impacting components inside the reactor pressure vessel during full power operation. The measuring system can be used without any interference in the regular reactor operation and without constructive changes of the existing in-core instrumentation. This provides a new and high-sensitive diagnosis tool to control and preserve the integrity of reactor internals.

In the low frequency range and under the influence of high radiation inside the reactor core a non-linear sensitivity of the sensor was observed, which has to be considered in determination of the neutron mechanical scale factors.

REFERENCES

[1] Runkel, Laggiard, Stegemann, Heidemann, Blaser, Schmid and Reinmann, "In-Core Measurements of Reactor Internals Vibrations by Use of Accelerometers and Neutron Detectors", Proceedings of Specialists Meeting on In-Core Instrumentation and Reactor Core Assessment, Mito-shi, Japan, 1996, OECD publication, 1997.

[2] Stolle, "Schwingungsuntersuchungen an Instrumentierungslanzen von Siedewasserreaktoren", Diplom-work IKPH, 1997.

[3] Stegemann, Runkel, Fiedler and Heidemann, "Ergebnisbericht zu Schwingungsmessungen innerhalb der Fahrkammerrohre von LVD-Lanzen des Blockes C der Kernkraftwerke Gundremmingen, 11.BE-Zyklus", Internal IKPH/KGB Report, No.: KGB/BC/BEZ.11/05.97, 1997.

[4] Laggiard, Fiedler, Runkel, Starke, Stegemann, Lukas, and Sommer, "Vibration Measurements in PWR Obrigheim by Use of In-Core Accelerometers", *Progress in Nuclear Energy*, Vol. 29, 1995.

[5] Laggiard, Runkel, Stegemann, Fiedler, "Determination of Vibration Amplitudes and Neutron-Mechanical Scale Factors Using In-Core Accelerometers in NPP Obrigheim", Proceedings of SMORN VII, Avignon 1995, OECD publication, 1996.

[6] Runkel, Laggiard, Stegemann, Fiedler, Heidemann, Mies, Oed, Weiß and Altstadt, "Application of Noise Analysis in Two BWR Units of Nuclear Power Plants Gundremmingen", Proceedings of SMORN VII, Avignon 1995, OECD publication, 1996.

[7] Altstadt and Weiß, "Numerische Untersuchungen zum mechanischen Schwingungsverhalten einer nassen LVD-Lanze", report produced for IKPH, 1993.

[8] Laggiard, Runkel and Stegemann, "One-Dimensional Bimodal Model of Vibration and Impacting of Instrument Tubes in a BWR", *Nuclear Science and Engineering*, 115, 62, 1993.

[9] Laggiard, Runkel and Stegemann, "Three-Dimensional Model of Vibration and Impacting of Instrument Tubes in a BWR and Transfer from Mechanical to Neutron Noise", *Nuclear Science and Engineering*, 120, 124, 1995.

DEVELOPMENT OF ADVANCED DIGITAL ROD POSITION INDICATION SYSTEM

Koki Inagaki, Hironobu Shinohara
Mitsubishi Heavy Industries, Ltd.
Kobe, Japan

Satoru Yasue, Masaru Tamuro
Mitsubishi Electric Corporation
Kobe, Japan

Abstract

The replacement of the existing analogue type rod position indication (ARPI) system by the DRPI system in existing plants is a problem in that the current DRPI system requires a large air-conditioned space in the containment vessel for maximum reliability. As a means of solving this problem we have developed an advanced DRPI system.

Our advanced DRPI system applies smaller signal processing cards using HIC technology, as well as smaller diameter cables to reduce the total cabinet size in the containment vessel. Additionally, we separate the detection resistances as large heat sources from the signal processing cards and apply hybrid heat resistant resin (HHR) to the card on which the resistances are installed to eliminate the air conditioning equipment.

After the completion of environmental tests we plan to apply the advanced DRPI system to domestic PWR plants.

Introduction

In all the existing PWR plants in Japan, the rod position indication (RPI) system is installed as part of the standard design. In nine of these plants, the analogue type RPI (ARPI) has been applied. The utilities of ARPI plants face the potential necessity of replacing ARPI with the digital rod position indication (DRPI) system, which aims to improve measuring accuracy, and to reduce the load of maintenance work.

Compared with ARPI, DRPI has several advantages:

- Higher accuracy even in the case of control rod motion.
- Higher reliability realised by redundant system configuration.
- No regular adjustment of analogue circuits.

On the other hand, the replacement of ARPI with conventional DRPI in existing plants has some problems as follows:

1) Replacement would require a large space for the cabinets inside of the containment vessel. In some plants, however, there is not sufficient room. The conventional DRPI needs cabinets installed in the containment vessel for detector signal conditioning and for data transmission to outside of the containment vessel.

2) Replacement would demand a large amount of work and proper financing. The data cabinet requires special air conditioning for higher reliability and a longer lifetime, which in turn needs air cooling through a chilled water system from the outside of the containment vessel through the additional containment penetration.

Based upon the above-mentioned situation, we decided to develop an advanced DRPI, which has smaller signal processing cards and a greater heat tolerance for cabinets in the containment vessel.

Conventional DRPI

The conventional DRPI has been applied to fourteen domestic plants, and has exhibited a good performance and a high stability.

System description

DRPI monitors the actual rod position of each control rod and actuates alarms in the main control room in case of a rod misalignment and/or rod drop event.

The system configuration is shown in Figure 1, compared with that of the advanced DRPI introduced in this paper. The system is mainly composed of detector assemblies, two data cabinets in the containment vessel, a control cabinet outside the vessel, and an indication unit on the main control board. The system has basically redundant configuration.

A detector assembly is mounted above the control rod drive mechanism (CRDM) coils on each CRDM housing. It consists of 42 discrete coils arranged with an interval of six steps and mounted

Figure 1.System configuration of the conventional DRPI system and the advanced one

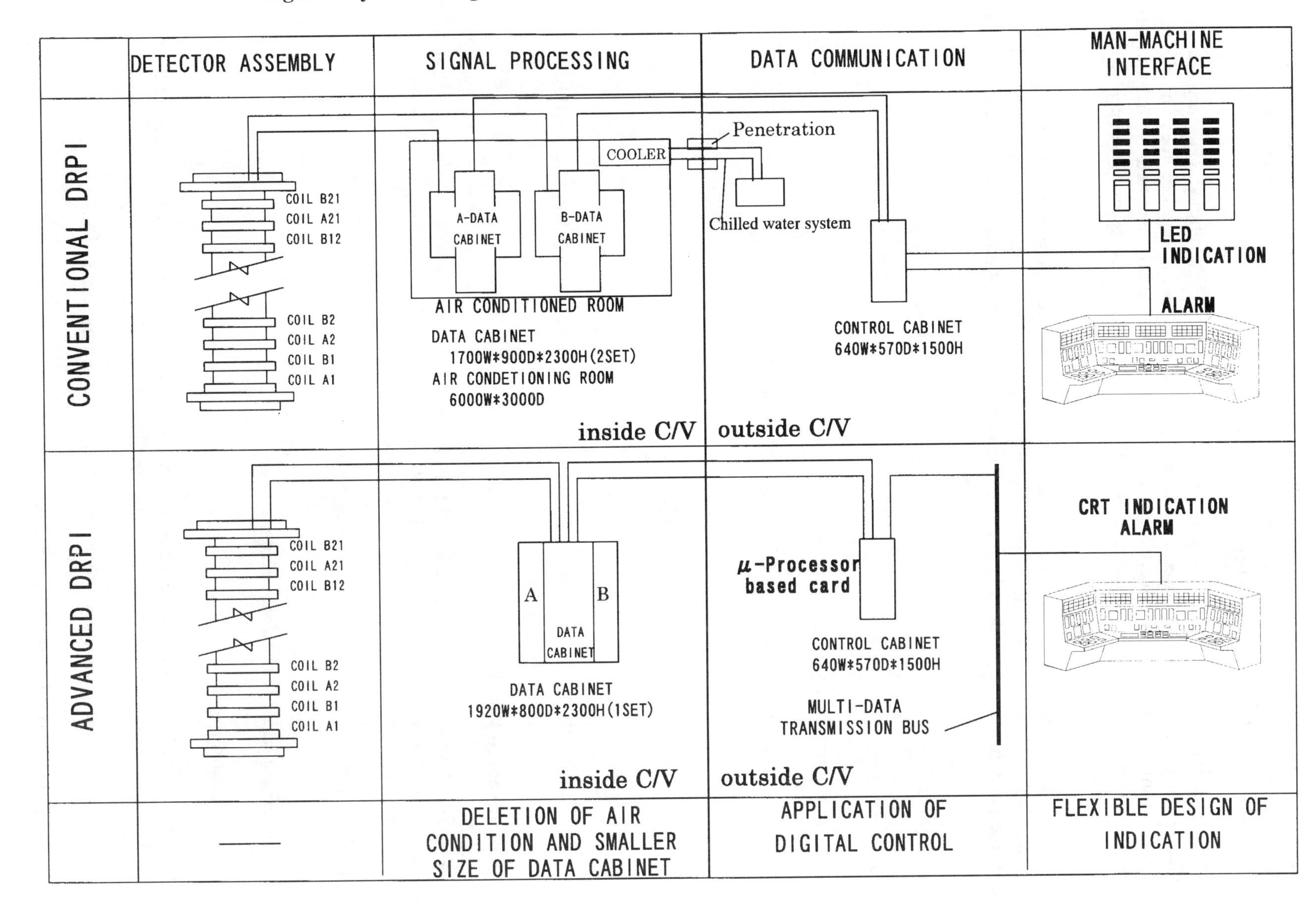

concentrically with a rod drive mechanism housing. The coils are separated into two groups, and each coil in one group is axially arranged above one coil in the other group. The detector senses the entry and presence of a rod drive shaft through its centreline.

The data cabinets are located in a room air conditioned by the chilled water system in the containment vessel, which is supposed to ensure high reliability and a long lifetime. The signals are processed from the detectors, converted to digital signals, and are transmitted by a multiplexing method to the exterior of the containment vessel. One data cabinet is assigned to each of the two groups.

The control cabinet directs the signal transmission of the overall system and transmits the rod position signal to the plant computer system and the indication unit on the main control board. It also detects rod position misalignment and rod drop, both of which actuate alarms in the main control room.

Main features

- Accuracy:
  - Within four steps (electronics).
  - Within six steps totalling all the normal conditions.
- Data cabinet:
  - Size: 1 700 W × 900 D × 2 300 H.
  - Number: 2.
  - Environment: temperature 10-35°C (requires air conditioning), humidity 0-100%.

Design of advanced DRPI

We have developed the advanced DRPI as a means of more feasibly replacing ARPI in the existing plants. The new design is applied to the signal processing card in the data cabinet in the containment vessel, cables and connectors and control cabinet.

The details of the new design are presented hereafter.

Signal processing cards in the data cabinet

Card size reduction

In order to reduce the size of the signal processing card in the data cabinet, we applied the hybrid IC (HIC) technology developed in our company (see Figure 2).

As a result the size of the signal processing card is reduced by some 75%.

- Card size: 160 × 100 (ref. conventional: 330 × 180).

Figure 2. HIC for the signal processing cards in data cabinet

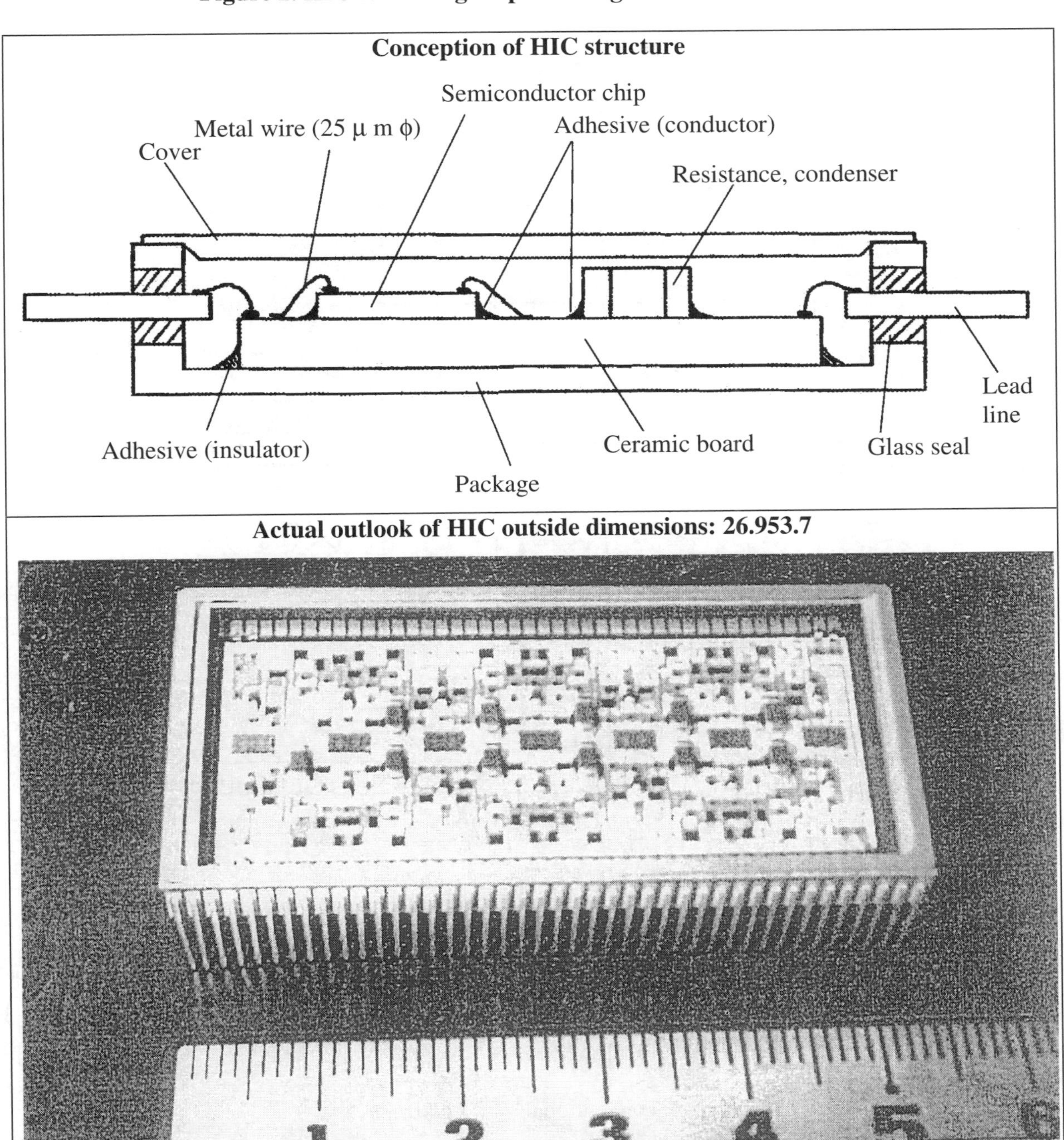

Tolerance of high temperature

We designed a new special card "resistance card" as illustrated in Figure 3. Detection resistances are separated from the signal processing cards because they are the main heat source. They are loaded in a resistance card of which the printed wiring board is made of hybrid heat resistant resin (HHR). HHR is available at the temperature of 145, while glass epoxy is not over 105.

Figure 3. Separation of the detection resistance from the signal processing card

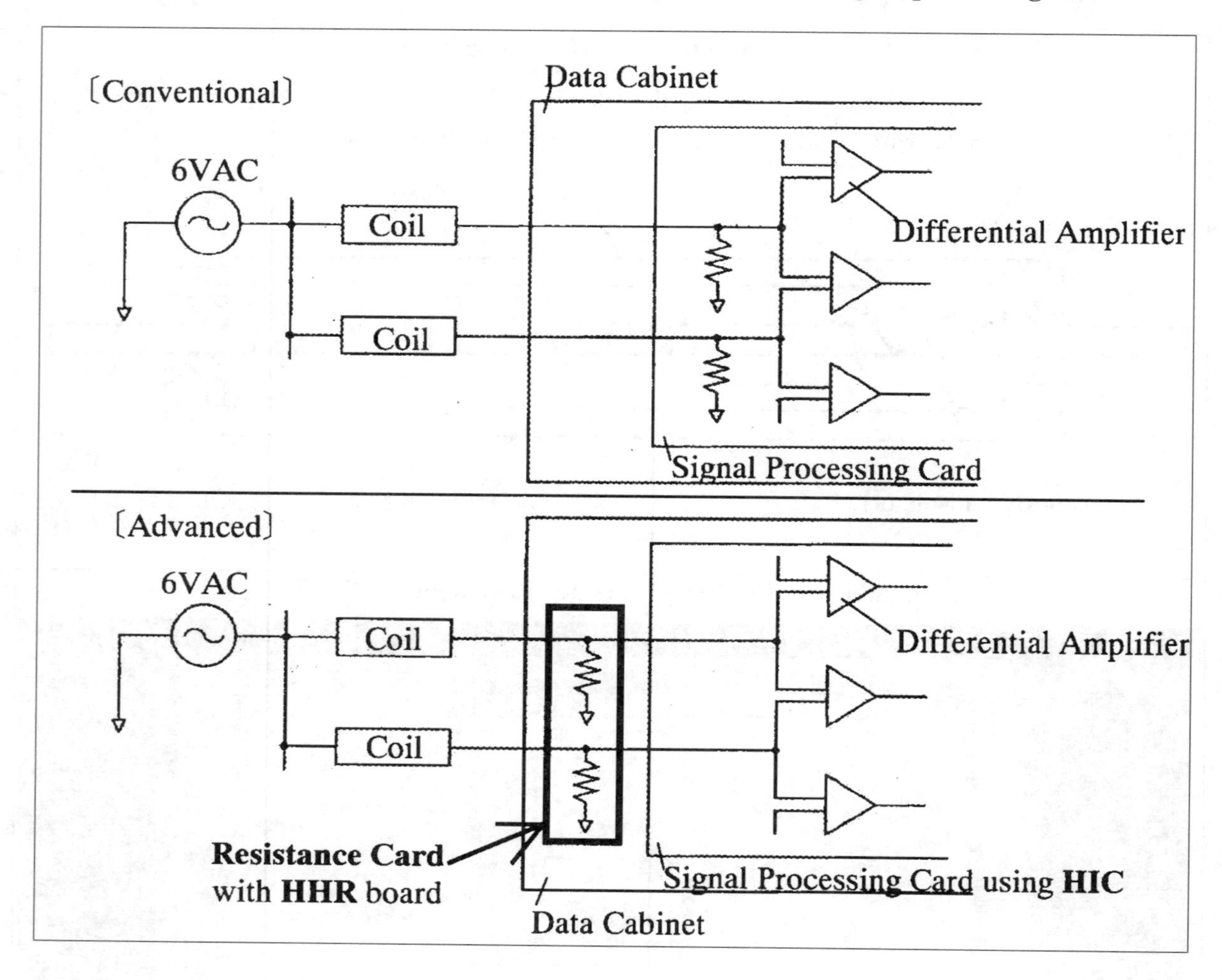

Cable and connector between detector and data cabinet

In order to reduce the size of the data cabinet we also developed a new cable and a connector linked to data shown as Figures 4 and 5.

- Cable size: ϕ15.5 (ref. conventional: ϕ22).
- Cable weight: 405 kg/km (ref. conventional: 650 kg/km).
- Connector size: $\phi 38.1 \times 31.5$ (fixed side – ref. conventional: $\phi 46.5 \times 48$), $\phi 44.5 \times 31.3$ (movable side – ref. conventional: $\phi 50 \times 130$).

Data cabinet

The rack structure of the data cabinet is designed so that heat generated in the rack may be discharged well through natural air mixing.

Figure 4. Comparison of cable specification of the advanced cable and the conventional one

	ADVANCED		CONVENTIONAL
Sectional view of cable connected into data cabinet	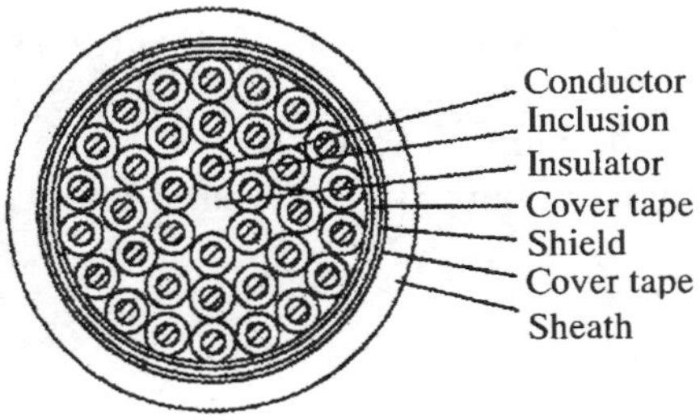 Conductor, Inclusion, Insulator, Cover tape, Shield, Cover tape, Sheath Example of 36 cores		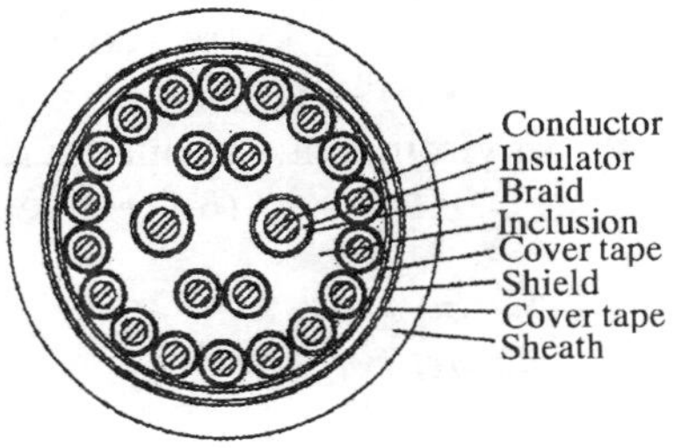 Conductor, Insulator, Braid, Inclusion, Cover tape, Shield, Cover tape, Sheath
Conductor	Tin plating mild copper wire		Tin plating mild copper wire
Insulator	Fire-retarding bridge polyethylene		Fire-retarding EP rubber nylon braid
Shield	Tin plating mild copper braid		Tin plating mild copper braid
Sheath	Fire-retarding low temperature special heat resisting vinyl		Fire-retarding chloro-sulfonated polyethylene
Size, no. of cores	0.5 mm^2 × 36 cores	0.5 mm^2 × 46 cores	0.5 mm^2 × 22 cores + 3.5mm^2 × 2 cores
Out diameter (mm)	15.5	17.0	22
Estimated weight (kg/km)	405	500	650

Figure 5. Comparison of advanced connectors and conventional one

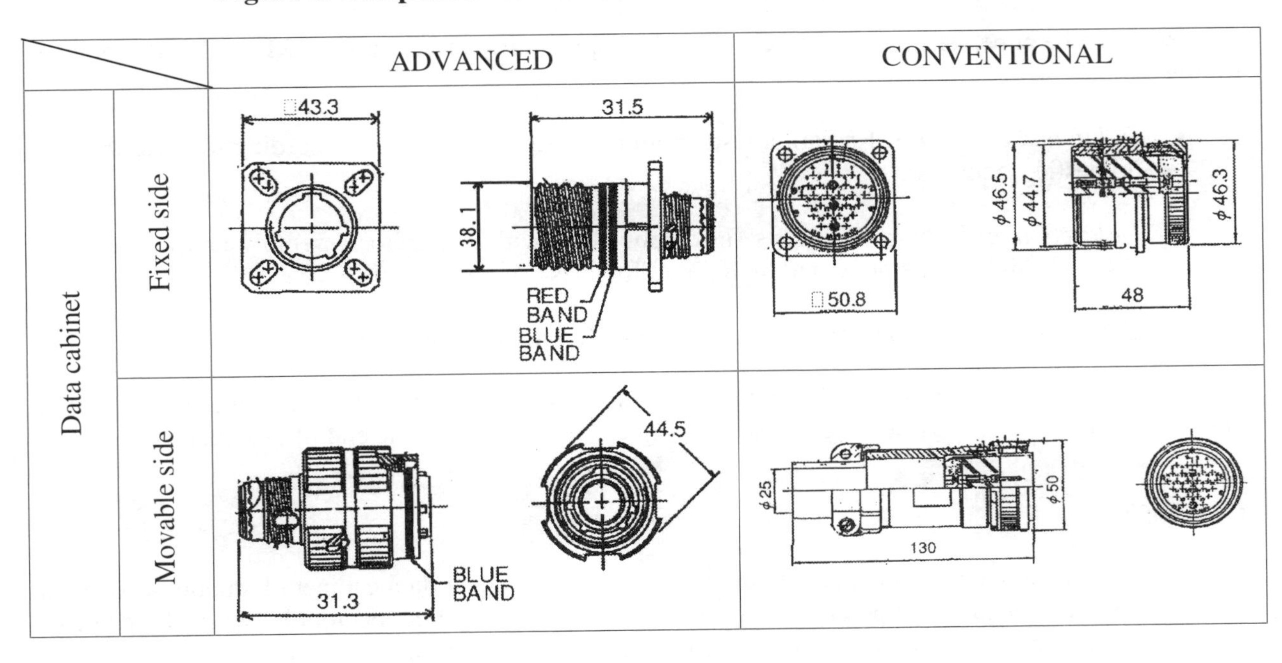

Based upon this structure and design improvements mentioned in previous sections, the data cabinet can be redesigned to be smaller and more tolerant of high temperature as follows:

- Cabinet size: 1 920 W × 800 D × 2300 H (one set)
  (ref. conventional: 1 700 W × 900 D × 2 300 H (two sets)

- Environment: temperature 0-50°C, humidity 0-95%
  (ref. conventional: temperature 10-35°C, humidity 0-100%)

Control cabinet

We apply the μ-processor based control cabinet to improve its reliability and design flexibility of the interface with the other instrumentation in the plant such as the plant control computer system (PCCS).

Advanced DRPI and feasibility check

System description

The system configuration of the advanced DRPI is illustrated in Figure 1, together with that of the conventional DRPI.

The application of the advanced DRPI can delete the need for a large, air conditioned space for the data cabinet, and increases design flexibility regarding interface with other instrumentation in the plant such as a PCCS and the reactor control system.

Tests of advanced DRPI system for PWR plants

We have been performing tests to evaluate the applicability of advanced DRPI to PWR plants as follows:

- *Cable and connector.* Radiation tests, high temperature tests, etc. (according to IEEE383 and domestic standards).

- *Data cabinet.* Performance tests (in combination with a detector assembly and mock-up of CRDM), high temperature and high humidity tests (ongoing).

- *Control cabinet.* Data communication tests with other instrumentation.

All the results of the tests which we have made so far indicate that the installation of the advanced DRPI in PWR plants is feasible. We intend to complete all the tests by the end of this year.

Conclusion

We have developed the advanced DRPI with a newly designed data cabinet of smaller size and a greater tolerance for high temperature, and also smaller cables and connectors between the detector and the data cabinet. We are now completing tests evaluating the feasibility of the application of the advanced DRPI in PWR plants. We are currently planning to apply the advanced DRPI to the existing RPI in domestic PWR plants.

EXPERIENCE WITH FIXED PLATINUM IN-CORE DETECTORS*

Richard J. Cacciapouti, Joseph P. Gorski
Duke Engineering & Services
400 Donald Lynch Boulevard
Marlborough, MA 01752

Abstract

Uniform sets of analyses were performed at nearly 40 exposure points over four cycles of operation with two independent in-core detector systems. Full in-core analyses for each set of data collected with both movable fission chambers and fixed self-powered platinum detectors show comparable results for peaking values. Statistics of predicted to measured signal differences are good. Compared to Cycle 1, the axial or three-dimensional component of uncertainty is unchanged after over ten years (six cycles) of operation. Over the same period, the radial uncertainty has decreased slightly. The uncertainty value used in technical specification surveillance has remained constant. The results show the use of self-powered platinum detectors to be a complete and independent system with the accuracy and functionality expected of an in-core detector system.

* This paper was not presented orally at the workshop.

Background

A fixed in-core detector system was designed and developed at Seabrook Station [1] to determine in-core power distributions. Seabrook Station is a four loop Westinghouse plant operating at 3 411 MWt and containing 193 assemblies. Unlike most Westinghouse plants, Seabrook Station contains two complete and independent in-core detector systems. The first is a movable in-core detector system, which uses movable fission chambers typical of Westinghouse plants similar to Seabrook Station. The second detector system employs self-powered fixed platinum detectors. Both of these systems were installed during plant construction.

Description of movable in-core detector system

The movable in-core detector system uses 58 reactor core instrument thimbles. One or more of six movable fission chambers traverse each thimble. The measurement of in-core power requires the six movable fission chambers to be passed through the core at least 12 times. As the detector is passed through the core, the signals are collected and saved on the main plant computer as a neutron flux trace. Each detailed axial trace consists of 61 relative axial neutron flux measurements. These traces, which collectively make up a flux map, are then processed with analytical predictions of detector reaction rates by INCORE-3 [2] to infer the measured power distribution and corresponding local peaking factors. The results are then compared to established limits to ensure that the core is operating within the limits of the technical specifications.

Description of fixed in-core detector system

The fixed detectors used at Seabrook Station are self-powered with platinum emitters and yield a signal proportional to the incident gamma and neutron flux. The fixed in-core detector system consists of 58 detector strings. Each string contains five self-powered platinum detectors for a total of 290 detectors in the core. These strings are an integral part of the instrument thimble. They are located in the same radial core locations as the movable in-core detector system. Each detector consists of a 13.5 inch long platinum emitter within the core and is connected to its associated lead wire. A compensation lead wire which is identical to the emitter lead runs parallel to the emitter lead within the sheath of each detector to correct for gamma induced background current. The emitter and leads are all packed in an Al_2O_3 dielectric insulator and bound in an Inconel sheath. The wires for a detector string form a helix around a central Inconel tube and are then bound by an Inconel sheath. The central Inconel tube is the path used by the movable fission chamber. The fixed in-core detectors are spaced along the thimble so that they fall in the mid-regions of the core between fuel assembly grids.

The data acquisition system developed at Seabrook Station [1] consists of the fixed in-core detector data acquisition software and two trains of front end multiplexing instrumentation. Each train reads 145 of the platinum detector channels. The signal developed within the platinum emitter is determined as the emitter and its lead signal less its compensation lead signal. Cross channel calibration is essentially avoided since only two analogue to digital measurement devices, one per train, are used to develop all 290 signals. Each channel loop is terminated with a 20KΩ precision resistor, which minimises detector leakage current and improves channel response time by maintaining a small resistance capacitance time constant.

The system hardware has been configured in such a manner that less than 0.08% of the detector signal is system noise. Signal common mode rejection is accomplished by maintaining a single common ground for each detector channel. The reactor ground is connected to each channel shield, which envelops the entire detector loop, including the multiplexer and analogue to digital instruments. Digital filtering is accomplished in the monitoring instrumentation by averaging 32 samples from each channel every minute. This filtering removes any residual AC component and results in a signal to noise ration of 8×10^{-4} at full power conditions.

For the first three cycles of operation, technical specification surveillance was provided by the movable in-core detector system. Data was also collected with the fixed in-core detector system for comparison and to determine accuracy, reproducibility and signal degradation. To use the fixed detectors for technical specification surveillance, the system qualification was submitted to the US Nuclear Regulatory Commission [3] for approval.

Power shape determination

The gamma and neutron interactions result in an axial signal that is not directly representative of power. The method used for determining power from this system begins with an assumption that the ability to predict the detector's measured signal from a neutronics calculation is equal to the ability of the same calculation to predict the in-core power distribution. This implies any differences between predicted and measured detector signals can be applied to local power predictions to infer the measured power [4].

The generation of a three-dimensional measured power distribution involves a combination of measured signals and analytical signal to power conversion factors. The fixed in-core detectors provide continuous signal data, which is collected and stored once per minute. The power distribution and predicted signals are generated with SIMULATE-3 [5]. The SIMULATE-3 model of Seabrook Station consists of four nodes per assembly radially and 24 nodes per assembly axially. When a measured power distribution is required, the SIMULATE-3 model is updated to the current plant conditions. Using these conditions, SIMULATE-3 calculates the power distribution and the detector constants. The detector constants include both the neutron and gamma responses [6] for the platinum detectors.

The Fixed Detector In-Core Code (FINC) was developed by Yankee Atomic Electric Company to infer the three-dimensional power distribution. FINC performs a cubic spline fit of the predicted and measured signals to axially expand the five original measured and predicted signals to 24 equal axial intervals (nodes). This is consistent with the axial resolution of the neutronics code model. The signals are assumed to be zero at an extrapolated distance above and below the bottom of the core, reducing the differences between prediction and measurement in these areas.

From these mathematically created axial detector signals, measured to predicted signal ratios are determined for use in the inferred power distribution calculation. Thus, the ratio of the measured to predicted detector signal for all 24 axial nodes in all 58 instrumented locations are generated. These ratios represent the local differences between the predicted and measured power in the instrumented locations in the core. Once the detector measured to predicted signal ratios have been determined, the full core measured power distribution is generated.

Since not all fuel assemblies in the core contain detectors, a system of determining power in uninstrumented locations is required. The FINC code uses a proportional weighting method to couple instrumented and uninstrumented assemblies in radial power distribution calculations. These weights are applied as given in the following equation:

$$P_{jk}^{meas} = P_{jk}^{pred} * \frac{\sum_{i}^{I} w_i \left(S_i^{meas} / S_i^{pred} \right)}{\sum_{i}^{I} w_i}$$

where: P^{meas} is measured power at location jk
P^{pred} is predicted power at location jk
w is weighting factor between I and jk
S^{meas} is measured detector signal at location I
S^{pred} is predicted detector signal at location I

This method of using detector ratios to modify the local predicted power distribution is applied in each of the 24 axial planes defined in the SIMULATE-3 model. The predicted power of axial nodes near a detector will be modified by the detector ratio determined for that axial node and the radial weighing scheme. The predicted axial power distribution for each individual assembly is modified by the local detector ratio. This means that the axial power shape in uninstrumented assemblies is derived from the predicted axial power shape in the uninstrumented assembly modified by local measurements from local instrumented locations.

Core operational data consistent with current operational conditions are used to update the SIMULATE-3 predictive model. Model update calculations of detector constants are performed very quickly on high-powered workstation computers at Seabrook Station. Thus, detector responses and in-core power distributions can be predicted for these conditions and used directly with the measurement data.

Fixed and movable detector results comparisons

During normal operation of the plant, an in-core detector analysis is performed to determine the in-core power distribution on a monthly basis. The purpose of this analysis is to demonstrate that the maximum peaking factors, as determined by the in-core power distribution, are less than the limits assumed in the safety analysis. Nearly 40 in-core power distributions have been processed by both the fixed in-core detector system and the movable in-core detector system for the same conditions. Data collected from both of these systems are compared in this work to show that both systems are reporting similar results for the same core conditions.

The primary parameters of concern for technical specification surveillance are the axial peak power in any pin, F_q, and the integrated peak power in any pin, F_{Δ_h}. Each of these values have been compared for each surveillance made with both the fixed in-core detector system and the movable in-core detector system. Results for Cycles 1-4 are presented in Tables 1-4.

The results provided in Tables 1-4 for display a deviation in F_q between the movable and fixed in-core detector systems. As cycle burn-up increases, the fixed in-core detector system predicts a lower value of F_q than that determined from the movable in-core detector system. All other data is in good agreement and confirms the accuracy of the fixed in-core detector system at determining the required surveillance parameters.

The measured value of F_q can be separated into its radial and axial components F_{Δ_h} and F_z. As shown in Tables 1-4, the F_{Δ_h} data from the two measurement systems is comparable for all four cycles. Therefore, the F_z values do not agree between the systems. The deviation is due to the methodological differences used to analyse the data. Axial power distributions using the movable in-core detector system are biased by the ^{235}U fission spectrum using a single plane model in INCORE-3 to analyse the data. The methodology used in the analysis of fixed in-core detector system data considers fissions from all sources as explained below.

The movable in-core detector system uses a ^{235}U fission chamber detector to measure the neutron flux axially through the core in each of the instrumented locations. The ^{235}U fission chamber produces a current proportional to the fissions generated from the incident neutron flux on a ^{235}U element. Thus, the movable in-core detector system measures the fission rate of ^{235}U in the core as a function of axial core position. At the beginning of the cycle, the fresh fuel dominates the core axial power shape and the ^{235}U fission rate shape is nearly the same as the axial power shape. However, as the cycle burn-up increases, the contribution from other nuclides become more dominant. The axial power shape within the core also changes from the classic cosine shape to a double humped shape. The double humped shape results from the depletion of the fuel in the central regions of the core and less depletion in regions above and below the centre of the core. The bottom of the core has a higher moderator density producing a softer spectrum, due to lower moderator temperature. The ^{235}U fission chamber is more sensitive to the softer spectrum at the bottom of the core than the harder spectrum near the top of the core. Thus, the axial power shape generated by the ^{235}U fission chamber will be more bottom peaked than the actual power shape.

From the data presented in Tables 1-4, Cycles 2 and 4 exhibit the deviation in F_q with burn-up, while Cycles 1 and 3 do not appear to exhibit this deviation. Cycle 1 was a fresh core and most all fissions were from ^{235}U. Even by the end of the cycle the ^{235}U fissions dominated the axial power shape. In Cycle 2, essentially two-thirds of the core contained burned fuel from Cycle 1 and burn-up dependence on F_q was observed near the end of cycle. In Cycle 3, the peak F_q values do not appear to exhibit a trend near the end of cycle. However, in Cycle 3, the peak F_q location is not the same as the peak F_{Δ_h} location. The F_{Δ_h} in the peak F_q location was measured higher with the fixed in-core detector system than that measured by the movable in-core detector system. Thus, the decrease in F_z was compensated by an increase in F_{Δ_h}. Cycle 4 showed the deviation and as expected the peak F_{Δ_h} value was in the same location as the peak F_q for most of the cycle. Although the peak F_q locations determined by each system were not the same, they are very near one another and have essentially the same axial power shape.

The results demonstrate that, as the core depletes, the peak F_q from the movable in-core detector system using a single plane model in INCORE-3 code is usually greater than that given by the fixed in-core detector system using the FINC code. The peak F_q from the movable in-core detector system is consistent with the ^{235}U axial fission rate shape. The peak F_q from the fixed in-core detector system is consistent with the axial power shape derived from all isotopes.

The single plane model for INCORE-3 used by Seabrook Station for this analysis is not the latest in use at other Westinghouse plants with movable in-core detector systems. A multi-plane model used by other Westinghouse plants compensates for ^{235}U reaction rate shape.

Technical specification surveillance

In the fourth cycle of operation, and after NRC approval, in-core power distribution surveillance was performed with the fixed in-core detector system. To aid the plant reactor engineer staff, a reactor analysis workstation was developed by Yankee Atomic Electric Company to process the data as needed. The workstation contains all software required to generate in-core constants and to develop power distributions from the platinum detector signals. A graphical user interface was developed based on specifications provided by the reactor engineering staff.

Conclusion

After more than ten years (six cycles) of operation, the fixed in-core detector system has continued to demonstrate the same accuracy as in the first cycle. No detector failures or signal strength degradation has been seen. The raw millivolt signals given by the fixed detectors are about the same at the end of Cycle 6 as during Cycle 1 measurements. The results show the fixed in-core detector system using self-powered platinum detectors to be a complete and independent system with the accuracy and functionality expected from an in-core detector system.

REFERENCES

[1] Joseph P. Gorski and Alan G. Merrill, "In-Core Power Monitoring Using Platinum In-Core Detectors at Seabrook Station", Advances in Mathematics, Computations and Reactor Physics, International Topical Meeting, 28 April-2 May 1991, Pittsburgh, PA, USA.

[2] A.J. Harris and H.A. Jones, "The INCORE-3 Program", WCAP-8402, March 1975.

[3] Joseph P. Gorski, "Seabrook Station Unit 1 Fixed In-Core Detector System Analysis", YAEC-1855-P-A, October 1992 (Yankee Atomic Electric Company Proprietary).

[4] Joseph P. Gorski, "Highly Detailed Axial Power Shape Generation from Fixed In-Core Detector Systems", Proceedings of the 1994 Topical Meeting on Advances in Reactor Physics, 11-15 April 1994, Knoxville, Tennessee.

[5] K.S. Smith, K.R. Rempe and D.M. Ver Planck, "SIMULATE-3: Advanced Three-Dimensional Two-Group Reactor Analysis Code, Methodology", STUDSVIK/NFA-89-04, November 1989.

[6] Dominic G. Napolitano and Donald R. Harris, "Sensitivity of Seabrook Station's In-Core Platinum Detectors", Advances in Mathematics, Computations and Reactor Physics, International Topical Meeting, 28 April-2 May 1991, Pittsburgh, PA, USA.

Table 1. Cycle 1 results

Date	Exposure MWd/MTU	Fixed detector system		Movable detector system	
		Maximum $F_{\Delta h}$	Maximum F_q	Maximum $F_{\Delta h}$	Maximum F_q
08/29/90	1 945	1.376	1.995	1.361	1.949
09/26/90	2 950	1.355	1.879	1.325	1.853
10/10/90	3 468	1.336	1.801	1.316	1.788
11/08/90	4 369	1.312	1.731	1.316	1.741
12/05/90	4 850	1.313	1.704	1.309	1.712
01/04/91	5 997	1.299	1.667	1.291	1.662
02/05/91	7 214	1.297	1.640	1.283	1.632
03/18/91	8 473	1.297	1.630	1.289	1.627
04/16/91	9 266	1.289	1.611	1.278	1.621
05/20/91	10 560	1.279	1.575	1.266	1.577
06/18/91	11 570	1.272	1.564	1.261	1.582

Table 2. Cycle 2 results

Date	Exposure MWd/MTU	Fixed detector system		Movable detector system	
		Maximum $F_{\Delta h}$	Maximum F_q	Maximum $F_{\Delta h}$	Maximum F_q
11/01/91	415	1.473	1.842	1.442	1.832
11/08/91	682	1.468	1.901	1.433	1.892
12/04/91	1 680	1.468	1.848	1.436	1.838
01/08/92	2 966	1.464	1.768	1.429	1.767
02/04/92	3 996	1.454	1.749	1.424	1.744
03/04/92	5 101	1.444	1.767	1.420	1.786
04/01/92	6 169	1.436	1.774	1.423	1.792
05/05/92	7 466	1.428	1.758	1.413	1.781
06/02/92	8 536	1.419	1.734	1.406	1.769
07/06/92	9 840	1.407	1.705	1.409	1.767
08/07/92	11 060	1.395	1.674	1.399	1.739

Table 3. Cycle 3 results

Date	Exposure MWd/MTU	Fixed detector system		Movable detector system	
		Maximum $F_{\Delta h}$	Maximum F_q	Maximum $F_{\Delta h}$	Maximum F_q
11/25/92	277	1.432	1.870	1.443	1.865
12/22/92	1 099	1.420	1.921	1.426	1.890
01/28/93	2 206	1.435	1.954	1.444	1.943
02/23/93	3 189	1.437	1.948	1.453	1.925
03/23/93	4 259	1.439	1.894	1.447	1.910
04/22/93	5 402	1.448	1.849	1.443	1.874
05/26/93	6 577	1.454	1.809	1.440	1.822
06/23/93	7 649	1.454	1.787	1.440	1.802
07/26/93	8 909	1.451	1.777	1.448	1.787
08/24/93	9 881	1.449	1.751	1.437	1.755
10/14/93	11 211	1.442	1.748	1.455	1.749
12/10/93	13 200	1.432	1.757	1.426	1.767

Table 4. Cycle 4 results

Date	Exposure MWd/MTU	Fixed detector system		Movable detector system	
		Maximum $F_{\Delta h}$	Maximum F_q	Maximum $F_{\Delta h}$	Maximum F_q
11/02/94	3 499	1.443	1.855	1.441	1.868
12/08/94	4 869	1.443	1.808	1.428	1.855
05/03/95	10 439	1.397	1.676	1.404	1.721
08/31/95	14 403	1.363	1.646	1.375	1.683

SESSION III

Improved Core Models in Core Monitoring

Chairs: A. Wells and T. Lefvert

IMPACT OF ADVANCED BWR CORE PHYSICS METHOD ON BWR CORE MONITORING

Hoju Moon and Al Wells
Siemens Power Corporation
2101 Horn Rapids Rd.
Richland, WA 99352, USA
Tel.: (509) 375–8265
E-mail: Hoju_Moon@nfuel.com

Abstract

Siemens Power Corporation recently initiated development of POWERPLEX™–III for delivery to the Grand Gulf Nuclear Power Station. The main change introduced in POWERPLEX™-III as compared to its predecessor POWERPLEX™-II is the incorporation of the advanced BWR core simulator MICROBURN-B2. A number of issues were identified and evaluated relating to the implementation of MICROBURN-B2 and its impact on core monitoring. MICROBURN-B2 demands about three to five times more memory and two to three times more computing time than its predecessor MICROBURN-B in POWERPLEX™-II. POWERPLEX™-III will improve thermal margin prediction accuracy and provide more accurate plant operating conditions to operators than POWERPLEX™-II due to its improved accuracy in predicted TIP values and critical k-effectives. The most significant advantage of POWERPLEX™-III is its capability to monitor a relaxed rod sequence exchange operation.

Introduction

The monitoring of a BWR core under normal operating conditions is achieved by combining measured local power range monitoring (LPRM) signals with calculated LPRM signals and calculated bundle/nodal/pin power distributions. The calculation of LPRM signals and power distributions is performed by a core physics method embodied in a core simulator code. The evolution of the BWR core monitoring system within Siemens Power Corporation (SPC) has taken place in phase with the evolution of the BWR core physics method and that of the computer technology as shown in Figure 1.

Figure 1. Evolution of Siemens Power Corporation BWR core monitoring system

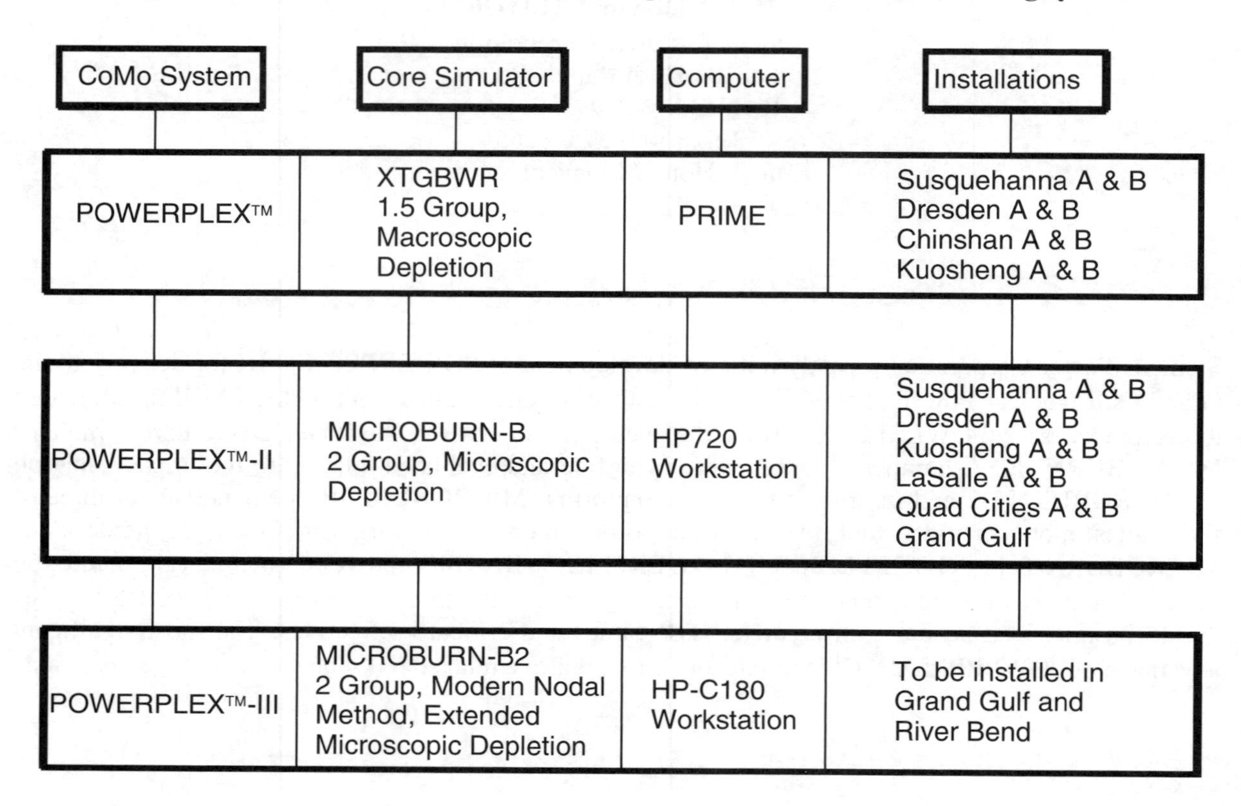

SPC recently initiated the development of POWERPLEX™-III for delivery to the Grand Gulf Nuclear Power Station operated by the Entergy Corporation. The main change introduced to POWERPLEX™-III as compared to its predecessor POWERPLEX™-II is the incorporation of the advanced BWR core simulator MICROBURN-B2. A number of issues were identified and evaluated relating to the implementation of MICROBURN-B2 and its impact on the core monitoring. The most important issues to be discussed in this paper are listed as follows:

1. Computing time, memory requirements.

2. Monitored thermal margin increase/decrease due to the use of advanced physics method.

3. Reliability of predicted control rod pattern and/or power/flow as guidance for operators.

4. Relaxed rod sequence exchange operation capability.

Plant characteristics

The Grand Gulf reactor is a BWR/6 jet pump plant containing 800 fuel bundles with a rated thermal output of 3 833 MWth. Each fuel bundle is modelled with one radial node and 25 axial nodes. The reactor normally operates in a spectral shift mode where the core flow is maintained at about 80% of the nominal flow until near the end of a fuel cycle at which time the core flow is increased to a maximum value. The current fuel cycle length is 18 months. Future cycle length, however, could be as long as 24 months. The reactor core contains 44 traversing in-core probes (TIPs) and LPRM locations, as well as 193 control blade locations. Both the TIPs and the LPRMs are neutron sensitive fission chamber detectors.

Core simulator evolution

The core simulators MICROBURN-B and MICROBURN-B2 are compared below.

	MICROBURN-B	MICROBURN-B2
Solution method	Two-group coarse mesh FDM	Two-group modern nodal method
Cross-section method	Microscopic cross-sections	Microscopic cross-sections
Fuel depletion method	Microscopic depletion	Extended microscopic depletion
Thermal-hydraulics method	XCOBRA	XCOBRA/THRP
Neutron tip response method	Gap flux method	Corner flux reconstruction
Gamma tip response method	Gamma reaction vs. nodal power	Gamma reaction vs. nodal/corner power
Local pin power method	Infinite medium lattice	Pin power reconstruction
Computing time per solution	1 minute on HP–735	2.5 minutes on HP–735
Computer memory per 50 fuel types	30 Mbytes	90 Mbytes/high disk swap 120 Mbytes/low disk swap

Axial power monitoring

The measured axial power distribution is determined based on the following TIP-to-power conversion equation:

$$P^m_{i,k} = P^c_{i,k} \sum_j W_{i,j,k} \frac{T^m_{j,k}}{T^c_{j,k}}$$

where: $P^m_{i,k}$ = measured nodal power

$P^c_{i,k}$ = calculated nodal power

$W_{i,j,k}$ = weighting factor for contribution from TIPs surrounding the node at (i,k)

$T^m_{i,k}$ = measured TIP

$T^c_{i,k}$ = calculated TIP

When the core physics method and the TIP calculation method are changed, a question is raised about whether or not the above TIP-to-power conversion will appropriately determine the nodal power density. Especially given that MICROBURN-B2 calculates TIP values based on corner flux values, the consistency of TIPs with nodal average powers is questioned. This question was investigated, and the results are displayed in Figure 2. The figure shows axial distributions of TIPs and nodal power densities calculated by MICROBURN-B2 and MICROBURN-B. The measured axial power distributions were determined from the measured axial TIP distribution and the calculated axial TIP and power distributions from each code. The calculated axial TIP and power distributions of MICROBURN-B2 are visibly different from those of MICROBURN-B. However, the final measured axial power distribution determined from MICROBURN-B2 data is virtually the same as the final measured power distribution determined from MICROBURN-B data. This is conclusive evidence that the axial power monitoring is not affected by the change in core simulator method, in particular a change in the TIP/LPRM calculation method.

Figure 2. Impact of axial TIP modelling on the measured axial power distribution

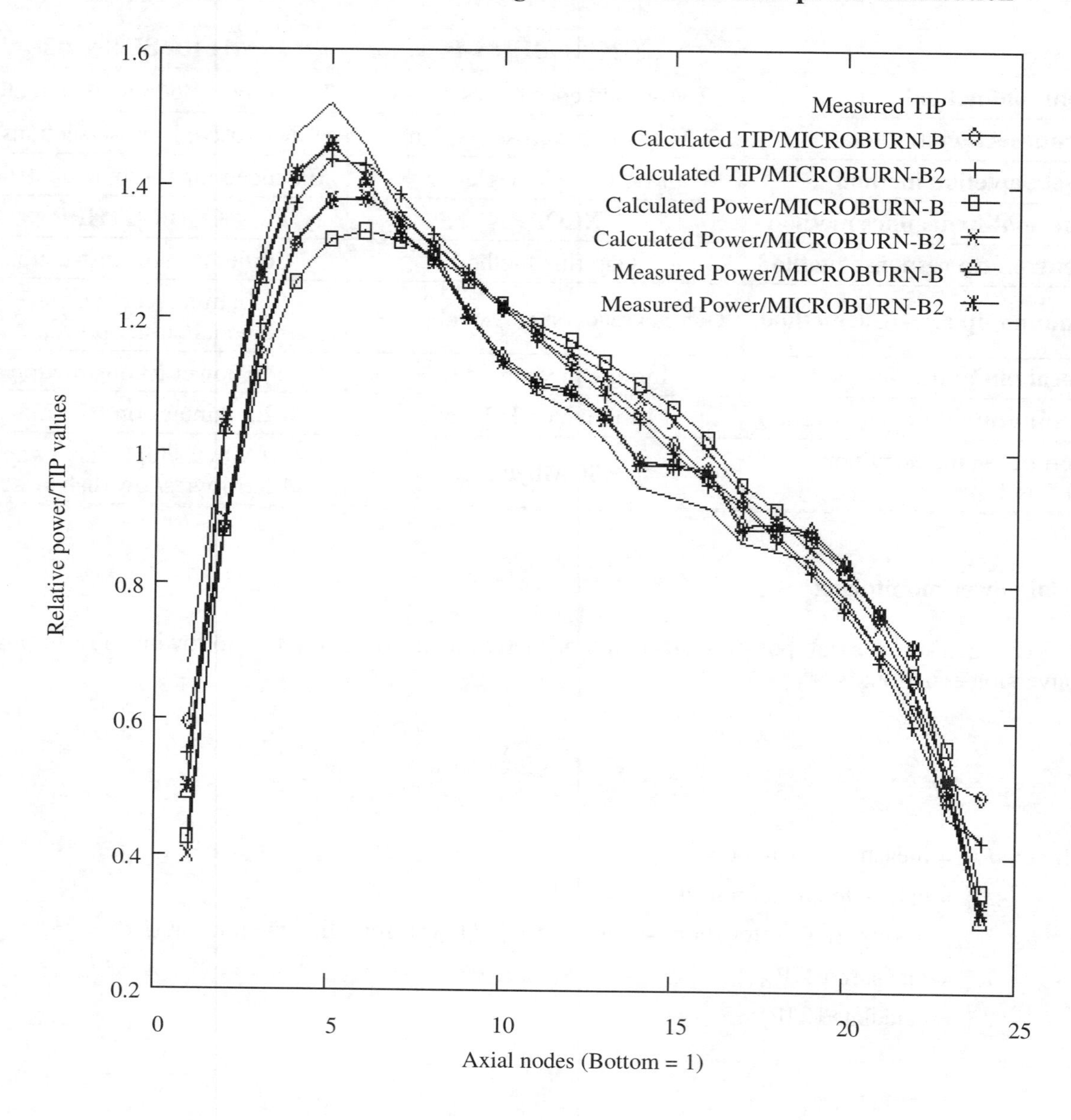

Radial power monitoring

Radial power monitoring relies on LPRM detectors, each of which is surrounded by four fuel bundles. In the Grand Gulf BWR, there are 44 LPRM detector locations next to which 176 fuel bundles reside. These fuel bundles are thus monitored somewhat more closely than other bundles which are further away from the LPRM locations. If the reactor is loaded in a diagonally symmetric pattern and the control rods are inserted likewise in a diagonally symmetric pattern, then the radial power measurement can utilise the LPRMs measured in diagonally symmetric locations for those bundles which are not immediately next to LPRMs. However, control rod patterns may not be diagonally symmetric depending on the rod sequence employed at any given time even if the loading pattern is diagonally symmetric. This situation is the same for all BWRs world-wide.

Due to the fact that each LPRM is surrounded by four fuel bundles, the monitored (measured) radial power of a high reactivity fuel bundle may turn out to be lower than its real value. This problem is illustrated in Figure 3 using a typical four-bundle cell composed of a fresh, a once burned, and two twice burned bundles. The calculated LPRM values for this cell by two different core simulators can be the same even if the calculated bundle powers are different. Since the calculated TIP values are the same, the amount of update per measured LPRM is the same. Thus the updated (measured) radial bundle power distributions will be as much different as the calculated bundle power distributions are.

Figure 3. Illustration of radial power monitoring based on two different core simulators

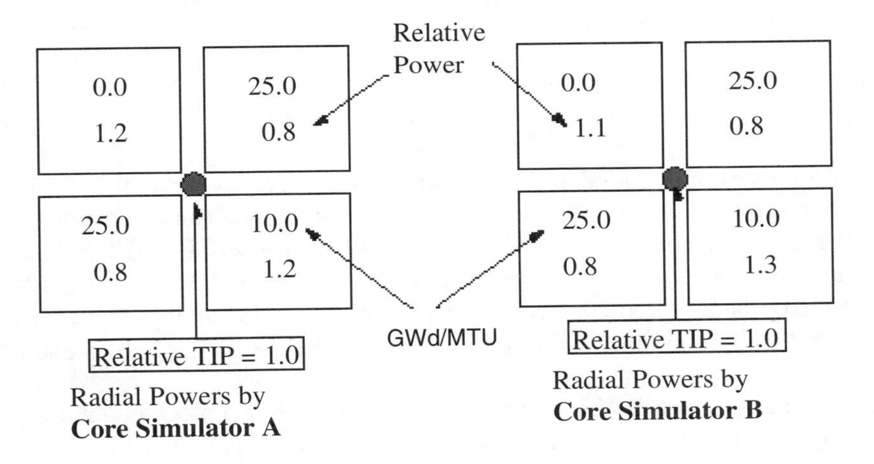

In essence, the radial power monitoring relies on the calculated radial power distribution in all BWR core monitoring systems. This implies that the quality of the core physics method is crucial to the accurate measurement of the radial power distribution. It has been observed that MICROBURN-B2 produces more accurate radial power distributions and sometimes significantly different depending on core loading pattern and fuel bundle types than its predecessor MICROBURN-B. An example of such a difference is provided in Figure 4. The figure shows that the maximum radial peaking factor by MICROBURN-B2 is about 4% larger than the radial peaking factor in the same bundle calculated by MICROBURN-B. Also, MICROBURN-B estimates the adjacent bundle to be the highest power bundle. The maximum power bundle of MICROBURN-B2 is a fresh Atrium™-10 of SPC, whereas that of MICROBURN-B2 is a once burned 9 × 9–2 bundle. This type of discrepancy is unusual and is only found in highly heterogeneous core loading patterns containing advanced mechanical design fuel bundles. It indicates the basic cross-section library difference as well as a thermal-hydraulic modelling difference to be the contributing factors. Its impact can be a 4-5% difference in the MCPR margin because the MCPR value is very sensitive to the radial peaking factor.

Figure 4. Example of radial power distribution differences

MICROBURN–B2, BOC of BWR-A Cycle 9

	1	2	3	4	5	6	7	8	9	10	11	12	13	14	15
1	1.141	1.117	1.250	1.032	1.165	1.059	1.139	0.823	0.817	0.935	1.134	0.962	1.034	0.859	0.476
2	1.141	1.117	1.250	1.032	1.165	1.059	1.139	0.823	0.817	0.935	1.134	0.962	1.034	0.859	0.476
3	1.119	1.278	1.210	1.233	1.191	1.235	1.048	1.154	0.959	1.158	1.141	1.153	1.018	0.860	0.463
4	1.251	1.220	1.224	1.003	1.180	1.201	1.255	1.219	1.218	1.023	1.220	1.134	1.054	0.835	0.463
5	1.033	1.235	1.012	0.758	0.850	1.193	1.212	1.205	1.152	1.249	1.203	1.175	0.878	0.827	0.442
6	1.170	1.188	1.180	0.850	0.875	1.005	1.259	1.045	1.206	1.071	1.241	1.129	1.014	0.790	0.427
7	1.075	1.238	1.189	1.193	1.010	1.242	***1.230***	1.286	1.228	1.260	1.183	1.134	0.816	0.756	0.387
8	1.147	1.052	1.259	1.213	1.260	1.225	***1.300***	1.104	1.271	1.054	1.186	1.052	0.916	0.693	0.354
9	0.830	1.160	1.221	1.208	1.044	1.280	1.094	1.258	1.080	1.190	1.091	0.994	0.800	0.483	
10	0.822	0.959	1.221	1.152	1.203	1.220	1.267	1.079	1.163	0.970	1.000	0.822	0.517		
11	0.944	1.162	1.028	1.249	1.078	1.256	1.048	1.188	0.968	0.996	0.825	0.524	0.335		
12	1.137	1.139	1.222	1.199	1.241	1.174	1.185	1.090	1.002	0.829	0.548				
13	0.964	1.155	1.133	1.178	1.120	1.134	1.052	0.995	0.828	0.525					
14	1.040	1.024	1.058	0.890	1.019	0.819	0.916	0.801	0.516	0.336					
15	0.867	0.864	0.832	0.831	0.798	0.759	0.694	0.482							
16	0.481	0.466	0.462	0.442	0.431	0.389	0.357								

MICROBURN–B, BOC of BWR–A Cycle 9

	1	2	3	4	5	6	7	8	9	10	11	12	13	14	15
1	1.103	1.099	1.198	1.021	1.108	1.073	1.108	.812	.746	.954	1.111	.988	1.035	.937	.549
2	1.103	1.099	1.197	1.021	1.109	1.072	1.108	.812	.746	.955	1.111	.989	1.035	.937	.549
3	1.099	1.224	1.241	1.187	1.227	1.203	1.078	1.126	.980	1.125	1.189	1.140	1.077	.891	.532
4	1.198	1.248	1.185	1.012	1.154	1.246	1.230	1.253	1.178	1.041	1.197	1.183	1.051	.913	.531
5	1.022	1.188	1.017	.713	.832	1.166	1.248	1.157	1.170	1.205	1.242	1.156	.911	.842	.500
6	1.111	1.221	1.151	.830	.799	1.029	1.216	1.046	1.146	1.078	1.208	1.170	.995	.851	.478
7	1.086	1.202	1.229	1.165	1.033	1.209	***1.266***	1.240	1.261	1.215	1.217	1.108	.835	.751	.417
8	1.114	1.079	1.231	1.246	1.214	1.258	***1.249***	1.106	1.213	1.064	1.151	1.085	.889	.717	.369
9	.818	1.128	1.252	1.156	1.038	1.233	1.090	1.249	1.073	1.134	1.115	.967	.815	.480	
10	.748	.976	1.178	1.169	1.143	1.250	1.207	1.073	1.158	.972	.965	.843	.518		
11	.960	1.125	1.041	1.204	1.082	1.208	1.052	1.130	.966	.947	.836	.520	.368		
12	1.111	1.183	1.195	1.235	1.204	1.205	1.143	1.111	.964	.838	.530				
13	.987	1.139	1.177	1.154	1.155	1.101	1.077	.965	.848	.524					
14	1.037	1.080	1.051	.917	.994	.829	.884	.814	.518	.373					
15	.942	.892	.906	.842	.855	.749	.715	.479							
16	.551	.533	.528	.495	.482	.417	.370								

An example of maximum radial power peaking factor trend over a full cycle is presented in Figure 5. Here the reactor is loaded with fuel bundles of a single mechanical design. MICROBURN-B2 calculated radial peaking factors are about 3-4% larger than the MICROBURN-B calculated values. The impact of such a radial peaking factor difference is seen in the calculated MCPR value shown in Figure 5. MICROBURN-B2 predicts about 3-5% smaller MCPR margins than MICROBURN-B. The POWERPLEX™-II core monitoring system follows the MICROBURN-B2 trend of MCPRs.

The same information for a BWR loaded with a mixture of two mechanical designs (9 × 9–2 and Atrium™-10) is provided in Figure 6. Here, MICROBURN-B2 calculated radial peaking factors are again about 3-4% larger than the MICROBURN-B calculated values. However, the POWERPLEX™-II calculated MCPRs follow more closely the MICROBURN-B prediction than the MICROBURN-B2 prediction during the period of BOC-MOC. This indicates that the core monitoring system did not see the increased power in those high power locations predicted by MICROBURN-B2. However, MICROBURN-B2 has been proven to be more representative of the real reactor condition than MICROBURN-B through many other BWR fuel cycle simulations. Thus Figure 6 is considered to provide evidence that the radial power distribution monitoring relies more on the core physics method than the measured LPRM signals.

Reliability of predicted control rod pattern

Plant operators are required to follow as closely as possible the predicted (suggested) control rod patterns provided by fuel cycle designers. This is because the control rod pattern at earlier cycle exposures affects the availability of predicted thermal margins at later cycle exposures and the cycle energy.

Figure 5. Radial power peaking and MCPR trends in BWRs of single assembly type loading

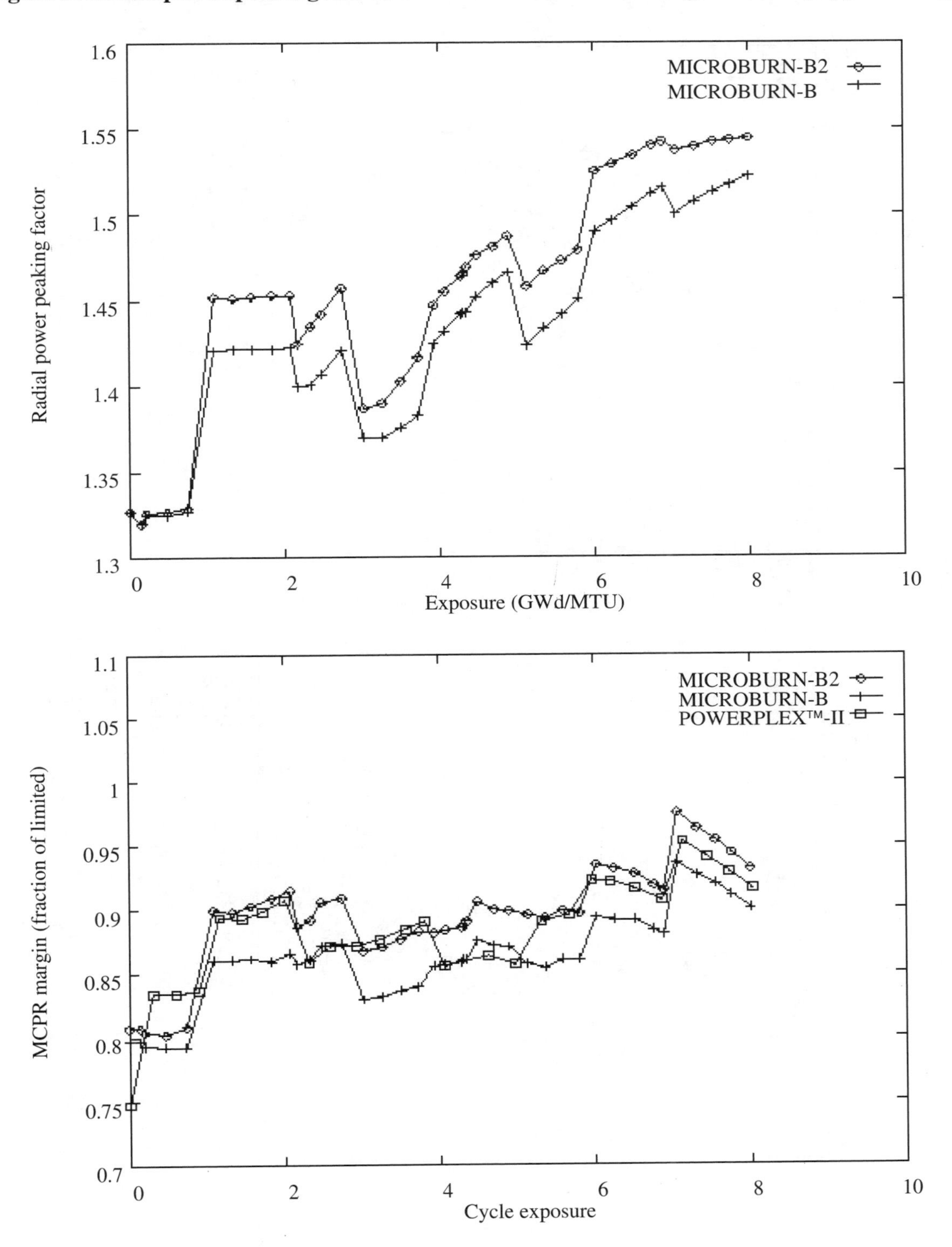

Figure 6. Radial power peaking and MCPR trends in BWRs of highly heterogeneous loading

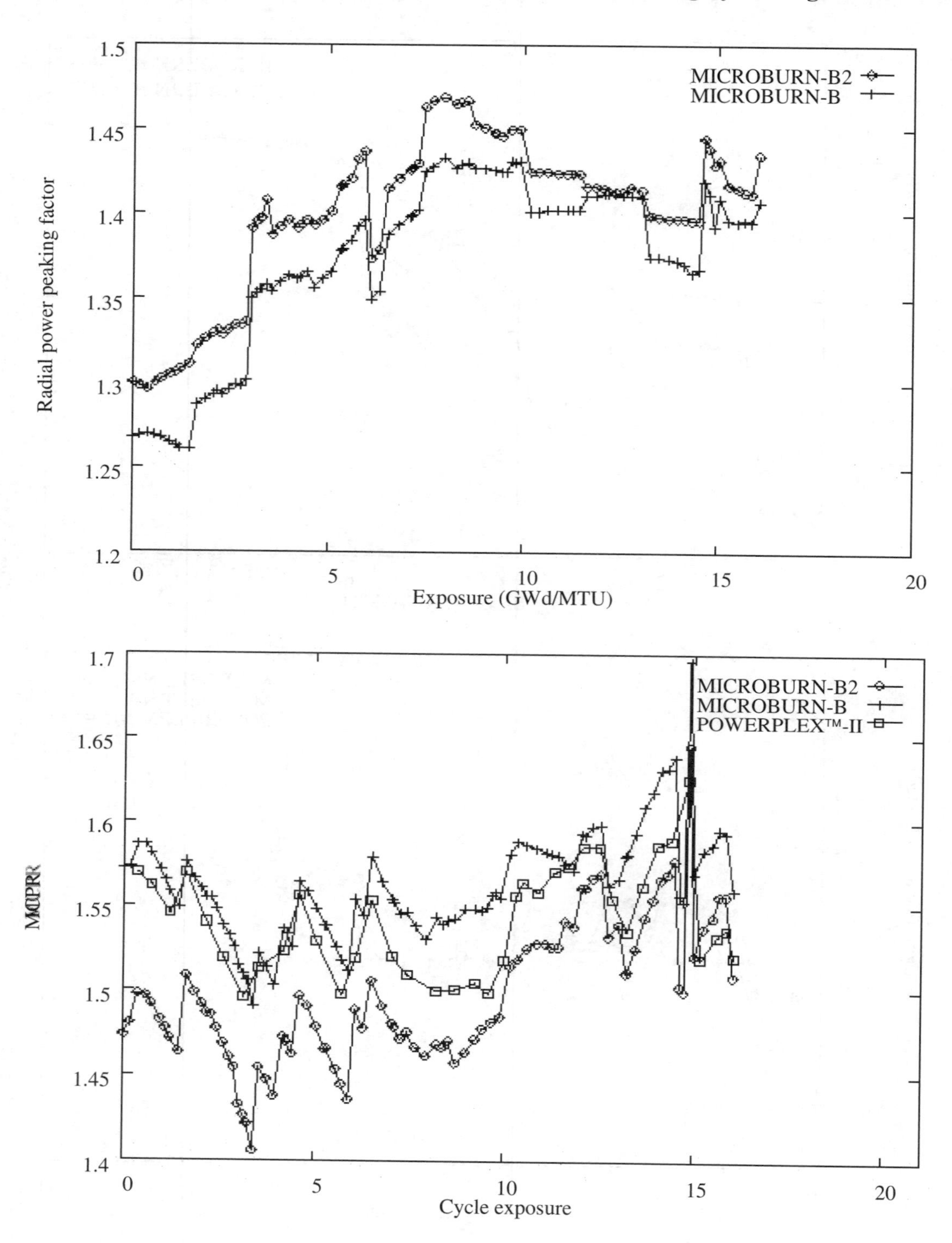

On a quick turn around basis, core power/flow transients (start-up, load excursion, system test, etc.) require control rod patterns predicted ahead of the actual manoeuvre. For this purpose, plant operators may depend on the prediction of core monitoring systems. Thus, for both short-term and for long-term operations, the predictive accuracy of a core monitoring system is important. Typically, the critical k-effective of a BWR core simulator may exhibit a dependency on the core void fraction or the core inlet sub-cooling (enthalpy). The agreement of predicted axial power distributions with the measured axial power distributions usually depends on the cycle exposure and the core loading pattern. MICROBURN-B2 reduces the uncertainty in the predicted hot critical k-effectives and axial power distributions compared to MICROBURN-B [1]. Thus it is anticipated that the new core monitoring system POWERPLEX™-III will provide more reliable control rod patterns to plant operators.

Relaxed rod sequence (RRS) exchange monitoring

The relaxed rod sequence (RRS) exchange operation is a new operating strategy [2] which employs a smaller number of control rod pattern changes than the conventional rod sequence exchange strategy. The RRS rod sequence exchange is illustrated in Figure 7 in comparison with the other two existing strategies.

Figure 7. Three types of control rod sequence exchanges

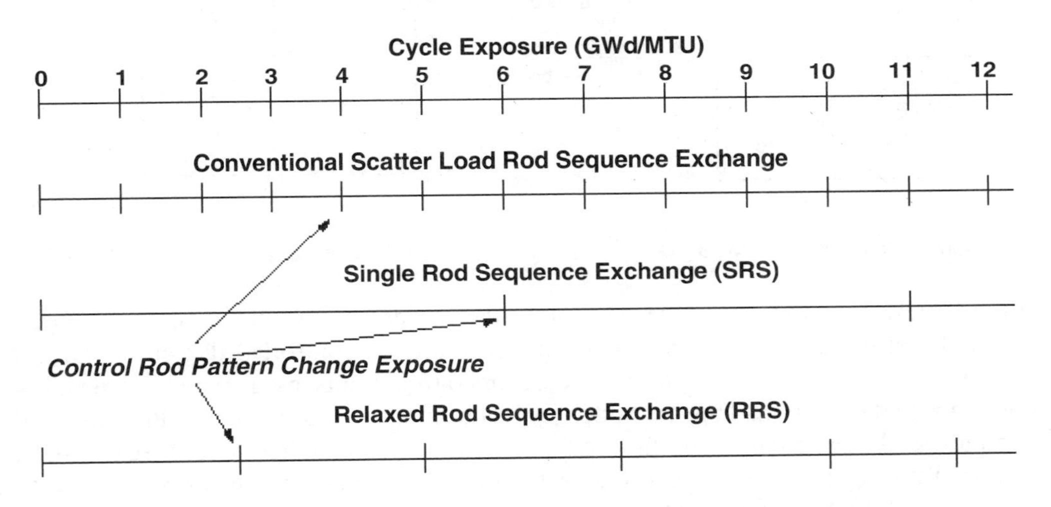

The advantage of RRS is that it reduces the operating cost of a BWR while preserving the loading pattern flexibility of the conventional rod sequence exchange. On the other hand, the single rod sequence (SRS) core has a limitation in selecting fuel bundles to be loaded in control cells. The implementation of RRS requires a special consideration of potential power peaking during and following a rod sequence exchange. The power ramp rate in fuel rods adjacent to the control blade to be withdrawn has to be closely monitored. Recently, there have been several reports of fuel rod failures that occurred in control cells following control blade withdrawals. Except for those cases which were related to certain manufacturing defects, these fuel rod failures are related to the high power ramp rate and the final high power peaking in regions adjacent to a withdrawn control blade. This issue is illustrated in Figure 8.

To effectively deal with the power peaking increase in an RRS core, an extension of the SRS power peaking penalty methodology could be employed. In fact, this is the path taken by plant operators facing this issue using their existing core monitoring systems. In the SRS power peaking

Figure 8. Illustration of RRS power peaking

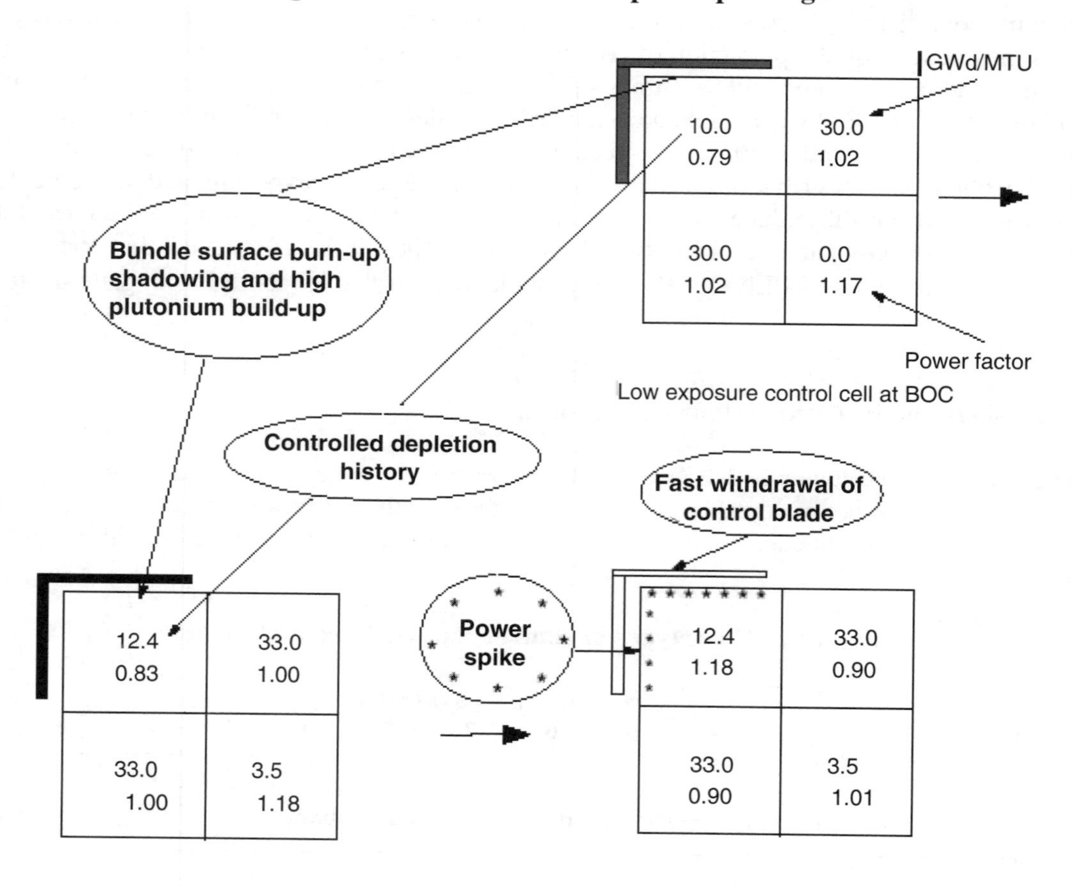

methodology, a few fuel rods near the control blade corner are examined in an infinite medium condition (single lattice depletion) for peaking increase after a controlled depletion. The calculated incremental increase is then applied to SRS core operating limits as a penalty depending on the duration of controlled depletion. This methodology has been applied to control cells composed of high exposure bundles. A comparison of this conventional penalty method and MICROBURN-B2 is summarised in Table 1. This table suggests that the penalty method should not be applied to control cells containing fresh fuel bundles and to low exposure lattices under cyclic control.

Table 1. RRS methodologies

Issues	Local peaking penalty method	MICROBURN-B2
Controlled depletion of twice burned bundles	Allowed	Allowed
Controlled depletion of once burned bundles	Allowed with an increase in uncertainty	Allowed
Controlled depletion of fresh fuel bundles	Not allowed	Allowed
Cyclic controlled depletion	Not modelled	Modelled
Controlled depletion reactivity effect	Difficult to model	Modelled

The inability of the penalty method to deal with a fresh bundle is related to the non-linear depletion of gadolinium. The fast exponential depletion of gadolinium during the early life of a fuel lattice makes it difficult to correlate the local peaking factor increase as a function of controlled depletion length. A cyclic control of a lattice is typical of BWR operations. An example of local peaking factor trend over cyclic control of a fuel lattice is shown in Figure 9. This figure shows that, contrary to the conventional notion, the local peaking factor of a previously controlled lattice does not return to the level of uncontrolled lattice at the same exposure. This indicates that the spectral history effect of a controlled depletion is not erased by a subsequent uncontrolled depletion. MICROBURN-B2 contains a local peaking method qualified [3] for modern bundle designs to effectively handle this situation. The MICROBURN-B2 result shown in Figure 9 was obtained using the standard SPC procedure, which is a single bundle uncontrolled depletion followed by instantaneous controlled solutions for selected exposure points. A controlled depletion branch in the lattice calculation is not necessary for MICROBURN-B2 due to its microscopic depletion method. This provides the flexibility needed for simulating a cyclic controlled depletion.

Figure 9. Effect of cyclic control on local peaking factors

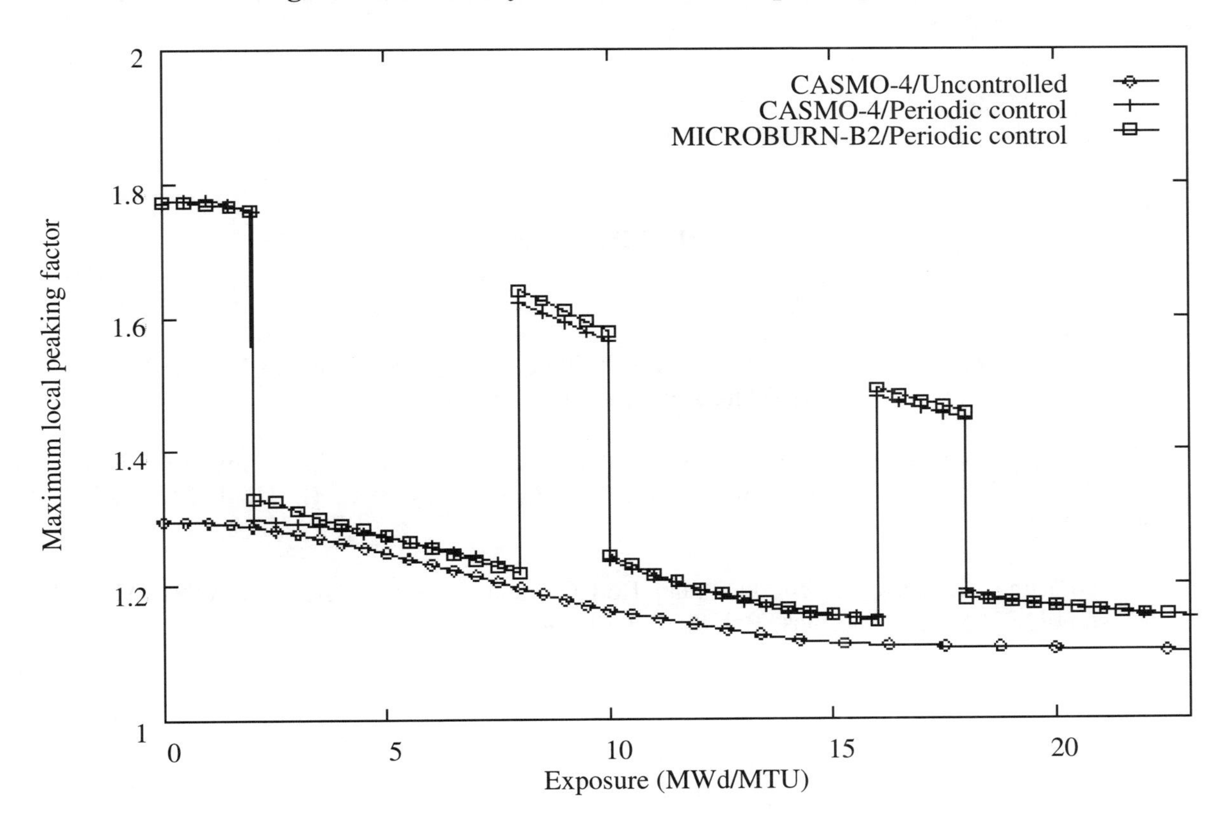

One last issue relating to the RRS operation is that a lattice with a previously controlled depletion has a reactivity potential different from a lattice with no previous controlled depletion at the same exposure and the same void history. Again, the microscopic depletion method provides a simple solution to this issue. In summary, a controlled depletion of a fresh lattice and a cyclic control of a lattice are complicated technical issues for plant operators trying to implement an RRS based on their current core monitoring systems. POWERPLEX™-III provides an excellent technical solution to these issues.

Summary and conclusion

The new BWR core monitoring system POWERPLEX™-III will benefit from the advanced BWR core simulator MICROBURN-B2 in key areas like radial power distribution monitoring, reliability of predicted control rod patterns, and relaxed rod sequence exchange core monitoring. However, it will require a computer with two to three times faster floating point speed and three to five times more memory than the current POWERPLEX™-II core monitoring system.

The controlled depletion of fresh bundles and the cyclic control of a fuel lattice in an RRS core make it unfeasible to rely on the conventional local peaking penalty method for monitoring purposes. The advanced capability of MICROBURN-B2 makes it simple and accurate to deal with an RRS core. POWERPLEX™-III is thus well suited for monitoring this advanced operating strategy.

The radial power distribution monitoring is largely dependent on the predictive capability of the core monitoring system rather than measured LPRM signals. Due to this reason, POWERPLEX™-III will improve the accuracy of the monitored (measured) radial power distribution, thus better ensuring the integrity of fuel rods.

REFERENCES

[1] St. Misu, H. Moon, "The Siemens 3-D Steady State BWR Core Simulator MICROBURN-B2," Proc. Int. Conf. Nuc. Sci. and Tech., Vol. 2, Long Island, NY (Oct. 1998).

[2] H. Moon, "Advanced Core Design Using MICROBURN-B2 BWR Core Simulator," Proc. Am. Nucl. Soc. Topical Meeting – Adv. in Nuclear Fuel Management II, TR–107728–V1, EPRI (March 1997).

[3] H. Spierling, St. Misu, H. Moon, "Fuel Rod Gamma Scan of a Siemens ATRIUM-10 Fuel Assembly," ANS Transactions, Vol. 80 (June 1999).

CORE MONITORING BASED ON ADVANCED NODAL METHODS: EXPERIENCE AND PLANS FOR FURTHER IMPROVEMENTS AND DEVELOPMENT

Alejandro O. Noël, Lorne J. Covington
Studsvik Scandpower

Alf Nilsson*, Daniel Greiner
Kernkraftwerk Leibstadt

Abstract

The core surveillance methods implemented in the 3-D core simulators SIMULATE-3 and PRESTO-2 will be described, with emphasis on the adaptation of calculated results to measured detector signals. Furthermore, experience from 10 cycles of on-line operation of PRESTO-2 in Leibstadt is presented.

* Present affiliation: Barsebäck Kraft AB

Introduction

The core simulators SIMULATE-3 [1,2] and PRESTO-2 [3], both employing advanced nodal methods, are today in use in a large number of BWR and PWR plants world-wide. Besides standard off-line use for core design, core tracking and operational support, the simulators are also used for on-line core monitoring in some plants. Several projects for implementing the simulators for core monitoring are in progress.

PRESTO has been in use on-line in the Brunsbüttel [4] and the Leibstadt [5] plants for more than a decade. SIMULATE-3 has been enhanced in some of the following areas to support BWR on-line core monitoring:

- A LPRM and TIP based adaptive power model.
- Pin power based PCI monitoring.
- Enhanced data control.

This on-line version of SIMULATE-3 is currently in use at four utilities and nine BWR reactors as a means of upgrading the on-line monitoring systems.

This paper addresses the models implemented in the simulator to calculate an inferred adapted power model using detector signals. The methods used allow thermal limits to be calculated using both predicted power shapes and power shapes that are adapted to measured TIP and LPRM readings. The paper will present the approach used in rejection and substitution of LPRM and TIP measurements, axial and radial expansion of the adapted power shape and available feedback mechanism for the adapted power shape.

The experience gained from 10 cycles of on-line core supervision at Kernkraftwerk Leibstadt is presented. Of special interest here is the actual power uprate in the plant which imposed special demands on the accuracy of the simulator and resulted in an upgrade of core physics methods, which since the last cycle is based on HELIOS/PRESTO-2.

Adaptive methods

Adaptive methods, as applied in SIMULATE-3 and PRESTO-2, are described below. The basic approach in both simulators is to accurately calculate detector signals and compare with corresponding measured values. The deviation between predicted and measured detector values is employed primarily to adapt the local power used in prediction of thermal limits in the plant. Minor differences exist in how the detector signals are being utilised but larger differences may also exist in the strategy for adaptation chosen by different users. A spectrum of methods exists, varying from using the adapted solution only as a back-up and check for the accuracy in the calculation, to the other extreme of feeding the adapted values back into the basic neutronics and depletion calculations in the simulator.

SIMULATE-3 adaptive model

Detector signals

Part of the SIMULATE-3 calculation sequence for each snapshot is to provide computed values for each of the traversing in-core probes (TIPs) and local power range monitors (LPRMs). These detector signals are computed in SIMULATE-3 by using the detailed flux reconstruction module, which is also used for the detailed pin-by-pin power distribution calculation. If the SIMULATE-3 calculation is performed in fractional core geometry, TIP and LPRM computed signals are folded out into full core geometry so that computed signals can be compared with measured signals.

TIP calculation

While the details of the flux reconstruction methods are described elsewhere, it is important to understand how the detector signals are computed. It is assumed that the detector is represented by a trace amount (or an explicit detector) of a known isotope (e.g. ^{235}U) or material (steel) and that the cross-sections for the detector have been generated in lattice code (CASMO) and placed into the neutronics data library.

The five basic steps in the detector calculation are:

- Reconstruction of the average fast and thermal neutron flux in the narrow-narrow gap of each assembly n, for each axial plane k:

$$\Phi^k_{n,1}, \Phi^k_{n,2}$$

- Interpolation of two-energy group, microscopic detector cross-sections corresponding to local conditions (e.g. bundle exposure, void history, control rod history, instantaneous void and fuel temperature) of each assembly n, for each axial plane k:

$$\sigma^k_{n,1}, \sigma^k_{n,2}$$

- Calculation of the detector reaction rate in the narrow-narrow gap of each assembly n, for each axial plane k:

$$TIPC^k_n = \Phi^k_{n,1}\sigma^k_{n,1} + \Phi^k_{n,2}\sigma^k_{n,2}$$

- Normalisation of the TIP signals to an average value of 1.0 over all axial planes and all narrow-narrow gaps. This normalisation is chosen so that the computed TIP distributions are independent of the TIPs that are in service, and it facilitates meaningful comparison of computed TIP signals between snapshots which might have different numbers of functional TIPs.

- Averaging of the TIP signals for the four assemblies n which surround the actual TIP location l in axial plane k:

$$TIPC^k_l = \frac{1}{4}\sum_{n=1}^{k} TIPC^k_n$$

LPRM calculation

The LPRM calculation is nearly identical to that of the TIP, with two minor differences. First, the detector type for the LPRM can be different (e.g. ^{235}U) from that of the TIP (e.g. gamma detector) which simply changes the microscopic cross-section for the detector in LPRM level z:

$$LPRMC_n^z = \Phi_{n,1}^z \tilde{\sigma}_{n,1}^z + \Phi_{n,2}^z \tilde{\sigma}_{n,2}^z$$

Secondly, the average fluxes at the LPRM locations are computed by performing an axial integration of the flux shapes (fourth-order polynomials) over the axial location of the LPRM as shown in the figure.

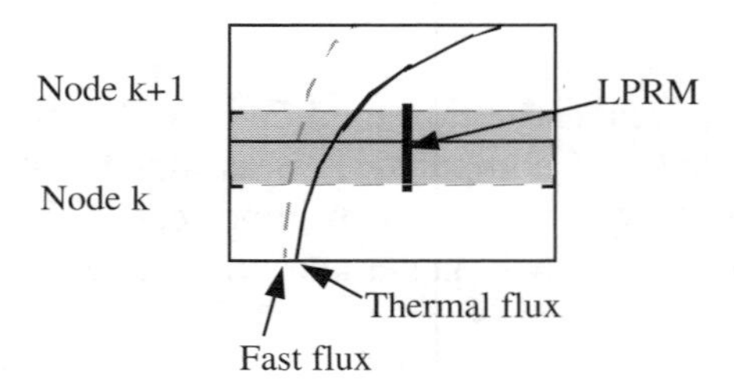

Note that the integration of the LPRM fluxes may require evaluation of the flux contributions from two adjacent nodes if the LPRM spans a nodal boundary. Finally, the LPRM signals for the four assemblies n which surround the LPRM location l in axial level z are averaged to obtain:

$$LPRMC_l^z = \frac{1}{4}\sum_{n=1}^{z} LPRMC_n^z$$

LPRM-to-TIP calibration

The LPRM signals are very useful in detecting changes in flux distribution within the reactor core. However, the LPRM signals change even if the flux is constant, because of detector isotopic depletion, instrumentation drift, detector failures, etc. In order to compensate for these (hopefully small) changes, the LPRMs are assumed to be calibrated with respect to TIP measurements on a periodic (at least monthly) basis.

TIP rejection and substitution

When the TIP measurements are performed, it is sometimes not possible to physically insert the TIP into each measurable location. On rare occasions, the TIP trace itself may have some undetected problem, and the measured TIP signal may not be reliable. When this happens, there will be no available TIP signal with which to perform the LPRM calibration. Under such circumstances, the calculated TIP signal can be used to substitute for the missing measured values.

First, when measured TIP signals are provided in 48 axial nodes, the signals are collapsed to 24 nodal values by simple averaging. Secondly, the 24 axial node measured TIP signals (TIPM) are all globally normalised so that the average measured TIP matches the average calculated TIP. Subsequently, the TIP signal for a given radial location l is assumed to be unusable if it fails either of the two following tests in any axial plane k:

$$\frac{TIPM_l^k}{TIPC_l^k} > TIPMAX, \quad \frac{TIPM_l^k}{TIPC_l^k} > TIPMIN$$

If this should occur, the measured TIP values for all planes of the failed TIP are replaced by the computed values:

$$TIPM_l^k = TIPC_l^k,\ k = 1, k\max$$

In many cases the TIP measurement system is limited in stroke to an elevation below the top of the active fuel. Consequently, a pseudo-measured TIP signal(s) is substituted in the plane(s) where measured data is absent. They are computed by assuming that the computed axial TIP shape can be used for the planes with missing data:

$$TIPM_l^k = TIPM_l^{k-1} \times \frac{TIPC_l^k}{TIPC_l^{k-1}},\ k = ktop,\ k\ \max$$

After measured TIP rejection and substitution is completed, the values of TIPM are re-normalised so that the average of the measured values matches that of the computed TIPs.

LPRM rejection and substitution

The LPRM signals are also evaluated to determine if the signals are reliable. First, all measured LPRM signals (LPRMM) are globally normalised so that the average measured LPRMs match the average calculated LPRMs. Subsequently, the LPRM signal for a given radial location l and axial level z is assumed to be unusable if it fails any of the following tests:

- Measured values are realistic (i.e. no zeros or negatives). Measured TIP and/or LPRM signal is "very" much different from predicted signal. This gross rejection test determines if any signals are failed.
- Rejection occurs with an excessive change in the LPRM ratio. This test ensures that any electronic spikes in the measured data are not used in the adaptation process.
- Rejection occurs if the LPRM signal is outside the RMS error of the other detectors in the same axial plane. This test ensures no one detector in any particular plane has excessive errors due to detector drift.

If this should occur, the measured LPRM values for the failed plane(s) of the LPRM are synthesised from the computed values using the relation:

$$LPRMM_l^z = LPRMC_l^z \times LPRMR_l(z)$$

where LPRMR(z) is computed by performing a quadratic (or linear) polynomial fit to the remaining three (or two) LPRM signals in the LPRM string. The expansion constraints are that the polynomial reproduce the ratio of measured-to-calculated LPRM at the valid levels:

$$LPRMR_l(z) = \frac{LPRMM_l^z}{LPRMC_l^z}$$

When the polynomial is evaluated at a specific LPRM location, the value of LPRMR(z) is required to be within the maximum and minimum constraint values to ensure that excessive extrapolation is not introduced from the polynomial fitting. If more than two of the levels are failed, the calculated values of the LPRM are substituted for the failed measured values. After measured LPRM rejection and substitution is completed, the values of LPRMM values are re-normalised so that the average of the measured values matches that of the computed LPRMs.

LPRM calibration

The LPRM calibration is performed so that the LPRM signal corresponds to the signals measured with the TIPs at the LPRM locations. In practice, the gain adjustment factors (GAFs) for the LPRMs may not be re-calibrated unless there is a threshold level of difference between the measured TIP signal and the measured LPRM signal. Analysis of LPRM calibration data indicates that this operational threshold may be as large as 5.0%. A difference between LPRM and TIP of such magnitude would make the interpretation of the subsequent LPRM signals difficult to relate to the desired core power distribution.

Consequently, when the LPRM calibration is modelled in SIMULATE-3, the measured TIP signals (DET.TIP card data) are compared with the LPRM signals (DET.PRM), and an LPRM deviation distribution is retained for subsequent use. The deviation distribution is defined as:

$$LPRMDEV_l^z = \frac{TIPLPRM_l^z}{LPRMM_l^z}$$

where the TIP-inferred LPRM signal is computed as:

$$TIPLPRM_l^z = \left(\frac{TIPM_l^{k1} + TIPM_l^{k2}}{2}\right) \times LPRMSHP_l^z$$

and:

$$LPRMSHP_l^z = \frac{LPRMCTIP_l^z}{\left(\frac{TIPC_l^{k1} + TIPC_l^{k2}}{2}\right)}$$

where $k1$ and $k2$ are the two axial planes containing level z of the LPRM string. Note that the local shape factor accounts for the differences between the nodal and localised LPRM position of TIP signal. $k1$ and $k2$ may be the same plane if the LPRM does not span a nodal boundary. It is also very important that the LPRMCTIP quantity is defined as the computed reaction rate for a TIP-type detector at the LPRM location (not the LPRM detector type).

If the operational gain adjustment factors were perfect, all of the LPRM deviations would be unity at the time of the LPRM/TIP calibration.

LPRM-based adaptation of power distributions

The basis for the determination of "measured" power distributions is that the computed power shape (POWC) is assumed to differ from the measured power shape (POWM) in the same manner as the computed TIP signals differ from the measured TIP signals.

At the TIP calibration points, it is easy to determine the ratio of the computed TIP to measured TIP signal at each radial TIP position and in each axial node. This ratio is defined to be TIPRAT:

$$TIPRAT_l^k = \frac{TIPM_l^k}{TIPC_l^k}$$

Radial power distribution adaptation

Given the values of TIPRAT for each axial plane of each TIP, one must construct a radial interpolation scheme for each fuel assembly. This is achieved by combining the TIPRAT values with a set of radial weighting factors for each of the TIPs (A, B, C, D) within a span of five fuel assemblies from the assembly in question.

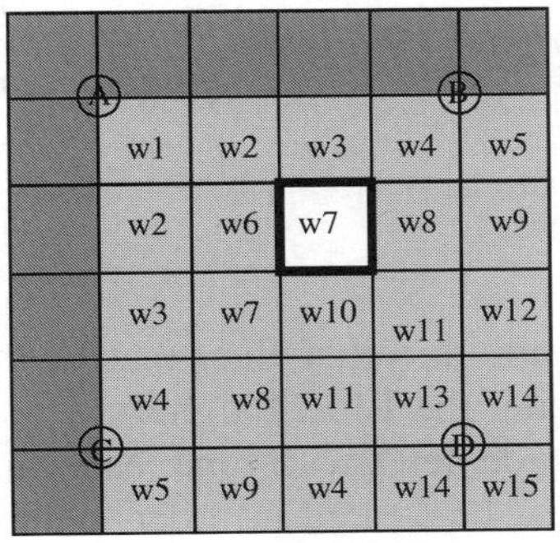

A B C D

w1	w2	w3	w4	w5
w2	w6	w7	w8	w9
w3	w7	w10	w11	w12
w4	w8	w11	w13	w14
w5	w9	w4	w14	w15

Note that the radial weighting factors in Figure 2 are relative to TIP A, and the weighting factors are assumed to be diagonally symmetric and the same for all detectors (the values have to be rotated before being used for the other TIPs of the figure).

The measured power distribution in each assembly n and axial plane k is computed as:

$$POWM_n^k = POWC_n^k \times \frac{\sum_{l=1}^{4} TIPRAT_l^k \times w_n^l}{\sum_{l=1}^{4} w_n^l}$$

The radial weighting factor array is user input and zero values are permitted. It should be noted that inclusion of zero values in the radial weighting array might imply that some fuel assemblies are not within reach of any LPRM/TIP location. If this occurs, the value of the adapted power shape is set equal to the un-adapted SIMULATE-3 predicted power in that fuel assembly. If the innermost two positions of the radial weighting array are non-zero, all fuel assemblies in the interior will be within reach of some LPRM/TIP, but at the outer edge of the core, there may be assemblies outside of the range of any LPRM/TIP. Note that the measured power distribution must be re-normalised to preserve the core-averaged value of 1.0 needed in the SIMULATE-3 depletion equations.

Pseudo-LPRM signal construction

Even though the LPRMs have been calibrated to produce the same signals as the TIPs (although they may have different detector types) the LPRMs cannot be compared directly to computed LPRM signals. First, the LPRM signals must be corrected for detector depletion effects, mis-calibration, and the differences between the types of detectors used for the TIP and LPRM signals. At each snapshot, the pseudo-LPRM signal is computed as:

$$LPRMP_l^z(E_i) = \frac{LPRMM_l^z(E_i)}{LPRMDEP_l^z(E_i)} \times LPRMDEV_l^z(E_0) \times \left[\frac{LPRMC_l^z(E_0)}{LPRMCTIP_l^z(E_0)} \right]$$

Note that the last factor in the above equation is simply the ratio of the LPRM signal computed with the LPRM-type detector to the LPRM signal computed with a TIP-type detector. The LPRMDEP factor is used to model the LPRM depletion effects between TIP calibrations, and it is defined in the following section. The pseudo-LPRM signal (LPRMP) is now directly comparable to the SIMULATE-3 computed LPRM signal, and it is used to define a deviation between the computed and measured LPRM signals as:

$$LPRMRAT_l^z(E_i) = \frac{LPRMP_l^z(E_i)}{LPRMC_l^z(E_i)}$$

LPRM depletion model

Once the reactor is at some condition other than that of the TIP calibration, it is necessary to synthesise the TIPRAT values by using the LPRM signals. This task is made more difficult because the fissionable material in the LPRMs depletes with time. Naturally, this depletion is not uniform in all of the LPRMs, and the depletion must be treated in order to separate the depletion effects from flux distribution changes in the LPRM signals. One of the LPRM depletion models in SIMULATE-3 allows the user to input the detector relative response curves versus detector fluence. These curves are then used to calculate the change in detector response since the last TIP calibration.

Axial shape adaptation

Given the pseudo-LPRM signals at each snapshot, an axial shape correction factor is computed for each axial plane of each LPRM/TIP string. The snapshot axial correction factor is computed as:

$$TIPRAT_l^k(E_i) = TIPRAT_l^k(E_0) \times \left[\frac{LPRMRAT(E_i)}{TIPRAT(E_0)}(z) \right]_l$$

where the bracketed quantity (the ratio of measured-to-computed LPRM signals divided by the ratio of measured-to-computed TIP signals at the LPRM/TIP calibration) is computed by performing a cubic polynomial fit for each LPRM string. The constraints on the polynomial expansion are the values of the function at the LPRM axial locations. The polynomial function is then evaluated for each axial plane to fill in the TIPRAT values for all axial planes (typically 25). When the polynomial of the above equation is evaluated for a specific plane, the value of the polynomial is required to be within the maximum and minimum constraint values to ensure that excessive extrapolation is not introduced from the polynomial fitting.

This definition of a planar axial shape for each TIP/LPTM string implies that the gross correction of the power distribution is determined by the error in the calculated LPRM signals, and the fine structure is determined by the error in the calculated TIP signals at the time of the LPRM/TIP calibration. Note that at the LPRM/TIP calibration point, the bracketed quantity is unity, and the TIPRAT values remain those obtained from the measured and computed TIPs.

If the error in the computed LPRM at any snapshot is identical to the errors of the TIPs at the LPRM/TIP calibration, the TIPRAT values are once again equal to those from the measured and calculated TIPs.

Iterative adaptation models

The power adaption models described above are very similar to other vendor models which are widely used models for on-line adaptions of measured power distributions and for determination of limits to operating margins. In principle, this type of adaptive model can be used exclusively for on-line adaptions without significant limitations. The adequacy, however, of any adaptive model is determined by the uncertainty in the predicted limits to thermal margins, and this uncertainty is difficult to determine.

It has been observed from gamma scan comparisons that in some cases the use of the full 3-D adapted power shapes can lead to worse bundle power predictions at EOC than those obtained with direct SIMULATE-3 calculations. The reason for the larger errors is primarily because the TIP/LPRM signals are not perfectly reliable for determining radial power distributions. For these reasons, SIMULATE-3 supports computation of adaptive power shapes in one of four different ways: 1) in full 3-D adaption; 2) in axial only adaption of each bundle; 3) in radial only adaption of each bundle; or 4) in core-averaged axial (1-D) adaption of all bundles.

One certainty is that the closer the predicted TIP and LPRM signals are to their measured analogues, the better the prediction, and the smaller the uncertainty in the predictions. Consequently, it is possible to use the adaptive power shapes for more than simply altering the power distribution in thermal limits calculation. In SIMULATE-3, the adaptive power shape can also be used to drive the iterative neutronics/thermal-hydraulics calculation and/or to perform the fuel assembly depletions.

PRESTO-2 adaptive model

The adaptive methods of PRESTO-2 are implemented in a code module outside the simulator kernel as an adaptive post-processor named ADAPT. ADAPT modifies the power, linear heat generation rate and thermal margin distributions to reflect the difference between measured and calculated detector signals. The correction function can be either based on the TIP differences only, or on the combination of TIP and LPRM detector differences. Once ADAPT has evaluated the modified (adapted) nodal power distribution, it derives all other thermal quantities distributions by means of the implemented plant-specific algorithms of PRESTO-2.

The adaptive methodology of ADAPT follows that described for SIMULATE-3 with the following differences:

- Detector signal validation: ADAPT assumes that the process computer delivers already validated TIP and LPRM signals. Any invalid detector signal shall be flagged as failed and is then excluded from the adaptive calculations.

- Pseudo-LPRM signal construction: ADAPT assumes that the input measured LPRM readings are already adjusted with the TIP calibration factors LPRMDEV and the depletion factors LPRMDEP. The process computer calculates the LPRMDEV factors, while the LPRMDEP factors can either be provided by the process computer or calculated by PRESTO-2. Note that no correction is applied to the pseudo-LPRMs (named as calibrated measured LPRMs $PRMMEA_z^l$ in ADAPT) to account for the different TIP and LPRM detector types.

- Predicted LPRM calculation: at the time of a TIP calibration, PRESTO-2 evaluates and stores an LPRM correction factor array defined as:

$$COFPRM_z^l = TIPC_z^l / LPRMC_{zl}$$

where $TIPC_z^l$ is the predicted LPRM detector signal using the TIP detector model. If the TIP and LPRM detectors are of the same kind (e.g. fission chambers), COFPRM reduces to 1.0 everywhere. Each time PRESTO-2 makes a new state point calculation, it calculates the predicted LPRM readings as:

$$PRMPRD_z^l = LPRMC_z^l * COFPRM_z^l$$

- Otherwise, the adaptive methods are identical to those discussed for SIMULATE-3. The deviation between calculated and measured LPRM signals, defined by LPRMRAT, is finally calculated in ADAPT as:

$$LPRMRAT_l^z = PRMMEA_l^z / PRMPRD_l^z$$

On-line experience of PRESTO at Leibstadt

Core supervision strategy

The core supervision strategy selected in the Leibstadt plant is to base the operation on purely calculated thermal margins without any adaptation as long as the simulator shows good accuracy compared to detector measurements. However, an adaptation method is still implemented and can be activated by the reactor engineers if insufficient accuracy is detected in the predictions.

The operating staff decided this strategy when it could not be demonstrated that adaptive calculations would actually reduce the uncertainty in the thermal margin calculations when the following factors were properly taken into account:

- Uncertainties in detector measurements.
- Uncertainties in detector models.
- Detector signals see contributions only from the four neighbouring fuel channels.
- Detector signals can not resolve the contribution from an individual channel (out of the four surrounding channels).

Even if the 3-D simulator delivers accurate results, the chosen strategy may still fail if:

- Input data to the simulator is not properly given.
- Process data contain errors, e.g. failed control rod position indications.

In order to assure the accuracy of simulator and to protect against errors of this type the following checks are systematically performed by the reactor engineers:

1. Periodic TIP comparisons: TIP/LPRM calibrations are performed at regular intervals not exceeding one month. The measured TIP detector readings are compared against those predicted by the simulator (PRESTO-2) to validate the on-line core supervision results.

2. LPRM surveillance: a snapshot of the differences between predicted and measured LPRM readings is saved after each TIP/LPRM calibration. These differences, and trends in differences, are then evaluated after each PRESTO-2 calculation. In the event the differences exceed pre-set limits, a warning is issued in the graphical user interface of the core supervision system, demanding an action from the reactor engineer.

3. LPRM signal tracking: given stable conditions, the LPRM readings are expected to remain stable within certain pre-set limits. Unexpected changes may be an indication of either a detector failure or other changes in core conditions, not properly tracked by the simulator.

In the event any of the above mentioned checks does not fulfil the acceptance criteria, the reactor engineer has a number of alternatives for action:

- Perform a new TIP/LPRM measurement and validate once more the results.
- Re-calibrate some LPRM detectors by means of a partial TIP/LPRM calibration.
- Ignore the warnings and/or bypass the detector(s) that triggered the warning if he concludes that they have failed (drifting detectors).
- Compare purely predicted vs. adapted thermal margins to decide whether there is a risk or not of exceeding the thermal limits.
- Finally, if the conclusion of all analyses is that the simulator fails to predict the actual core conditions with the expected accuracy and the adaptive method yields more conservative thermal margins; the reactor engineer may switch on the adaptive method. After setting the adaptive method switch on, the core supervision system shows in all displays the adaptive thermal margins instead of the purely predictive.

Note that the adaptive thermal margins are always available in the database. The effect of turning the adaptive method switch on is that all system edits and panels show the adapted margins instead of those purely predicted.

3-D simulator methods

A precondition for the KKL core supervision strategy is that, given correct input data, the 3-D simulator will deliver accurate thermal margin distributions. With the development of advanced nodal methods, KKL could further improve the accuracy of their on-line results by upgrading the core physics methods in 1998. The introduction of improved physics methods was also seen as an important step in the process of uprating the power of the plant, which has reached 112% in the present Cycle 16. Two of the improvements include:

- The introduction of the new transport theory lattice code HELIOS. Besides improving the accuracy of basic cross-section data, the use of HELIOS, with its generalised geometrical capabilities, also supported the development of a more accurate detector model based on pin power reconstruction.
- The introduction of the new 3-D simulator PRESTO-2. It employs a two-energy group advanced nodal code supporting intranodal flux reconstruction, pin power reconstruction and the more accurate detector model developed from HELIOS.

Once the new methods were adopted in 1997, KKL started an ambitious V&V programme that also led to the licensing of the new methods. In parallel the on-line core supervision system was being upgraded and qualified in a parallel operation mode. The new methods have been in on-line production use since BOC 15, September 1998. The implementation of the new methods and the upgrade of the core supervision system were accommodated with technical support from Studsvik Scandpower.

Results

Figures 1 and 2 show the on-line TIP comparison nodal resp. radial r.m.s. values for KKL Cycles 6 to 15. The data summarises the on-line results during the 10 years of core supervision results using the present methods. The dashed vertical line shows the boundary between the "old" core physics based on RECORD/PRESTO and the "new" physics methods with the introduction of HELIOS/PRESTO-2.

The introduction of the new methods clearly reduced the uncertainties in the TIP detector predictions. Although in some cycles the TIP comparisons resulted in nodal r.m.s. greater than the administrative limit of 7%, it was never necessary to switch the adaptive methods on, mainly because even with somewhat higher uncertainties the thermal margins were large enough to guarantee that the thermal limits would not be exceeded.

Figure 3 shows the trend of the thermal limits during Cycles 14 and 15. It can be observed that during Cycle 15 the MFLPD margin was much smaller than during Cycle 14, mainly due to the power uprate of the plant to 109% in this cycle. The introduction of the new methods, however, highly improved the accuracy in the TIP comparisons, which opens the possibility in the future to take credit for the new, more accurate methods.

Figure 1. KKL C06-C15 TIP nodal r.m.s.

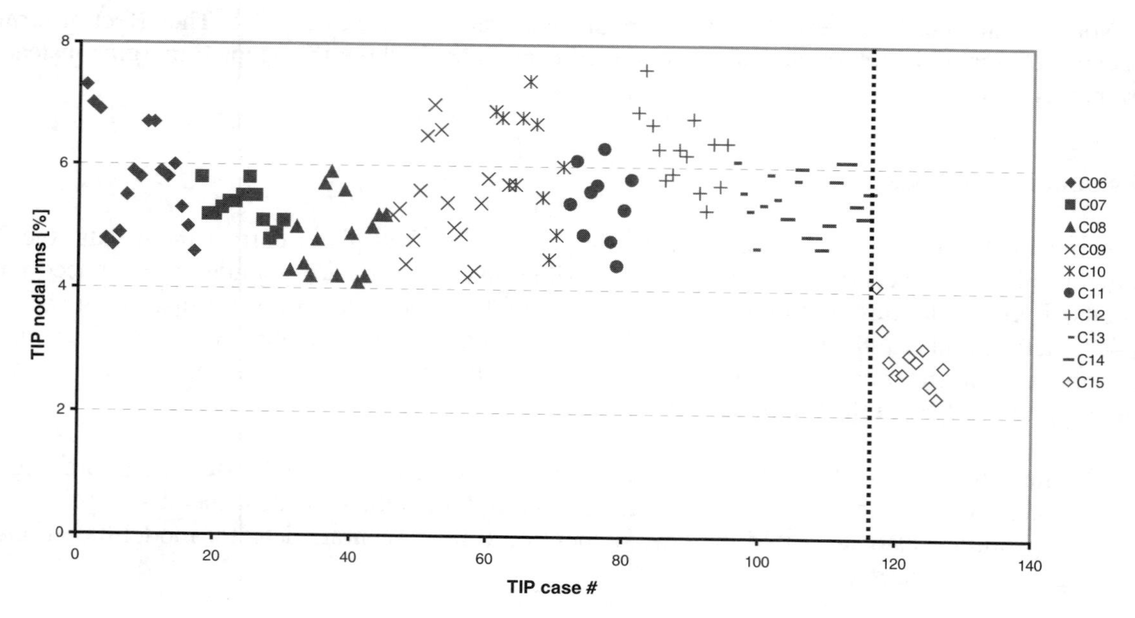

Figure 2. KKL C06-C15 TIP radial r.m.s.

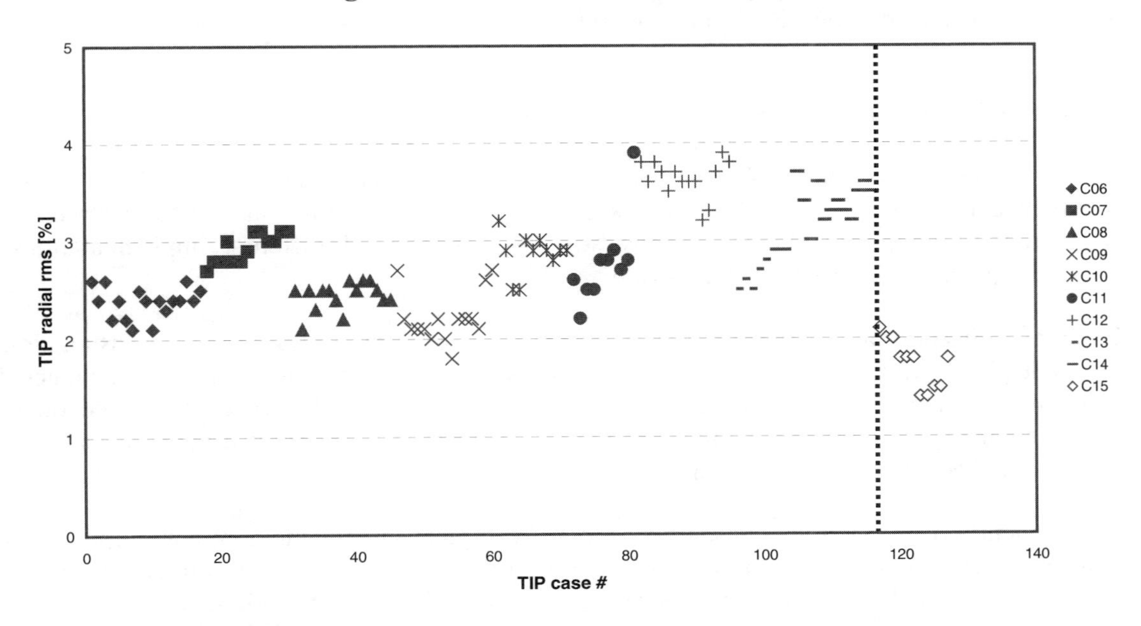

Figure 3. KKL on-line core supervision, thermal margins

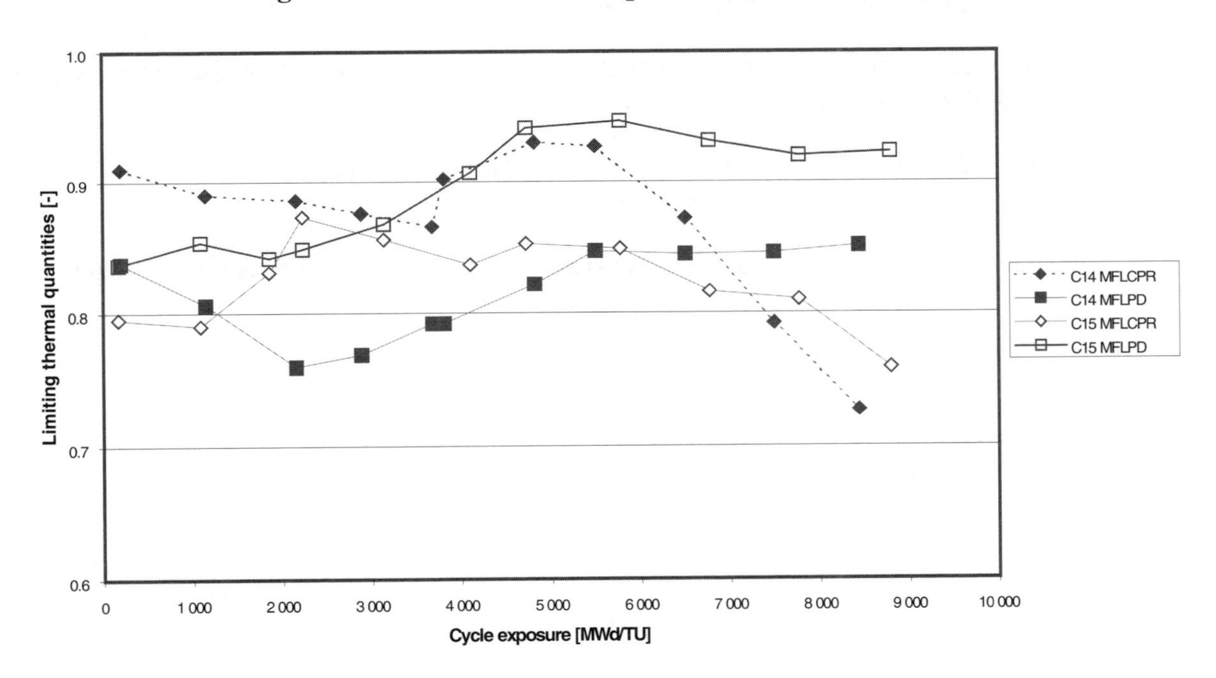

Conclusions from the KKL experience

The chosen strategy of core supervision based on purely calculated thermal limits has proven to be successful, and it has not yet been necessary to switch to adaptive methods. However, essential for this approach are accurate and well-proven simulation methods and careful tracking of the simulator predictions vs. detector measurements.

The introduction of the new physics methods in connection with the power uprate of the plant has highly reduced the uncertainties. KKL plans to continue the operation with this strategy.

Summary and future perspectives

Basic methods of adaptation of simulator results vs. detector measurements have been described as implemented in SIMULATE-3. This allows the user to make an optional choice of strategy for implementation on-line, as is being demonstrated by a variety of different implementations of SIMULATE-3 presently in progress world-wide.

Experience from an alternative strategy was presented from the PRESTO on-line installation at the Leibstadt plant. Although this core supervision system has all the algorithms implemented for adaptation to the detectors, the plant has been successfully operated for 10 cycles based on pure predictions of thermal limits. The methods of adaptation have only served as a back-up method but there has not yet been a need for activating that option.

Independently of the strategy chosen for adaptation to plant instrumentation, new improved core surveillance systems are expected to play an important role in the operation of both BWR and PWR plants in the future. With modernised I&C systems and the availability of modern computer hardware and software, there are prospects for developing reliable systems at reasonable costs. In the design of modern core surveillance systems the following features will be emphasised:

- Integration of off-line and on-line application of the simulator, i.e. the same simulator version will be used, the database is shared for off-line and on-line use, etc.
- Modular software structure, with clear separation between simulator, database, man-machine interface, etc. to assure easy maintainability and flexibility.
- High data security and a clear distinction between accessibility for different user categories.

REFERENCES

[1] K.S. Smith, *et. al.*, "SIMULATE-3 Methodology", STUDSVIK/SOA-89/04 (1989).

[2] K.R. Rempe, K.S. Smith and A.F. Henry, "SIMULATE-3 Pin Power Reconstruction Methodology and Benchmarking", *Nucl. Sci. Eng.*, 103, p. 334 (1989).

[3] S. Børresen, N.E. Patiño, "Methods of the Advanced Nodal Simulator PRESTO-2", Proc. ANS Topl. Meeting, Vol. 1, Myrtle Beach, South Carolina, 23-26 March 1997.

[4] S. Lundberg, W. van Teeffelen, J. Wenisch, "Core Supervision Methods and Future Improvements of the Core Master-PRESTO System at KKB", these proceedings.

[5] W. van Doesburg, S. Børresen, "Simulator Based Core Monitoring – Experience with PRESTO at KKL", *Trans. Amer. Nuc. Soc.* 75, p. 334 (1996).

JEF-2 CROSS-SECTION LIBRARY FOR CASMO-4: IMPACT ON CORE MONITORING OF OKG REACTORS

Per Claesson
Reactor Physicist
OKG AB
572 83 Oskarshamn, Sweden

Abstract

Core follow calculations with SIMULATE-3/CASMO-4 using JEF-2 cross-sections have been performed for the three Oskarshamn reactors. Comparisons have been made with the long used cross-section library ENDF/B4.

A stronger void coefficient is seen on the 2-D level with the JEF-2 cross-sections. TIP results are affected and the results are mixed for the three reactors. Cold k_∞ is higher with JEF-2 cross-sections, raising the cold k_{eff} level to the same level as for the hot k_{eff}, which remains unaffected. The burn-up bias for cold k_{eff}, which is seen with ENDF/B4, is not present with JEF-2. The impact on core monitoring is small although a greater confidence in the prediction of the shutdown margin is reached.

General

The three Oskarshamn BWR reactors have been analysed with the three-dimensional nodal code SIMULATE-3 6.02. Cross-sections to SIMULATE-3 have been generated with the multi-group two-dimensional code CASMO-4 1.28 using the JEF-2 cross-section library. Hot and cold k_{eff} have been calculated, as well as thermal margins and shutdown margin. Extensive TIP comparisons have also been made. Comparisons have been made with the long used cross-section library ENDF/B4 using CASMO-4 1.13.

The Oskarshamn reactors are ABB Atom built BWRs of 1 375, 1 800 and 3 300 MW thermal. They have 448, 444 and 700 bundles, and Oskarshamn-3 uses internal circulation pumps. The power density is 36, 44 and 49 kw/lt. The reactors run in annual cycles and use spectral shift, coast-down, control cells, monosequence, low leakage loading and sometimes feedwater temperature reduction.

The fuel in Oskarshamn-1 for the analysed cycles is Exxon 8*8, KWU 9*9, ABB Svea-64 and Svea-96S. For Oskarshamn-2 it is ABB 8*8 and Svea-64, KWU 9*9 and 10*10. For Oskarshamn-3 it is all ABB 8*8, Svea-64, Svea-100, Svea-96S and Svea-96 Optima.

CASMO model

CASMO is a multi-group two-dimensional transport theory code for burn-up calculations. Gadolinium depletion is performed automatically within CASMO-4, including the special ABB Atom zebra model.

A 40 neutron energy group library and 10 gamma energy group library have been used. The JEF-2 library contains more than 300 nuclides compared to 103 for ENDF/B4.

CASMO generated cell data to SIMULATE in a standard way using the S3C option, including history effects from void and control rods. Void history calculations were made at 0, 40 and 80% void.

SIMULATE model

SIMULATE-3 is an advanced three-dimensional two-group reactor analysis code. It is a nodal diffusion model that uses discontinuity factors. SIMULATE-3 also includes a reflector model and a model for pin power reconstruction which have been used in all calculations.

There is also a model for fuel temperature as a function of burn-up. These temperatures are calculated with Interpin-CS and implemented in the input.

The bypass flow is calculated explicitly. The heat balance model in SIMULATE is used except for cases with reduced feedwater temperature. This model is validated versus measured temperature values from the plant.

The control rod handle is modelled with 5 cm extra length of 30% of normal absorption and the reactivity worth of the spacer is taken into account through this model in SIMULATE.

2-D results

Comparisons at the 2-D level show quite significant differences for k_{∞}. The largest difference is seen at high void content. This is valid for hot and cold cases, and also for cases controlled by control rod. The differences are very similar within the same fuel type but can differ slightly for another fuel type. The main trends, however, remain valid.

With JEF-2 cross-sections lower hot k_∞ at high void and somewhat higher k_∞ at low void is calculated. In other words a stronger void coefficient (more negative) is calculated with JEF-2 cross-sections (see Appendix 1).

While hot k_∞ is lower with JEF-2 cross-sections, at high void the cold k_∞ is higher.

k_∞ calculated with the JEF-2 library increases with burn-up compared to the k_∞ with ENDF/B4, especially for higher void contents. This larger conversion, which is seen with the JEF-2 cross-sections, is due to differences in ^{238}U resonance data.

The gadolinium burn-up rate is slightly slower with JEF-2 cross-sections, i.e. the k_∞ does not increase as fast as with ENDF/B4 in the gadolinium burn-up range. This is also valid for the ABB zebra concept in Svea-64.

Figure 1. k_∞ difference (JEF-2 – ENDF/B4) for a Svea-96S in Oskarshamn-3, uncontrolled hot and cold

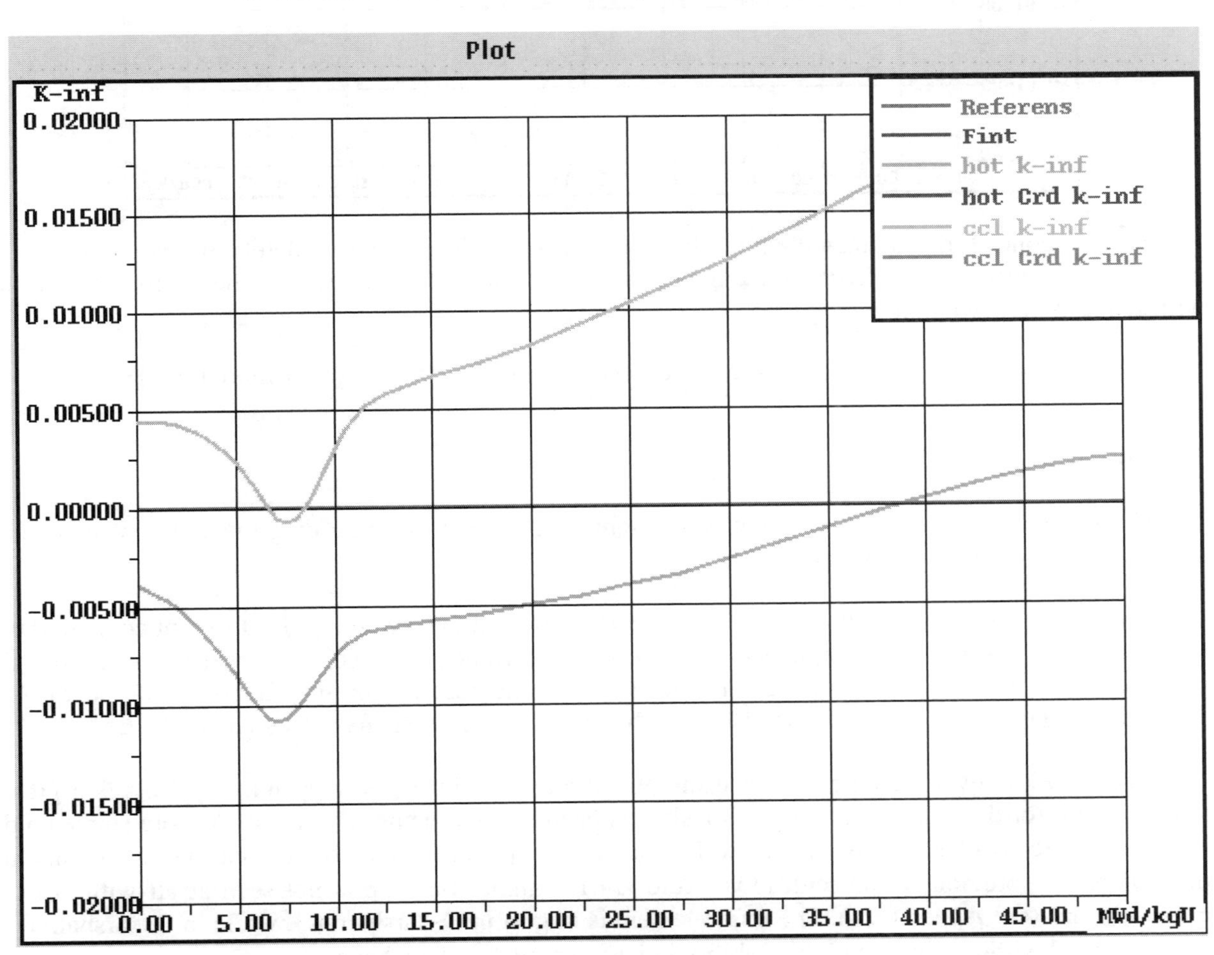

Figure 2. k$_{\infty}$ difference (JEF-2 – ENDF/B4) for a Svea-96S in Oskarshamn-3, uncontrolled hot and cold, controlled hot and cold

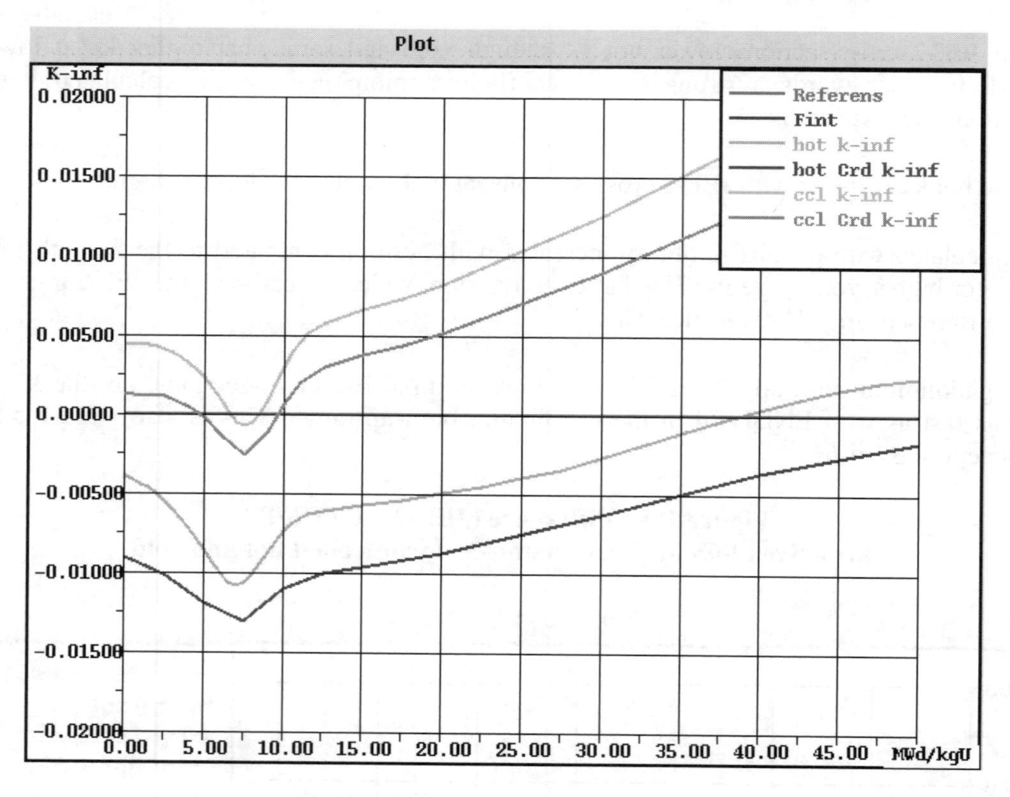

k$_{\infty}$ with control rod calculated with JEF-2 cross-sections shows approximately 300 pcm more of a negative difference (at 20 MWd/kgU) compared to uncontrolled. In other words, the control rod worth with JEF-2 is somewhat higher.

The internal power distribution is affected very little. Very few rods differ more than 1%.

3-D results

Comparisons on the 3-D level show quite significant differences regarding the axial power shape due to the stronger void coefficient.

The hot k_{eff} level is unchanged, however the shape during one cycle is very different compared to the ENDF/B4 results. This is explained with the stronger (more negative) void coefficient for JEF-2 cross-sections, which pushes the power downward in the reactor for the first part of the cycle. The reduced gadolinium burn-up rate also contributes to this different shape (see Appendix 2).

The cold k_{eff} level increased when using JEF-2 cross-sections (see Appendix 3). This is a great improvement for the cold level that previously has been too low using ENDF/B4. A more stable cold level in between cycles is also seen, as well as a smaller spread for the cold critical measurements in one cycle. The previous trend with lower cold k_{eff} for higher burn-up is not seen at all with JEF-2 cross-sections (see Appendix 4). The improvement is largest in Oskarshamn-3, while in Oskarshamn-1 the k_{eff} spread within a cycle and the stability in between cycles is unchanged.

The shutdown margins are unchanged for all three reactors.

Table 1. Hot and cold k_{eff} level for the three Oskarshamn reactors

k_{eff}	O1	O2	O3
Hot eofp	1.002	1.005	0.999
Cold	1.002	1.001	0.999

Table 2. Local critical measurements for Oskarshamn-3

Cycle	Number	Temp. °C	CASMO-4 1.13 SIMULATE-3 6.02		CASMO-4 1.28 JEF SIMULATE-3 6.02	
			Average	Std	Average	Std
1	14	25	0.99675	0.00062	0.99917	0.00094
2	6	48	1.00003	0.00114	1.00226	0.00147
3	6	70	0.99732	0.00081	1.00062	0.00065
4	7	20	0.99443	0.00098	0.99754	0.00089
4b	5	45	0.99388	0.00132	0.99804	0.00115
5	5	58	0.99388	0.00039	0.99736	0.00090
6	8	31	0.99412	0.00116	0.99845	0.00120
7	10	32	0.99422	0.00161	0.99875	0.00117
8	9	20	0.99302	0.00176	0.99830	0.00131
9	10	53.5-65.6	0.99126	0.00231	0.99784	0.00132
10	10	41-48	0.99261	0.00156	0.99866	0.00096
11	10	37	0.99394	0.00069	1.00007	0.00122
12	10	37-43	0.99294	0.00119	0.99929	0.00105
13	10	45	0.99179	0.00216	0.99831	0.00140
Average value ⇒			0.99430	0.00126	0.99888 ☺	0.00115 ☺

For Oskarshamn-3 the already very good TIP results is improved for most of the cycles with JEF-2 cross-sections. In particular for Cycle 14, which with ENDF/B4 was very poor, the nodal RMS went from 6% down to 4% (see Appendix 5).

For Oskarshamn-2 and Oskarshamn-1, however, the stronger void coefficient in JEF-2 cross-sections pushes the power too much to the bottom of the core in many cycles. This is often seen in the first part of the cycle. Consequently, the power will often overshoot in the top of the core at the end of the cycle. The TIP results for Oskarshamn-1 are therefore on the average unchanged, and for Oskarshamn-2 are a little worse with the JEF-2 cross-sections.

Impact on core monitoring

The impact on the core monitoring is overall very small. However, the absence of burn-up bias for the cold critical k_{eff} level gives a greater confidence in the prediction of the shutdown margin and in the prediction of the critical pattern at critical measurements.

Having the same hot and cold k_{eff} level also makes prediction of first global criticality more precise.

The stronger void coeffic ient will for some reactors (e.g. Oskarshamn-3) give better axial agreement for the calculated power but for others (e.g. Oskarshamn-2) it will lead to worse agreement.

Table 3. TIP results for Oskarshamn-3: SIMULATE-3 6.02 CASMO-4 1.13 with ENDF/B4 vs. CASMO-4 1.28 with JEF-2 cross-sections

Cycle	Number of TIP comparisons	Average RMS nodes ENDF/B4	Average RMS nodes JEF	Min-Max RMS nodes ENDF/B4	Min-Max RMS nodes JEF
1	10 neutron	0.065	0.087 ☹	0.050-0.090	0.056-0.115
2	3 neutron	0.098	0.101 ☹	0.090-0.103	0.099-0.102
	8 gamma	0.047	0.053 ☹	0.041-0.062	0.048-0.065
3	8 gamma	0.043	0.037 ☺	0.024-0.062	0.032-0.047
4	6 gamma	0.033	0.036 ☹	0.030-0.036	0.031-0.036
4b	6 gamma	0.050	0.059 ☹	0.035-0.087	0.041-0.089
5	13 gamma	0.042	0.046 ☹	0.034-0.057	0.036-0.067
6	15 gamma	0.042	0.041 ☺	0.034-0.065	0.034-0.062
7	16 gamma	0.045	0.039 ☺	0.034-0.071	0.030-0.051
8	13 gamma	0.041	0.036 ☺	0.032-0.054	0.029-0.045
9	15 gamma	0.032	0.031 ☺	0.023-0.042	0.023-0.049
10	15 gamma	0.034	0.028 ☺	0.026-0.042	0.024-0.032
11	21 gamma	0.032	0.033 ☹	0.024-0.039	0.025-0.044
12	24 gamma	0.034	0.032 ☺	0.030-0.043	0.026-0.040
13	18 gamma	0.038	0.031 ☺	0.036-0.045	0.029-0.036
14	15 gamma	0.048	0.036 ☺	0.038-0.063	0.031-0.042
Total	13 neutron	0.073 (neu)	0.090 (neu)	0.050-0.103 (neu)	0.056-0.115 (neu)
	193 gamma	0.039 (gam)	0.036 (gam) ☺	0.023-0.087 (gam)	0.023-0.089 (gam)

Appendix 1. k_∞ and void coefficient (pcm/%) vs. exposure

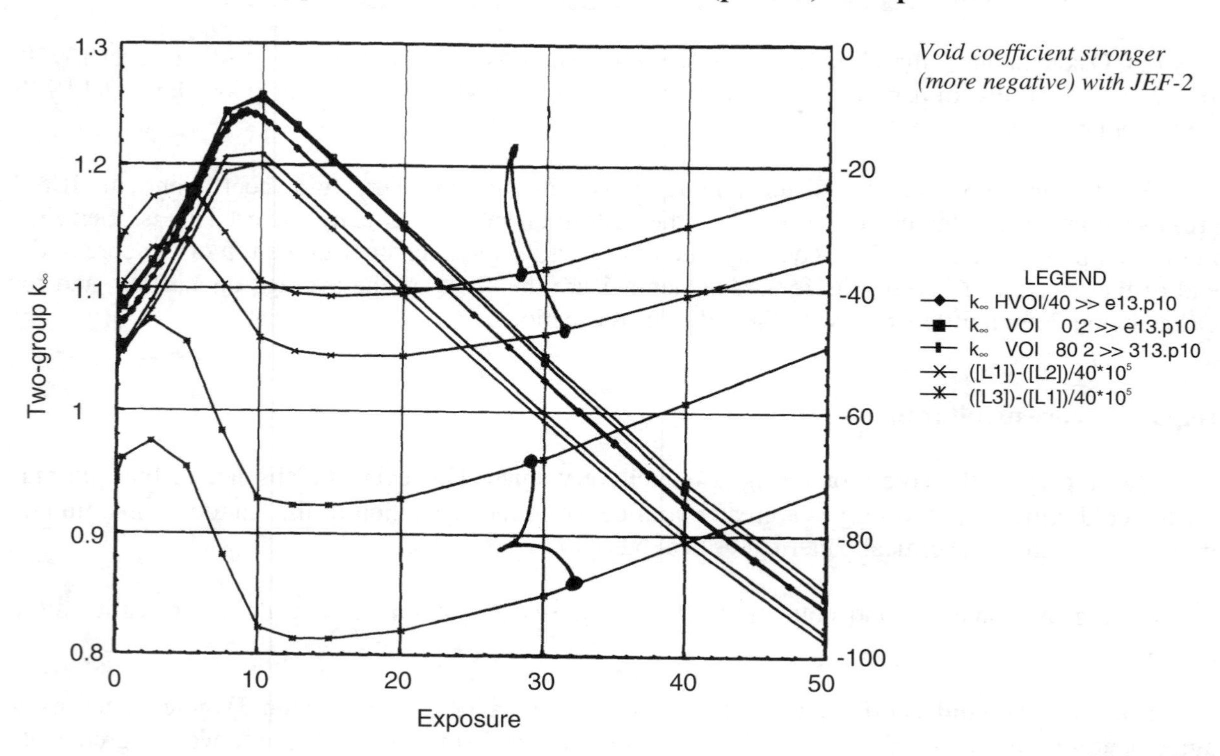

Appendix 2. Cold critical measurements CASMO-4 JEF and ENDF/B4 SIMULATE-3 for statepoints

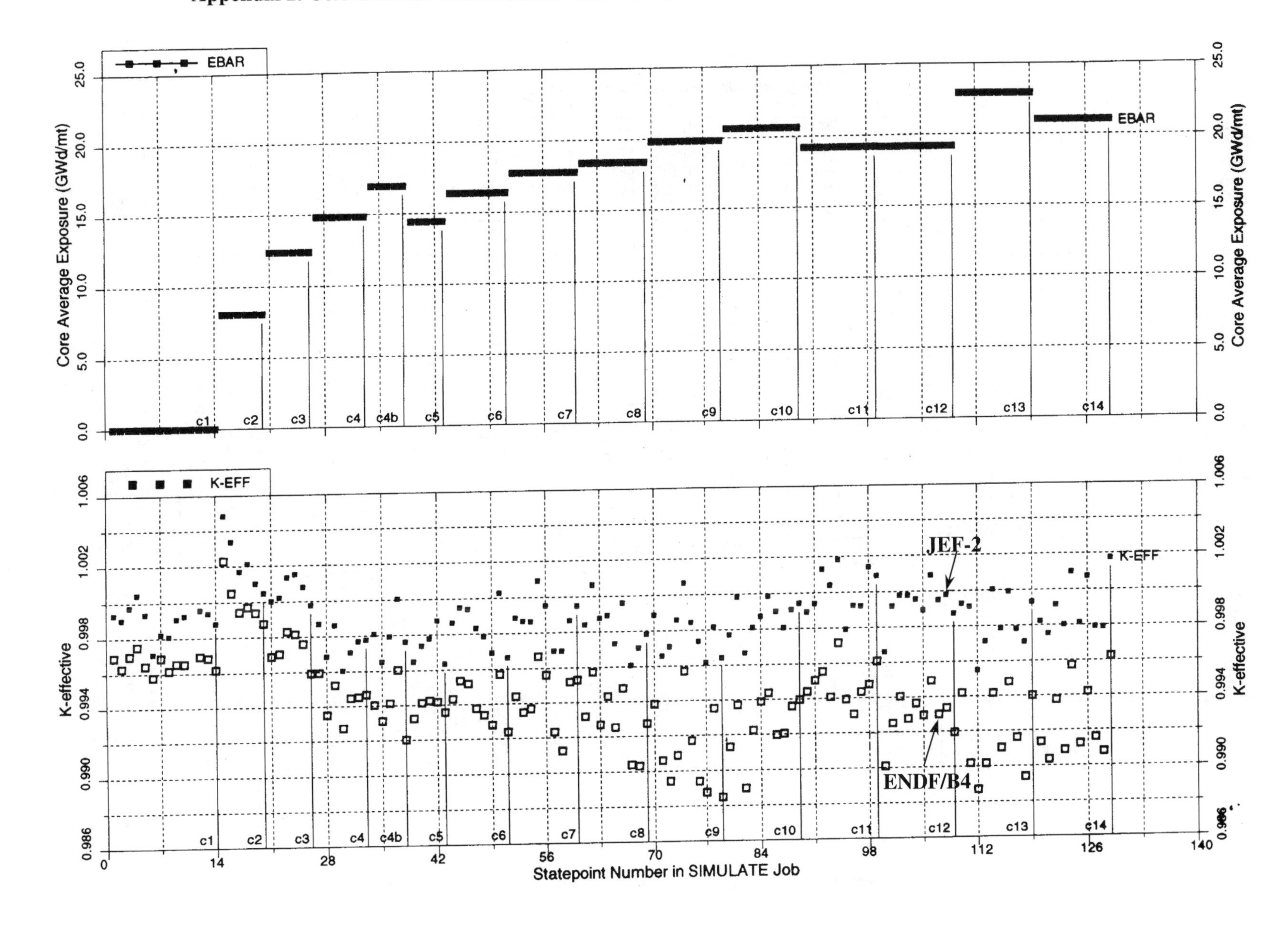

Appendix 3. Cold critical measurements CASMO-4 JEF and ENDF/B4 SIMULATE-3 vs. exposure

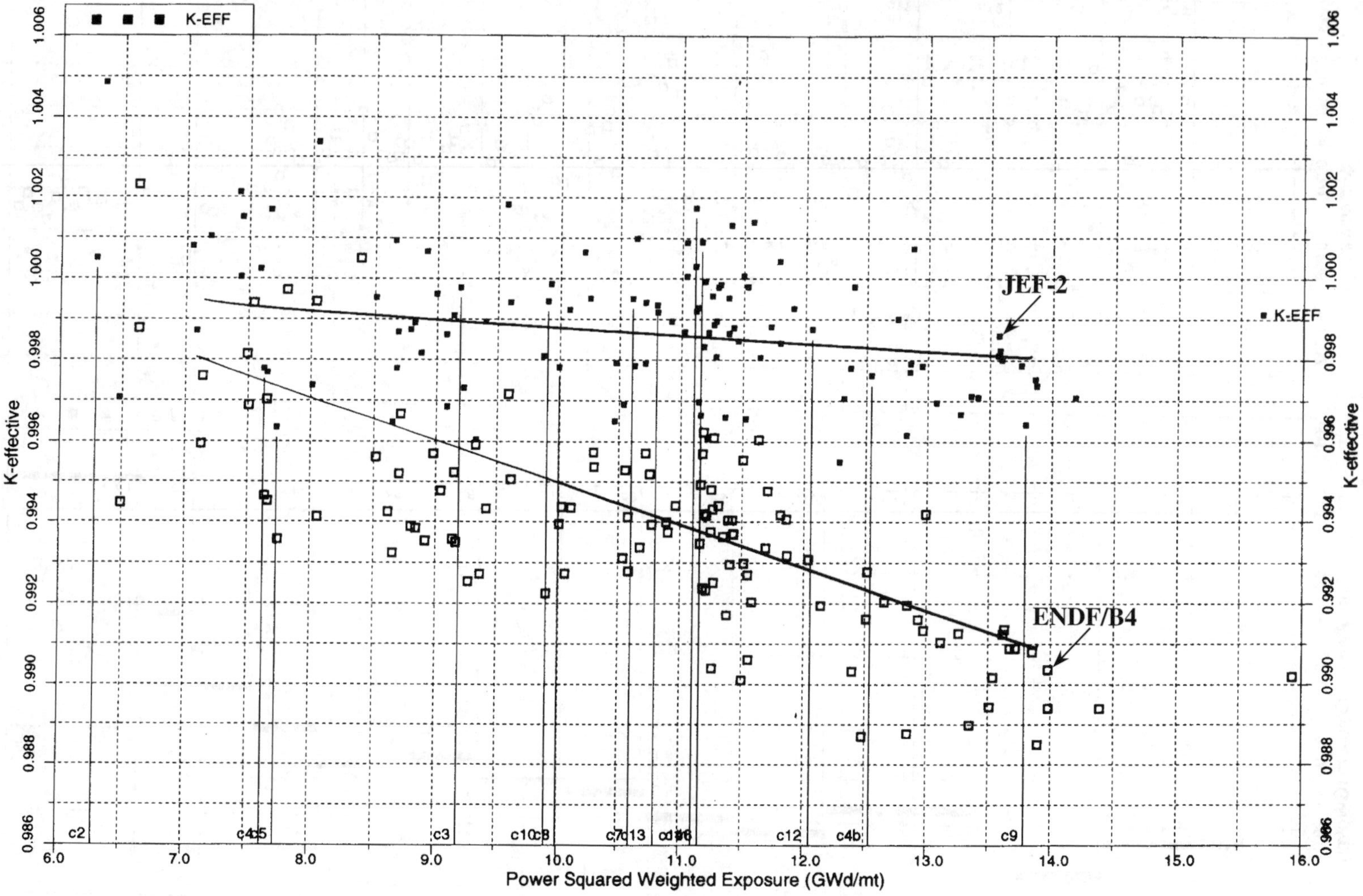

Appendix 4. Hot k_{eff} CASMO-4 JEF and ENDF/B4 SIMULATE-3 vs. integrated exposure

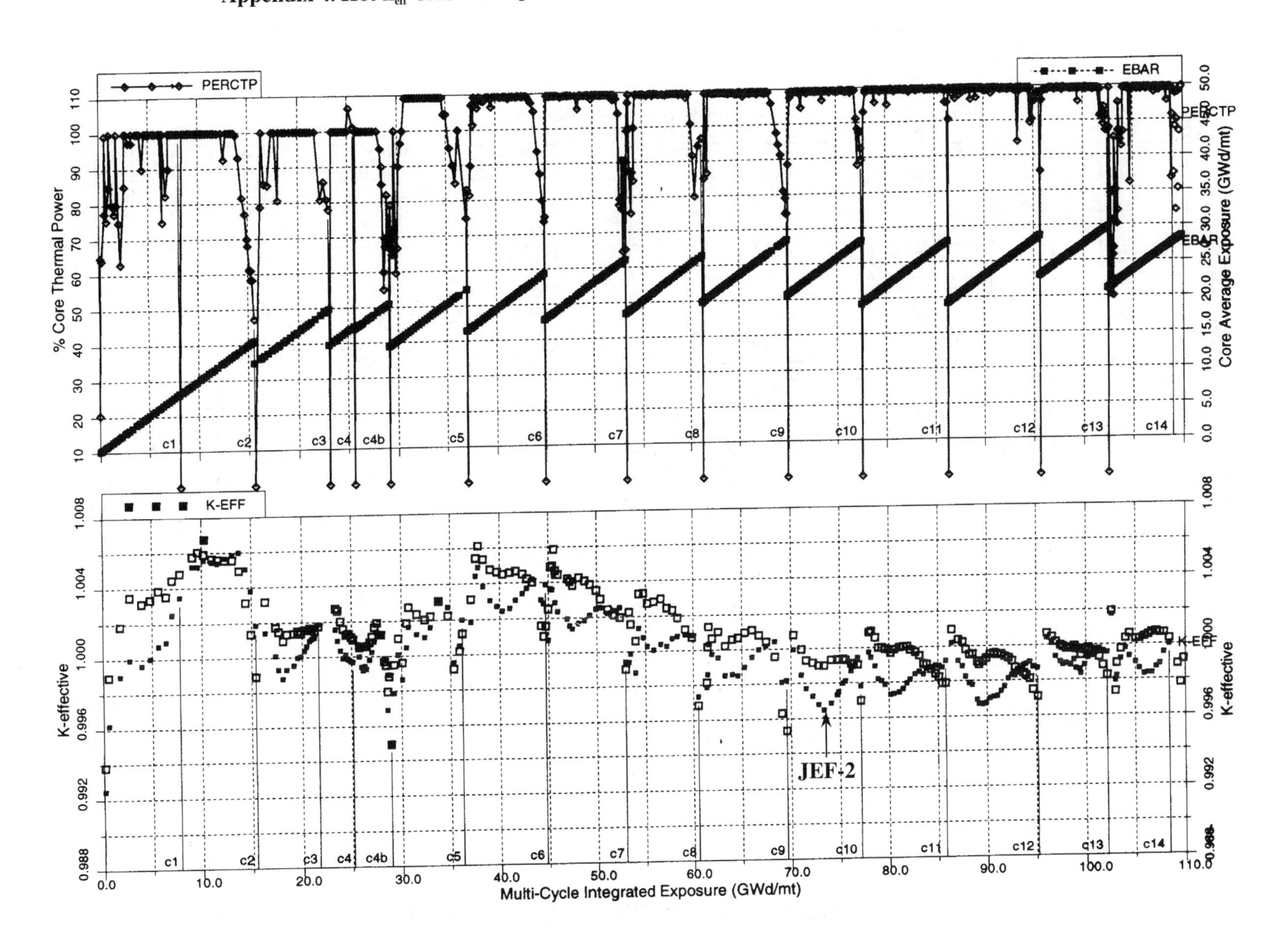

Appendix 5. TIP RMS deviation CASMO-4 JEF and ENDF/B4 SIMULATE-3 vs. integrated exposure

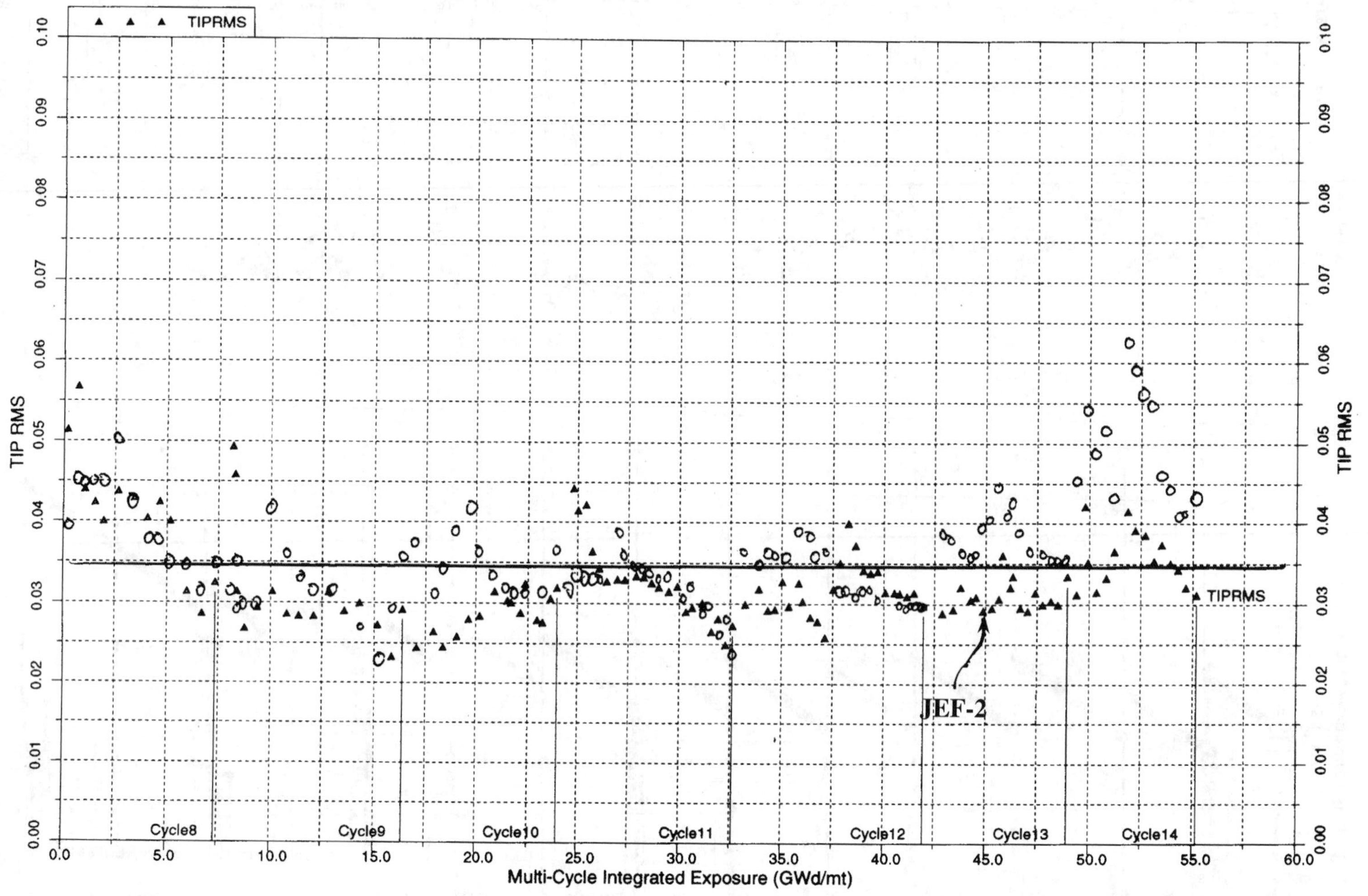

BWR CORE STABILITY PREDICTION ON-LINE WITH THE COMPUTER CODE MATSTAB

Thomas Smed, Pär Lansåker
Engineering Department
Forsmark Kraftgrupp AB
742 03 Östahmmar

Philipp Hänggi
Nuclear Engineering Laboratory
Swiss Federal Institute of Technology
8092 Zürich, Switzerland

Abstract

MATSTAB is a computer program for three-dimensional prediction of BWR core stability in the frequency domain. This tool has been developed, and is currently used, to perform core design and optimisation with regard to core stability. The requirement regarding the predicted decay ratio of the new core is one of the limiting factors, or key parameters, in core design. To be useful, the tool should be fast and simple to apply. The results must be delivered promptly and experts should not be required to interpret them.

Alternatively, the area of application for MATSTAB can be described as on-line monitoring using predictive tools. Core stability properties can be calculated for a number of presumptive reactor states, planned or unplanned.

A 3-D code operating in the frequency domain may be the best tool to use for the purposes just mentioned. Some strong advantages are that the results are given promptly, they require no post-processing and are directly amenable to graphic presentation of eigenvectors, etc.

Introduction

Core stability has been a key issue for BWR plants over the past few decades. In order to avoid instability a number of activities have been performed, both experimental and theoretical work. The aim has been to detect and suppress oscillations or simply to avoid them. In order to avoid oscillations one needs either to have great theoretical expertise or large operating margins. To operate with large margins is not practical and is also quite expensive. In order to avoid oscillations, the necessary knowledge and tools for the task are needed.

The use of stability monitors was intended to enhance observation possibilities. One property in particular, the ability to create an "early warning", was potentially very attractive. An early warning system based on analysis of measured time series and trends (as stability monitors are) has its limitations. The stability margin may, even in the best cases, be calculated for the present operating point, not for oncoming operating points. Some easy trend analysis and using rules of thumb may of course be used. In many cases the main interest is to predict the stability margin in the near future at an operating point different than the present. Typically in these cases the present state is very stable, and for some reason a transient behaviour or planned action tends to bring the reactor to an area susceptible to lower stability margins. The use of advanced system models – predictive tools – are required in these cases.

The technical demands on predictive tools of this kind are severe. A short computation time is highly desirable. The results, e.g. decay ratio, frequency and oscillating mode, should be possible to extract without any help from specialists.

The computer code MATSTAB was developed for a closely related problem. MATSTAB is used to predict and to optimise the reactor to core stability in the core design process. MATSTAB has been successfully used in production at Forsmark for the last couple of years.

Technical background

Computer methods and their capabilities

When it comes to core stability calculations, the most common methods used are time domain transient codes in 3-D or frequency domain in 1-D. The main advantage of frequency over time domain analyses is there is no need for time integration and thus no numerical error. In frequency domain the equations are linearised and solved directly and there is no need for interpretation of decaying oscillations in time series. In the iteration process of solving the frequency domain equations, the eigevalue as well as the eigenvector is extracted. The eigenvector gives amplitude and phase information that may be used for more specific evaluations. The main reason for using time domain, in spite of the drawbacks mentioned above, is to use 3-D modelling, which previously was not possible to perform in frequency domain. The use of 3-D modelling must be repeated many times to meet a level of desirable accuracy. The possibility of performing 3-D calculations in frequency domain with the computer code MATSTAB, however, extracts the best properties from both methods.

Frequency domain codes are well known to be very fast. A set of equations is only solved once instead of every time step in time domain. On the other hand it needs much more computer memory: about a couple of hundred megabytes of primary memory. The computation time is typically 10 minutes for cases that would take about three hours in time domain.

Only time domain codes can predict non-linear effects such as limit cycles and their absolute amplitudes.

Computer code MATSTAB

The primary objectives for developing MATSTAB and using a new method in calculating the margin to the onset of instability in BWRs were fast execution time without loosing prediction accuracy and to provide the capability to determine core-wide as well as out-of-phase oscillations.

The fast executing time is achieved by evaluating the core and system dynamics in the frequency domain. This procedure is based on the assumption that the dynamic behaviour of the power void feedback mechanism is linear for small deviations around steady state operating conditions and can therefore be described by a system matrix A_s. This approach is widely accepted as long as the studies are restricted to operating points below the stability limit. Indeed, when determining the distance to instability from noise measurements of power reactors, linear concepts give the best results (e.g. ARMA, AR).

The eigenvalues of the system matrix A_s describe the damping and the frequency of the different oscillation modes and hence the decay ratio of the dominating oscillation. The eigenvectors of A_s contain information about the shape of the oscillation and provide a very efficient method for sensitivity analysis.

From the number of available transient codes, RAMONA-3.9 has been chosen to be the base for the physical reactor model in MATSTAB. The reason for this choice was the good results produced with RAMONA in several international stability benchmarks and its wide acceptance. MATSTAB uses basically the same physical model as RAMONA-3.9, which covers a drift flux model using three equations to describe the mass balances for water and steam, and the energy and momentum balance equations for the mixture of the two phases. Special emphasis is put on the core modelling with full 3-D neutron kinetics with each fuel bundle represented.

Constitutive relationships are given for non-equilibrium vapour generation and condensation, unequal phase velocities as well as wall friction and heat transfer effects.

All together, MATSTAB, which is entirely programmed in MATLAB, is a very efficient tool to predict stability and to conduct sensitivity analysis for boiling water reactors.

The dynamic system

The continuous-time dynamical system consists of a set of n_x first-order differential equations of the form:

$$\frac{\partial}{\partial \tau}\mathbf{x}(\tau) = f(\mathbf{x}(\tau), \mathbf{y}(\tau)),\ \mathbf{x}(\tau) = \begin{pmatrix} x_1(\tau) \\ \vdots \\ x_{n_x}(\tau) \end{pmatrix},\ \mathbf{y}(\tau) = \begin{pmatrix} y_1(\tau) \\ \vdots \\ y_{n_y}(\tau) \end{pmatrix} \qquad (1)$$

To close the system (1), all parameters must be represented by the state variables. This leads to the n_y algebraic equations.

$$0 = g(\mathbf{x}(\tau), \mathbf{y}(\tau)) \qquad (2)$$

$\mathbf{x}(\tau)$ in Eq. (1) can be substituted by $\mathbf{x_0} + \Delta\mathbf{x}(\tau)$, where $\mathbf{x_0}$ is the steady state value and $\Delta\mathbf{x} := \Delta\mathbf{x}(\tau)$ is a small deviation around $\mathbf{x_0}$, $\mathbf{y}(\tau) = \mathbf{y_0} + \Delta\mathbf{y}$, respectively. A consecutive Taylor expansion of $f(\mathbf{x},\mathbf{y})$ and $g(\mathbf{x},\mathbf{y})$ yields:

$$\frac{\partial}{\partial\tau}\mathbf{x}(\tau) = f_0 + \frac{\partial f(\mathbf{x},\mathbf{y})}{\partial\mathbf{x}}\Delta\mathbf{x} + \frac{\partial f(\mathbf{x},\mathbf{y})}{\partial\mathbf{y}}\Delta\mathbf{y} + O(\Delta\mathbf{x}^2, \Delta\mathbf{y}^2, \Delta\mathbf{x}\Delta\mathbf{y}) \tag{3}$$

$$0 = g_0 + \frac{\partial g(\mathbf{x},\mathbf{y})}{\partial\mathbf{x}}\Delta\mathbf{x} + \frac{\partial g(\mathbf{x},\mathbf{y})}{\partial\mathbf{y}}\Delta\mathbf{y} + O(\Delta\mathbf{x}^2, \Delta\mathbf{y}^2, \Delta\mathbf{x}\Delta\mathbf{y})$$

Neglecting second order terms and using the fact that the left hand side of (1) is zero for the steady state $(\mathbf{x_0},\mathbf{y_0})$ leaves us with:

$$\frac{\partial}{\partial\tau}\Delta\mathbf{x} = \mathbf{A}\Delta\mathbf{x} + \mathbf{B}\Delta\mathbf{y} \tag{4}$$

$$0 = \mathbf{C}\Delta\mathbf{x} + \mathbf{D}\Delta\mathbf{y}$$

The resulting matrix $\left[\begin{array}{c|c}\mathbf{A} & \mathbf{B}\\ \hline \mathbf{C} & \mathbf{D}\end{array}\right]$ for the RAMONA model has the respectable size of about half a million equations. The system concludes to the equation:

$$\frac{\partial}{\partial\tau}\Delta\mathbf{x} = (\mathbf{A} - \mathbf{B}\mathbf{D}^{-1}\mathbf{C})\Delta\mathbf{x} = \mathbf{A}_s\Delta\mathbf{x} \tag{7}$$

Complex analysis

For convenience $\Delta\mathbf{x}(\tau)$ from (7) is renamed $\mathbf{x}(\tau)$. This yields the set of differential equations which is central for our study.

$$\frac{\partial}{\partial\tau}\mathbf{x}(\tau) = \mathbf{A}_s \cdot \mathbf{x}(\tau)$$

With the eigenvalue λ_i and the right eigenvector $\mathbf{e_i}$ and the left eigenvector $\mathbf{f_i}$:

$$\mathbf{A}_s\mathbf{e_i} = \lambda_i\mathbf{e_i}$$

$$\mathbf{f_i^T}\mathbf{A}_s = \lambda_i\mathbf{f_i^T}$$

One can show that:

$$\mathbf{x_i}(\tau) = \mathbf{e_i}e^{\lambda_i\tau}\left[\mathbf{f_i^T}\mathbf{x}(0)\right]$$

and the following interpretation of the eigenvectors and eigenvalues can be made. The right eigenvector, $\mathbf{e_i}$ describes the relative magnitude and phase of the participating states. Note that the factor $\mathbf{f_i}^\mathbf{T}\mathbf{x}(0)$ is depending on the initial state and does not affect the decay-ratio, magnitude and phase. $\mathbf{f_i}$ may be used for sensitivity studies and contain valuable information.

With the complex eigenvalue $\lambda_i = \sigma + j\omega$ and the solution of $\mathbf{x}(\bullet)$ can be written like:

$$\mathbf{x_i}(\tau) = ke^{\sigma\tau}(\cos\omega\tau + j\sin\omega\tau)$$

The stability criteria are described by the decay ratio, which is the ratio of two consecutive maxima of the impulse response.

Figure 1. Definition of decay ratio

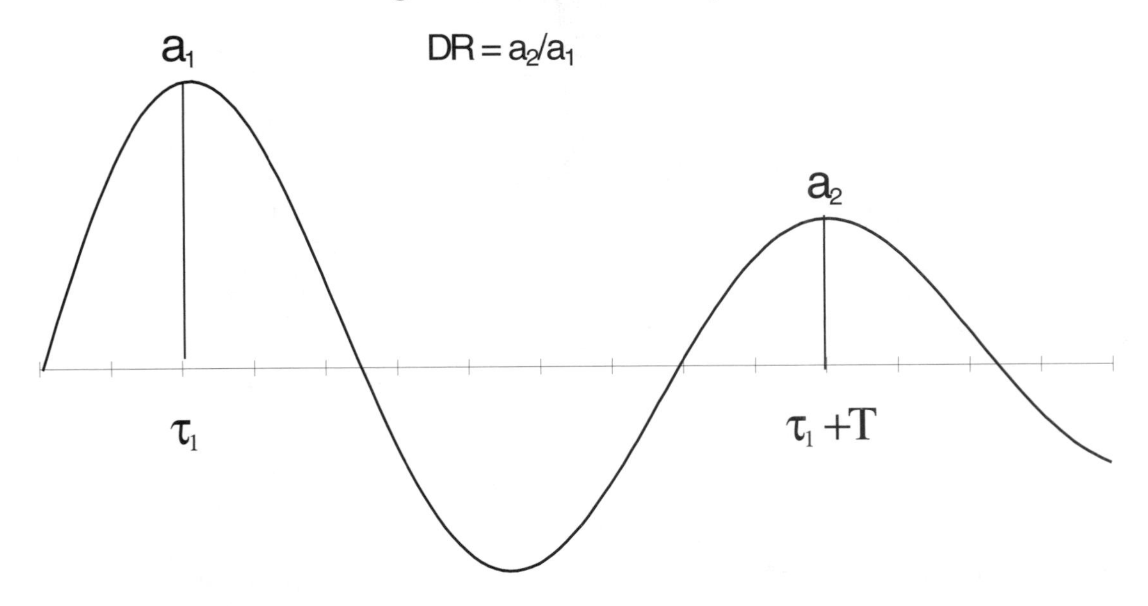

And the decay ratio becomes:

$$\underline{\underline{DR}} = \frac{a_2}{a_1} = \frac{ke^{\sigma(\tau_1+T)}\left(\cos(\omega(\tau_1+T)) + j\sin(\omega(\tau_1+T))\right)}{ke^{\sigma\tau_1}\left(\cos(\omega\tau_1) + j\sin(\omega\tau_1)\right) = \frac{e^{\sigma(\tau_1+T)}}{e^{\sigma\tau_1}}} = e^{\sigma T} = \underline{\underline{e^{\frac{\sigma 2\pi}{\omega}}}}$$

since cos(••) = cos(•• + T) where T is the damped oscillation period and $\omega = \frac{2\pi}{T}$.

Naturally, the eigenvalue and in particular the real part of it is the key parameter. Nevertheless, both the right and the left eigenvector contain important information on the shape of the mode.

Graphical presentation

Eigenvector components

The right eigenvector components are suitable to be presented in graphical format. The figure below shows the eigenvector components of the void and fast neutron flux for a channel in the core. The absolute value represents the relative amplitude of the oscillation and the phase value the phase shift between signals. Note that the neutron flux vectors are narrower than the void vectors and that void and neutron flux is shifted with about 180°.

Figure 2. Phase diagram showing right eigenvector components, phase and amplitude, for void and fast neutron flux in one core channel

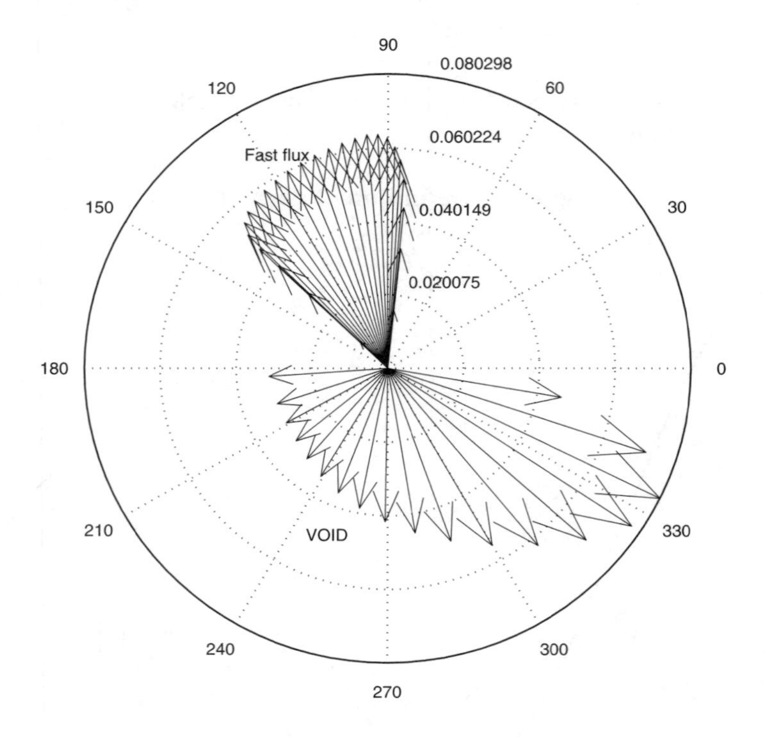

Presentation of eigenvectors of different variables

Figure 3(a) shows the magnitude of the eigenvector for fast neutron flux. Control rods are fairly well distributed in the core. Regions with high amplitudes are shown as white areas. In this case, it is difficult to say if the any adjustment in control rod sequence will affect the stability.

Figure 3(a). Absolute value of the right eigenvector for fast neutron flux

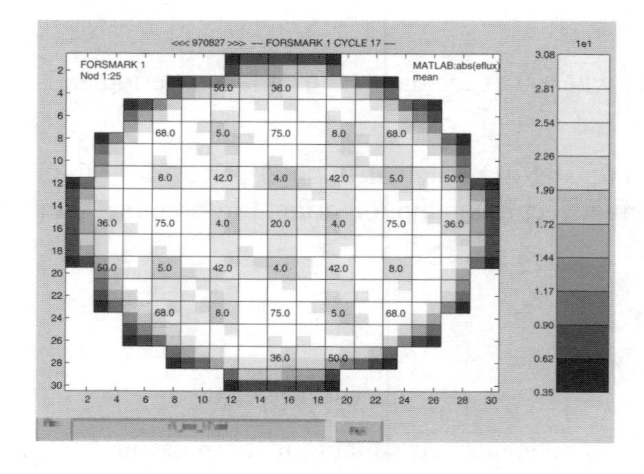

For the same case, when the void vector is displayed, a clearer picture is given. In Figure 3(b) one can see fairly small and isolated regions with higher oscillation amplitudes (shown as white areas). In this case, it is easy to see where a control rod adjustment would be most efficient to increase stability margin.

Figure 3(b). Absolute value of the right eigenvector for void

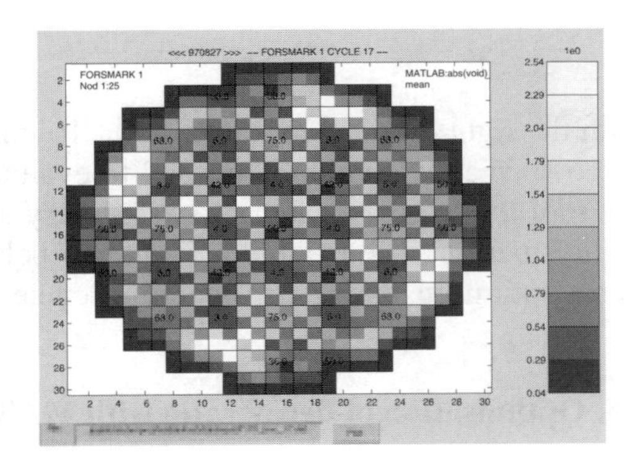

Presentation of absolute and phase values of eigenvectors

For the same case as above, the eigenvector for an out-of-phase, regional oscillation mode is displayed in Figure 4(a). The absolute values indicate the location of the symmetry line of the oscillation mode. In this case the regional mode was well damped. The decay ratio was 0.19 compared to the global mode decay ratio of 0.41.

Figure 4(a). Absolute values of eigenvector of an out-of-phase oscillation mode

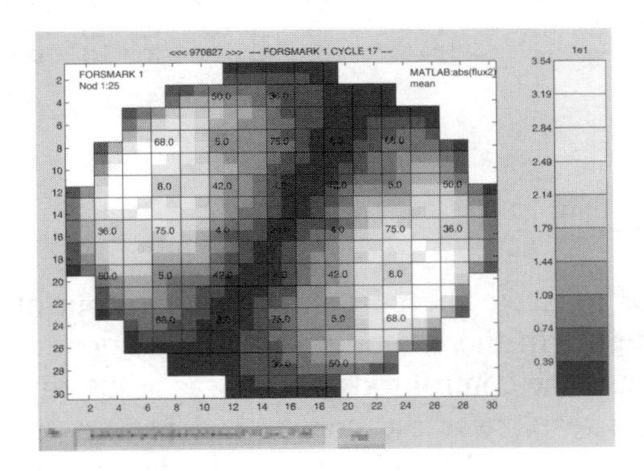

The phase plot of the eigenvector (see Figure 4(b)) will show a phase shift of 180° between the core halves. The phase diagram shows clearly what mode the oscillation may take. However in this case the regional mode would not be possible to detect in reality nor in time domain simulations.

Figure 4(b). Phase values of eigenvector for an out-of-phase mode

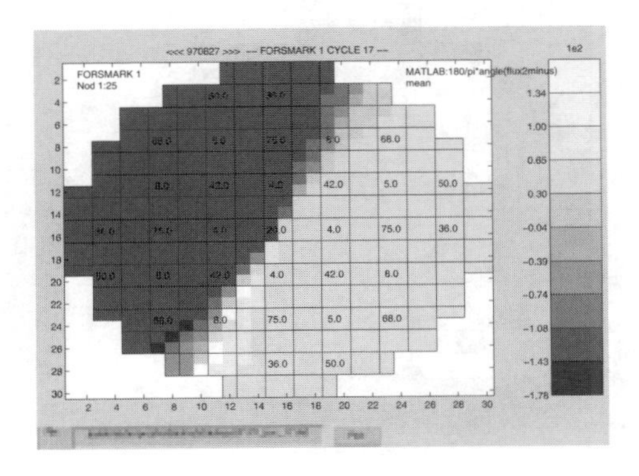

Practical experiences

Core design

Decay ratio calculation with the right eigenvector takes about 10 minutes. Since the decay ratio is extracted from the eigenvalue, no special evaluation is needed to evaluate the stability. Therefore the core engineer is able to use the tool in the core design process. Every step in an operating cycle can easily be checked against stability margin. The example (Figure 5) below for Forsmark 3 Cycle 15, shows how the stability may be optimised during an operating cycle using a matched control rod sequence.

Figure 5. Optimisation of decay ratio with MATSTAB

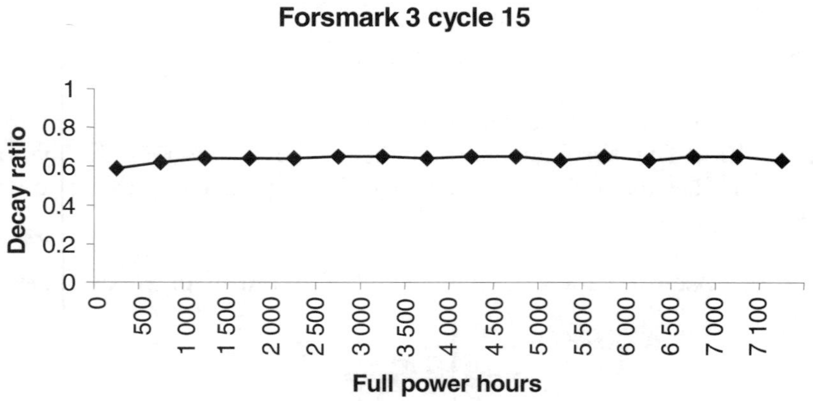

Forsmark 3 cycle 15

Decay ratio

0
0.2
0.4
0.6
0.8
1

0
500
1 000
1 500
2 000
2 500
3 000
3 500
4 000
4 500
5 000
5 500
6 000
6 500
7 000
7 100

Full power hours

Regional oscillation modes

The most common factor to make the core susceptible to out-of-phase oscillations is to make unfavourable control rod patterns. In the example below, on Forsmark 2 Cycle 15, one control rod stuck due to mechanical reasons. The control rod pattern was adjusted to that, rather than to shut down and perform maintenance. The result was the control rod pattern shown in Figure 6(a). A ring of control rods formed which affected the subcritical oscillation mode. In Figure 6(b) the void eigenvector reveal regions, off centre, with oscillation tendencies. For the global mode the decay-ratio was 0.40.

Figure 6(a). Eigenvector for neutron flux, global mode

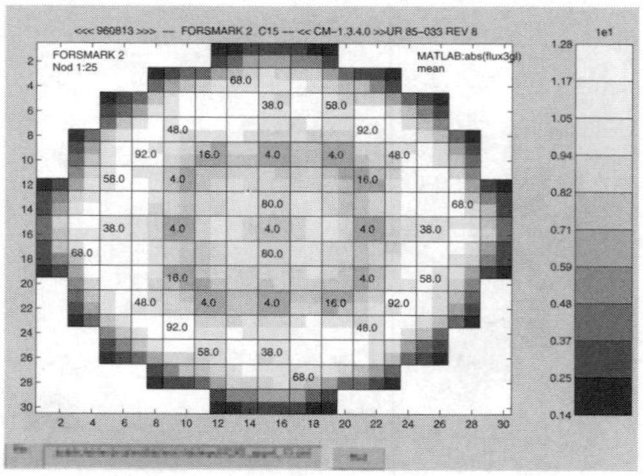

Figure 6(b). Eigenvector for void, global mode

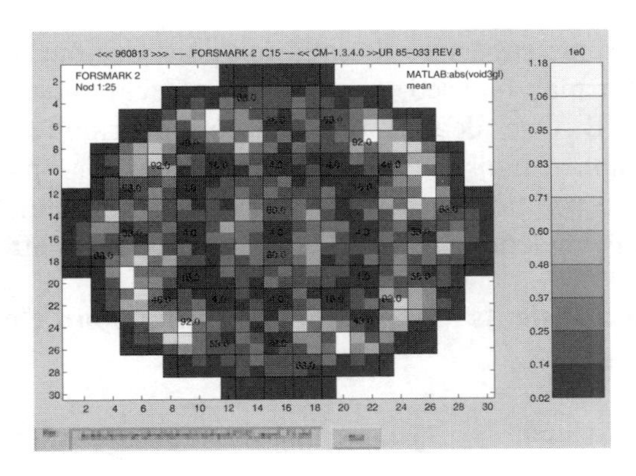

Below the most excitated region mode is displayed, one can clearly see the symmetry line of oscillation mode. The decay ratio exceeded the global mode. The decay ratio was 0.56. Although this is a rare occasion, the example shows the obvious risk for out-of-phase incidents due to lack of information in the core design process. There are few other alternative ways to extract information of damped regional modes. It may not easily be extracted from stability measurements and time domain simulations due to the fact that both need non-global excitations to override the global mode. MATSTAB on the other hand gives the information of any present risks for out-of-phase oscillations, even if the decay ratio for the subcritical modes is relatively low.

Figure 7(a). Eigenvector absolute values

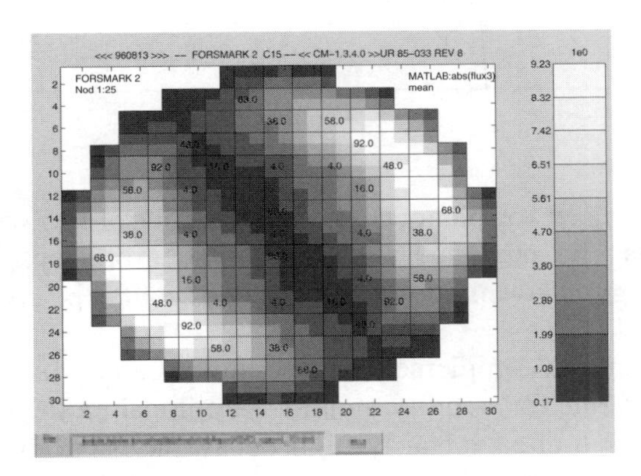

Figure 7(b). Eigenvector phase values

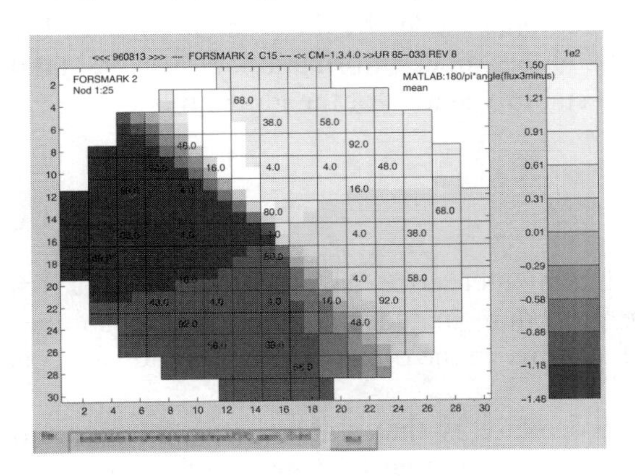

Proposed method

Present core monitor systems supervise thermal margins, dry-out, etc. As a complement to this, results from the core simulator may be used to extract core decay ratio and oscillation modes from a computer code like MATSTAB. One idea is to predict core distributions after a pump runback from full power or any other unplanned transients (see figure below). The predicted distributions are then used to calculate the dynamic solution in case of a specific transient. When high decay ratios are predicted, it is possible to perform actions to avoid an instability event after a transient.

Figure 8. Some possible transients to move the operating point closer to the stability limit

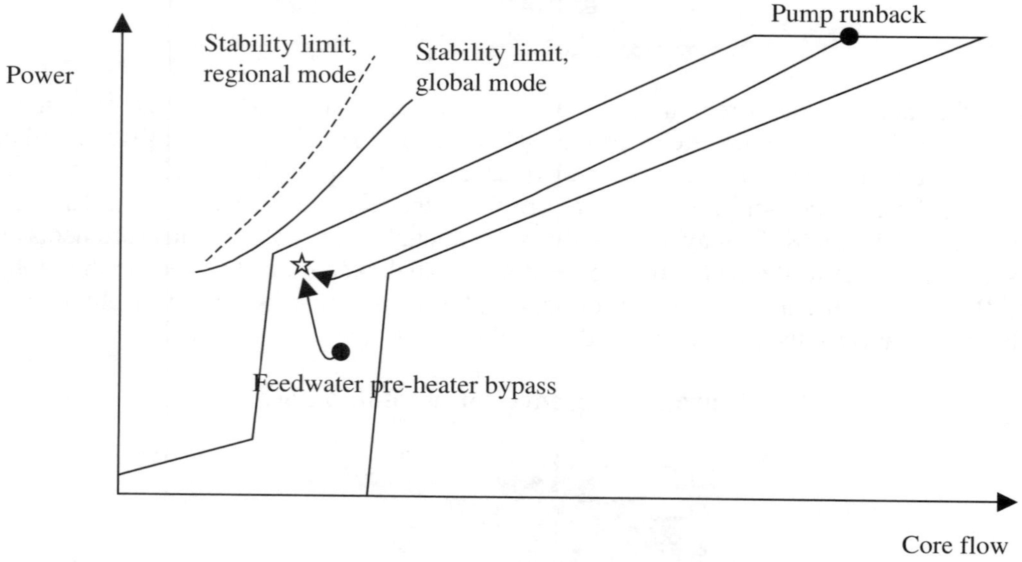

The method can also be used for planned actions like control rod withdrawal from a low power region. The decay ratio may be predicted a certain number of steps in the planned control rod sequence. Both global and regional oscillation modes may of course be predicted.

The method shall be seen as a complement to stability monitors. It features enhancements in core monitoring which are not possible for stability monitors. Regional mode detection is also very difficult with stability monitors and is limited to fairly high decay ratios, even with the most sophisticated methods. The reason for that is that the signal information in the PRM system dominates by presenting information on global characteristics, often very close in frequency. There is a possibility in some cases, however, to extract the regional modes by differentiation of LPRM data with data in symmetry positions, so as to suppress the global mode information. However, stability monitors may be the best method to evaluate the current stability status for the global mode.

Conclusions

The recently developed computer code MATSTAB offers a feasible procedure for on-line predictions of stability. The procedure includes simulation of a number of predetermined transients followed by an evaluation of stability in the obtained quasi-stationary operating point.

The advantage of this procedure is that stability can be monitored throughout the operating domain, a task that would require an enormous amount of off-line effort.

VNEM: VARIATIONAL NODAL EXPANSION METHOD FOR LIGHT WATER REACTOR CORE ANALYSIS

Makoto Tsuiki
Institute for Energy Technology, OECD Halden Reactor Project

Sverre Hval
Institute for Energy Technology, Kjeller

Abstract

A variational nodal expansion method (VNEM) has been developed as a next generation nodal neutronics method based on the neutron diffusion theory for three-dimensional core physics simulators of light water reactors. VNEM does not include the process of assembly homogenisation, therefore the local power distributions are obtained in a highly accurate manner without the process of "pin power reconstruction". In spite of its accuracy, the computing time required by VNEM is typically a few per cent of that of the reference, standard method. VNEM enhances the nodal neutronics part of today's standard LWR core monitoring method within the current established framework of the single assembly, infinite-lattice calculation, and the global nodal core analysis.

Introduction

An accurate neutronics analysis method of light water reactor (LWR) cores is needed for core monitoring systems to efficiently operate the core with a smaller margin to limiting parameters. It is also needed for in-core fuel management, and fuel/core design systems to optimise the core loading patterns, as well as for fuel designs of higher reliability. As the LWR core becomes more and more heterogeneous because of the usage of mixed oxide (MOX) fuel or fuel of much higher burn-up, a new higher order nodal method appears necessary.

With this background, a variational nodal expansion method (VNEM) for solving multi-group neutron diffusion equations was developed and implemented in a three-dimensional (3-D) nodal diffusion code. Results from test calculations proved that it gave more accurate power distributions in LWR cores [1,2,3]. The method does not include the process of homogenisation, therefore the intra-nodal local power distributions are obtained in a consistent way without "dehomogenisation", or "pin power reconstruction". The global quantities like the core eigenvalue and the nodal power distributions are also calculated with improved accuracy.

The purpose of this paper is to present the VNEM method and the results of its benchmarking against some test cases, including MOX fuel assemblies, in order to show that VNEM can be a next generation nodal neutronics method for LWR core monitoring with improved accuracy.

The variational method has been used to develop new nodal methods to avoid the process of homogenisation/dehomogenisation of nodes [4,5,6]. The most important feature that distinguishes the present method from previous approaches is that the ratio of the group neutron source distribution within a node, $S_g(r)$, in the whole core system to that within the single assembly infinite-lattice, $S_{inf,g}(r)$, is expanded with a system of orthogonal polynomials. Here sufficient accuracy is obtained by a few polynomials because $S_{inf,g}(r)$ includes local fluctuations of $S_g(r)$ caused by the assembly heterogeneity, so the ratio $S_g(r)/S_{inf,g}(r)$ is a smooth function of spatial variables. Therefore the present method enhances the neutronics part of today's standard LWR design codes without significant modification of the current established framework of the single assembly infinite-lattice calculation and the global nodal core analysis. Additional calculations are required only at the single assembly infinite-lattice level. The computing effort required for these additional calculations will be almost negligible compared to the whole lattice burn-up calculation. The nodal equation has only a few unknowns per node and can be solved within a reasonably short computing time.

The present method was implemented in a 3-D nodal diffusion code, and tested for benchmark cases. The maximum error of pin powers, nodal powers and the eigenvalue error were observed to be less than 1.8%, 1.5% and 0.05%, respectively. The computing time was estimated to be less than 3% of that of the reference code for practical LWR core applications.

Method

Fundamental assumption

The basic equations are the multi-group neutron diffusion equations, with the standard notations:

$$-\nabla D_g(r)\nabla\Phi_g(r)+\Sigma_g(r)\Phi_g(r)=S_g(r) \tag{1}$$

$$S_g(r) = (\chi_g/\lambda)\Sigma_{g'=1,G}\nu\Sigma_{f,g'}(r)\Phi_{g'}(r) + \Sigma_{g'=1,G}\Sigma_{g,g'}(r)\Phi_{g'}(r) \tag{2}$$

where g = 1, 2, ..., G indicates the neutron energy group and r = (x, y, z) is the spatial co-ordinate.

Let us assume that, within a node, the group neutron source (the rhs of Eq. (1)) for group g can be expanded by:

$$S_g(r) = \Sigma_{n=0,N} a_{n,g} S_{inf,g}(r) P_n(r) \tag{3}$$

and that the boundary value of the g-th group neutron flux on a node surface i, $\Gamma_g(s_i)$, can be expanded by:

$$\Gamma_g(s_i) = \Sigma_{m=0,M} b_{i,m,g} P_m(s_i) \tag{4}$$

where $S_{inf,g}(r)$ is the neutron source for the group g obtained by a single assembly infinite-lattice calculations and $P_n(r)$ is a three-dimensional Legendre polynomial of the n-th order normalised within a node. Index i (= 1, 2, 3, 4, 5, 6) indicates the left, right, front, back, bottom and top surfaces of the node, respectively, s_i is the co-ordinate on the node surface i, and $P_m(s_i)$ is a two-dimensional Legendre polynomial of m-th order normalised on the node surface i. $\Sigma_{n=0,N}$ indicates the sum with respect to the index n = 0 to N, and $\Sigma_{m=0,M}$, m = 0 to M. So N and M are the maximum orders of expansion for $S_g(r)$ and $\Gamma_g(s_i)$, respectively. The coefficients $a_{n,g}$ and $b_{i,m,g}$ will be determined later.

On the assumptions given above, the rigorous neutron fluxes solutions for Eqs. (1) within a node can be represented by:

$$\Phi_g(r) = \Sigma_{n=0,N} a_{n,g} \Phi_{n,g}(r) + \Sigma_{i=1,6}\Sigma_{m=0,M} b_{i,m,g} \Psi_{i,m,g}(r) \tag{5}$$

where $\Phi_{n,g}(r)$ is defined by the solution of the diffusion equation:

$$-\nabla D_g(r)\nabla\Phi_{n,g}(r) + \Sigma_g(r)\Phi_{n,g}(r) = S_{inf,g}(r)P_n(r), \quad n = 0,1,2,\ldots,N \tag{6}$$

with a boundary condition that $\Phi_{n,g}(s_i) = 0$ at all the node surfaces. Function $\Psi_{i,m,g}(r)$ is defined by the solution of:

$$-\nabla D_g(r)\nabla\Psi_{i,m,g}(r) + \Sigma_g(r)\Psi_{i,m,g}(r) = 0, \quad m = 0,1,2,\ldots,M \tag{7}$$

given a boundary condition that:

$$\begin{aligned} \Psi_{i,m,g}(r) &= P_m(s_i) && \text{on node surface i, and} \\ &= 0 && \text{on the other node surfaces} \end{aligned} \tag{8}$$

By substituting Eq. (5) into Eq. (1), and using Eqs. (6) and (7), one can prove that Eq. (5) becomes the rigorous solution of Eq. (1) from assumption (3). Also from assumption (4), and the boundary conditions on $\Phi_{n,g}(r)$ and $\Psi_{i,m,g}(r)$, one can prove that the function $\Phi_g(r)$ given by Eq. (5) is rigorously equal to the boundary value $\Gamma_g(s_i)$ on surface s_i. Therefore, the expansions (3) and (4) are the only assumptions, or approximations, made in the present method.

The expansion functions $\Phi_{n,g}(r)$ and $\Psi_{i,m,g}(r)$ are obtained by numerically solving Eqs. (6) and (7), respectively, within a single node given the neutron sources and the boundary conditions at the node surfaces. This can be done within a lattice physics code, or with an additional small code which runs after the lattice code has completed all the calculations needed.

Variational method

The expansion coefficients, $a_{n,g}$ and $b_{i,m,g}$ are determined by the variational principle. The linear combination, given by Eq. (5), is substituted into a functional of the neutron fluxes [7]:

$$J[\Phi_g] = \Sigma_{node} \int_{node} \begin{bmatrix} D_g(r)(\partial\Phi_g(r)/\partial x)^2 + D_g(r)(\partial\Phi_g(r)/\partial y)^2 + Dg(r)(\partial\Phi_g(r)/\partial z)^2 \\ +\Sigma_g(r)\Phi_g(r)^2 - 2\Phi_g(r)S_g(r) \end{bmatrix} d^3r \tag{9}$$

where Σ_{node} is the summation over all the nodes in the core, and $\int_{node} d^3r$ is the integration over node volume. It should be noted that, from the definition of the node:

$$\Sigma_{node} \int_{node} d^3r = f_{core} d^3r : \text{means the integration over the whole core}$$

The functional (9) is a quadratic form of the coefficients $a_{n,g}$ and $b_{i,m,g}$, and can be minimised if one sets the partial derivative with respect to these parameters equal to zero. Let $\Phi_g^e(r)$ be the exact solution of Eq. (1), then $J[\Phi_g^e]$ is the minimum value of the functional, and is also a constant. Therefore minimising $J[\Phi_g]$ is equivalent to minimising $J[\Phi_g] - J[\Phi_g^e]$, which becomes, after some partial integration:

$$J[\Phi_g] - J[\Phi_g^e] = \int_{core} \begin{bmatrix} D_g(r)(\partial\Phi_g(r)/\partial x - \partial\Phi_g^e(r)/\partial x)^2 + D_g(r)(\partial\Phi_g(r)/\partial y - \partial\Phi_g^e(r)/\partial y)^2 \\ +D_g(r)(\partial\Phi_g(r)/\partial z - \partial\Phi_g^e(r)/\partial z)^2 + \Sigma_g(r)(\Phi_g(r) - \Phi_g^e(r))^2 \end{bmatrix} d^3r \tag{10}$$

if $\Phi_g(r)$ takes the same value, or the same 1st derivative as $\Phi_g^e(r)$ at the core boundary. The values of coefficients, $a_{n,g}$ and $b_{i,m,g}$, required to minimise the functional (9) give the neutron flux solution "closest to the exact" in the sense that it gives the minimum of the "distance" defined by Eq. (10).

Nodal equations

If one substitutes Eq. (5) into Eq. (9), and differentiates the resulting equation with respect to the expansion coefficient $a_{n,g}$, and also substitutes Eq. (5) into the group neutron source $S_g(r)$, a set of linear equations to determine $a_{n,g}$ is obtained:

$$\begin{aligned} \Sigma_{n'=0,N} U_{n,n',g} a_{n',g} = (1/\lambda)\Sigma_{g'=1,G} \left\{ \Sigma_{n'=0,N} U^F_{(n,g),(n',g')} a_{n';g'} + \Sigma_{i=1,6} \Sigma_{m=0,M} V^F_{(n,g),(i,m,g')} b_{i,m,g'} \right\} \\ + \Sigma_{g'=1,G} \left\{ \Sigma_{n'=0,N} U^S_{(n,g),(n',g')} a_{n',g'} + \Sigma_{i=1,6} \Sigma_{m=0,M} V^S_{(n,g)(i,m,g')} b_{i,m,g'} \right\} \end{aligned} \tag{11}$$

for n = 0, ..., N, and g = 1, ..., G, where the coefficients are given by:

$$U_{n,n',g} = \int_{node} \Phi_{n,g}(r) S_{inf,g}(r) P_{n'}(r) d^3r \tag{12a}$$

$$U^F_{(n,g),(n',g')} = \int_{node} \Phi_{n,g}(r) \chi_g \nu\Sigma_{f,g'}(r) \Phi_{n',g'}(r) d^3r \tag{12b}$$

$$V^F_{(n,g),(i,m,g')} = \int_{node} \Phi_{n,g}(r) \chi_g \nu\Sigma_{f,g'}(r) \Psi_{i,m,g'}(r) d^3r \tag{12c}$$

$$U^S_{(n,g),(n',g')} = \int_{node} \Phi_{n,g}(r) \Sigma_{g,g'}(r) \Phi_{n',g'}(r) d^3r \tag{12d}$$

$$V^S_{(n,g),(i,m,g')} = \int_{node} \Phi_{n,g}(r) \Sigma_{g,g'}(r) \Psi_{i,m,g'}(r) d^3r \tag{12e}$$

Using the same procedure with respect to the expansion coefficient $b_{i,m,g}$, one obtains a set of linear equations of the form:

$$\begin{aligned} &\Sigma_{i'=1,6}\Sigma_{m'=0,M} C_{(i,m),(i',m'),g} b_{i',m',g}\Big|_{node\ 1} + \Sigma_{i'=1,6}\Sigma_{m'=0,M} C_{(i,m),(i',m'),g} b_{i',m',g}\Big|_{node\ 2} \\ &= (1/\lambda)\Sigma_{g'=1,G}\left\{\Sigma_{n'=0,N} B^F_{(i,m,g),(n',g')} a_{n';g'} + \Sigma_{i'=1,6}\Sigma_{m=0,M} C^F_{(i,m,g),(i',m',g')} b_{i',m',g'}\right\}_{node\ 1} \\ &+ \Sigma_{g'=1,G}\left\{\Sigma_{n'=0,N} B^S_{(i,m,g),(n',g')} a_{n',g'} + \Sigma_{i'=1,6}\Sigma_{m=0,M} C^S_{(i,m,g),(i',m',g')} b_{i',m',g'}\right\}_{node\ 1} \\ &+ \text{similar terms for node 2} \end{aligned} \tag{13}$$

for m = 0, ..., M, and g = 1, ..., G. Nodes 1 and 2 are adjacent nodes that share the common interface i. The coefficients are given by:

$$C_{(i,m),(i',m'),g} = \int_{surface} \Psi_{i,m,g}(r) D_g(r) \nabla\Psi_{i',m',g}(r) z\ ds \tag{14a}$$

$$B^F_{(i,m,g),(n',g')} = \int_{node} \Psi_{i,m,g}(r) \chi_g \nu\Sigma_{f,g'}(r) \Phi_{n',g'}(r) d^3r \tag{14b}$$

$$C^F_{(i,m,g),(i',m',g')} = \int_{node} \Psi_{i,m,g}(r) \chi_g \nu\Sigma_{f,g'}(r) \Psi_{i',m',g'}(r) d^3r \tag{14c}$$

$$B^S_{(i,m,g),(n',g')} = \int_{node} \Psi_{i,m,g}(r) \Sigma_{g,g'}(r) \Phi_{n',g'}(r) d^3r \tag{14d}$$

$$C^S_{(i,m,g),(i',m',g')} = \int_{node} \Psi_{i,m,g}(r) \Sigma_{g,g'}(r) \Psi_{i',m',g'}(r) d^3r \tag{14e}$$

where $\int_{surface} ds$ is the integration over the surface i of a node, and z is a unit vector which is outward normal to the surface i.

Power distribution

The power density distribution R(r) within a node can be given by, from Eq. (5):

$$\begin{aligned} R(r) &= \Sigma_{g=1,G}\Sigma_{f,g}(r)\Phi_g(r) = \Sigma_{g=1,G}\Sigma_{n,0,N} a_{n,g}\Sigma_{f,g}(r)\Phi_{n,g}(r) \\ &+ \Sigma_{g=1,G}\Sigma_{i=1,6}\Sigma_{m=0,M} b_{i,m,g}\Sigma_{f,g}(r)\Psi_{i,m,g}(r) \end{aligned} \tag{15}$$

After the coefficients $a_{n,g}$ and $b_{i,m,g}$ are obtained as in the previous section, the nodal power R_{node} can be calculated by:

$$R_{node} = \Sigma_{g=1,G}\Sigma_{n=0,N}U^{R}_{n,g}a_{n,g} + \Sigma_{g=1,G}\Sigma_{i=1,6}\Sigma_{m=0,M}V^{R}_{i,m,g}b_{i,m,g} \tag{16}$$

where:

$$U^{R}_{n,g} = \int_{node}\Sigma_{f,g}(r)\Phi_{n,g}(r)d^3r \tag{17a}$$

$$V^{R}_{i,m,g} = \int_{node}\Sigma_{f,g}(r)\Psi_{i,m,g}(r)d^3r \tag{17b}$$

Similarly, by using Eq. (15), one can obtain power density at any point within a node, or the average power density of any region within a node. For example, the "nodal pin power" of a fuel rod in a node may be defined as the average linear power density of the fuel rod in the node:

$$R_p = \int_p R(r)d^{3r}/H_{node} \tag{18}$$

where R_p is the nodal pin power, $\int_p$ is the integration over the volume of the fuel rod within the node, and H_{node} is the height of the node. Substituting Eq. (15) into Eq. (18), one obtains:

$$R_p = \Sigma_{g=1,G}\Sigma_{n=0,N}U^{R}_{p,n,g}a_{n,g} + \Sigma_{g=1,G}\Sigma_{i=1,6}\Sigma_{m=0,M}V^{R}_{p,i,m,g}b_{i,m,g} \tag{19}$$

where:

$$U^{R}_{p,n,g} = \int_p\Sigma_{f,g}(r)\Phi_{n,g}(r)d^3r/H_{node} \tag{20a}$$

$$V^{R}_{p,i,m,g} = \int_p\Sigma_{f,g}(r)\Psi_{i,m,g}(r)d^3r/H_{node} \tag{20b}$$

Benchmark results

HAFAS 2-D BWR benchmark problem

The present method was applied to the HAFAS two-dimensional BWR benchmark problem [8]. The results were compared to a finite-difference diffusion calculation with 22×22 space meshes per each node as the reference. For the present method calculations were performed with (N=4, M=5), and (N=2, M=2), where N and M are the order of expansion defined by Eqs. (3) and (4). One has 17 unknowns per (node, group) for (N=4, M=5), 9 unknowns for (N=2, M=2), and $22 \times 22 = 484$ unknowns in the reference calculation. The fuel region of each of the assemblies is divided into 16×16 fuel rod cells of equal volume. The pin power is defined to be the average power density in a fuel pin cell.

Table 1 summarises the results of the comparison, including the core eigenvalue comparison. The present method with (N=4, M=5) gives almost the same results as the reference. Even with (N=M=2), its agreement to the reference is much better than the known conventional nodal methods. The computing time of the present method was measured to be 2.9% of the reference calculation per each source iteration (N=M=2).

In Table 1, LOGOS and ANDEX are nodal diffusion codes based on an advanced modified one-group diffusion theory, and the modern nodal diffusion method (transverse integration, assembly discontinuity factors, etc.), respectively. The maximum nodal power error of VNEM is about 1/3 of these conventional methods even with N=M=2.

Table 1. Comparison summary for HAFAS benchmark problem

	Reference	**VNEM N=4, M=5**	**VNEM N=2, M=2**	**LOGOS** [9]	**ANDEX** [9]
k_{eff}	1.04337	-0.002%	-0.05%	-0.06%	-0.09%
Pmax[1]	1.741	0.03%	1.4%	2.8%	-3.6%
rms[2]	–	0.04%	0.9%	1.7%	2.2%
max[3]	–	0.1%	1.5%	-4.2%	-5.7%
Mmax[4]	1.629	0.4%	1.6%	Not available	
rms[5]	–	0.3%	0.8%		
max[6]	–	-0.6%	1.8%		

[1] Maximum nodal power in the core, errors in per cent
[2] Root mean square of nodal power errors in per cent
[3] Maximum nodal power error in per cent
[4] Core maximum of maximum nodal pin power (MPP), errors in per cent
[5] Root mean square of MPP errors in per cent
[6] Maximum MPP error in per cent

An actual 2-D BWR benchmark problem

The present method was also applied to actual two-dimensional BWR benchmark cases [1]. The core for these cases was taken from the horizontal cross-section of a 1 100 MW (electric) BWR core of a recent design, which was strongly heterogeneous because it was loaded with three kinds of fuel assemblies with different enrichment and containing a number of gadolinium-poisoned fuel rods.

A hot operating case (with 40% in-channel void fraction), and two cold critical cases (with the A- and the B-sequence control rod patterns), are used for comparison. Here the cold cases are to test the accuracy of the present method in calculating the cold critical eigenvalue. Reference calculations were performed by the ordinary few-group diffusion method using the same energy groups and space meshes as in the lattice physics code.

The results are summarised in Table 2. The present method calculations were performed with (N=4, M=5) and (N=2, M=2) as in the case of HAFAS benchmark. From these comparisons one can confirm the conclusion obtained from the results of HAFAS benchmark problem. The present method with (N=4, M=5) gives almost the same results as the reference. Even with (N=M=2) its agreement is quite satisfactory for practical purposes.

NEACRP 2-D PWR MOX benchmark

The present method was also tested against the NEACRP two-dimensional PWR MOX benchmark problems [3]. In the NEACRP specification [10] there are five benchmark problems, labelled C1 through C5. The first two, C1 and C2, concern infinite checkerboard geometry, and are not included here. The reason for this is simply that the appropriate boundary condition (the four-side periodic boundary condition) for this geometry is still not implemented in VNEM. The results are summarised in Table 3.

Table 2. Comparison summary for an actual BWR benchmark problem

(a) Hot operating case

	Reference	N=4,M=5	N=2,M=2
k_{eff}	1.04453	-0.002%	-0.006%
Pmax[1]	1.827	-0.3%	-0.1%
rms[2]	–	0.1%	0.1%
max[3]	–	-0.3%	0.2%
Mmax[4]	1.803	-0.4%	-0.2%
rms[5]	–	0.1%	0.2%
max[6]	–	-0.4%	0.9%

(b) Cold critical eigenvalues

	Reference	N=4, M=5	N=2, M=2
k_{eff}[A]	1.00009	-0.041%	-0.007%
k_{eff}[B]	1.00581	-0.023%	-0.054%

[1~6] See notes for Table 1
[A] A-sequence cold critical case
[B] B-sequence cold critical case

Table 3. Comparison summary for NEACRP PWR MOX benchmark problem

	C3	C4	C5
k_{eff}	1.01761	0.90550	0.93614
	0.002%	0.001%	0.000%
Pmax[1]	1.073	1.844	1.440
	-0.01%	-0.10%	-0.02%
rms[2]	0.01%	0.07%	0.01%
max[3]	0.01%	0.10%	-0.02%
Mmax[4]	1.212	2.303	1.714
	-0.30%	-0.80%	-0.00%
rms[5]	0.20%	0.31%	0.19%
max[6]	0.7%	1.0%	1.2%

[1~6] See notes for Table 1

In all three MOX benchmarks (C3~C5), the order of expansion is (N=4, M=5). It was observed that (N=2, M=2), which was practically sufficient for full-core benchmark cases shown in the previous sections, does not give good results for cases C4 and C5, in which the power distribution is strongly affected by zero-flux boundary, or the reflector. Also from these comparisons it can be said that the present method with (N=4, M=5) gives almost the same results as the reference.

Discussions and conclusions

It is difficult to determine the minimum order of expansion (N and M) to obtain a required accuracy theoretically. However, as far as light water reactor cores are concerned, it seems practical to find them by numerical test calculations. From the results of 2-D benchmark calculations shown above, it seems that N=M=2 will be satisfactory for almost all cases of practical interest in LWR analysis. This implies that, for 3-D cases, N=M=3 will be the practically acceptable minimum.

The result of a 3-D VNEM benchmark calculation [2] supports this inference. For N=M=3, the number of unknowns of VNEM will be less than a few per cent of that in the reference fine-mesh, finite-difference calculation. This means the computing time of VNEM will also be less than a few percent of the reference calculation, because the computing time is approximately proportional to the number of unknowns.

The number of the coefficients in the equations of VNEM is much larger than that in the conventional methods. This may cause a long computing time for the reproduction of the coefficients from a coefficients library. However:

1) The number of coefficients can be drastically reduced by using the symmetry of a node.

2) The off-diagonal coefficients (e.g. $U_{n,n',g}$ with $n \neq n'$) are much smaller than the diagonal ones (e.g. $U_{n,n,g}$), because of the orthogonality of Legendre polynomials, so the latter do not require very detailed reproducing procedures.

3) Pin power coefficients can be reduced by the "candidates method", in which only the pin power coefficients for small number of fuel rods (the candidates) that have the possibility of becoming the peak rod, are reproduced, and the pin power calculation is limited to these candidate rods.

Because of these factors, the computing time for the reproduction of the VNEM coefficients can be within a reasonable limit. The burn-up history effect on the VNEM coefficients can be included in the same way as used in the conventional nodal methods.

From the above consideration and the fact that the present method is significantly faster than the reference, it can be concluded that the present method is a promising basis for the development of a new 3-D nodal method for LWR cores.

REFERENCES

[1] M. Tsuiki, *et al.*, "A Variational Nodal Expansion Method for the Solution of Multi-Group Neutron Diffusion Equations", Proc. Int. Conf. on the Physics of Nuclear Science and Technology, Long Island, New York, 1998.

[2] M. Tsuiki, "A Variational Nodal Expansion Method for Three-Dimensional Cartesian Geometry", Proc. Int. Conf. on Mathematics and Computation, Reactor Physics and Environmental Analysis in Nuclear Applications, Madrid, Spain, 1999.

[3] S. Hval, *et al.*, "Benchmark Calculations using the VNEM Code on PWR-Lattices", Proc. Reactor Physics Calculations in the Nordic Countries, Goeteborg, Sweden, 1999.

[4] P. Ellison, *et al.*, "A Coarse Mesh Finite Element Method for Boiling Water Reactor Analysis", Proc. Topl. Mtg. Advances in Reactor Computations, Salt Lake City, Utah, USA, 1983.

[5] T.H. Fanning, *et al.*, "Variational Nodal Transport Methods with Heterogeneous Nodes", *Nucl. Sci. Eng.*, 127, 154, 1997.

[6] E. Nichita, *et al.*, "A Finite Element Method for Boiling Water Reactors", Proc. Int. Conf. on the Physics of Nuclear Science and Technology, Long Island, New York, USA, 1998.

[7] I. Dilber, *et al.*, "Variational Nodal Methods for Neutron Transport", *Nucl. Sci. Eng.*, 91, 132, 1985.

[8] K. Smith, *et al.*, "The Determination of Homogenized Diffusion Theory Parameters for Coarse Mesh Nodal Analysis", Proc. Topl. Mtg. Advances in Reactor Physics and Shielding, Sun Valley, Idaho, USA, 1980.

[9] T. Iwamoto, *et al.*, "New Nodal Diffusion and Pin Power Calculation Method Based on Modified One Group Scheme", Proc. Topl. Mtg. Advances in Reactor Physics, Charleston, S.C., USA, 1992.

[10] C. Cavarec, *et al.*, "Benchmark Calculations of Power Distribution within Assemblies", NEACRP-L-341, 1994.

SESSION IV

Improved Core Monitoring Systems, Design and Operating Experience

Chairs: Y. Shimazu and S. Andersson

SOME ASPECTS OF THE NEW CORE MONITORING SYSTEM AT NPP DUKOVANY AND FIRST EXPERIENCE

M. Pecka
CHEMCOMEX Praha

J. Svarny
Skoda Nuclear Machinery Plzen

J. Kment
NPP Dukovany

Abstract

Methods of experimental data interpretation and on-line calculation performed by the SCORPIO-VVER core surveillance system as well as results obtained from the system during the trial operation are discussed. Interpretation of measurements and 3-D power distribution determination are emphasised. A brief description of SCORPIO-VVER FA sub-channel characteristics methodology is given. Various methods of SPND signal conversion to local power, which are currently tested within the SCORPIO system, are described and results of their testing are presented. The SPND detectors delivered by IST, USA, loaded for the first time last year, are treated as well. SCORPIO-VVER predictive functions are inspired by HRP SCORPIO implementations at several PWRs (Ringhals, Sizewell B) and are tailored in order to meet the specific needs of VVER-440. A short description of the SCORPIO predictive capabilities is given, and results of predictions in comparison with real data are presented.

The SCORPIO-VVER system

The SCORPIO-VVER core surveillance system was built in co-operation with IFE Halden, NRI Rez, Skoda Nuclear Machinery Plzen and Chemcomex Praha. The first version of the system is designed for VVER-440/V-213 reactors and has been installed at Unit 1 of NPP Dukovany since March 1998. The system provides a support for operators and reactor physicists, and in the near future it will replace the old computing system VK3.

The following functions are provided by the modules of the system: communication with data acquisition units, interpretation of input signals, calculation of basic reactor and primary circuit parameters, 3-D power distribution determination, limit checking and thermal margin calculation, PCI margin calculation and predictive functions (including power distribution calculation and limit checking). Interactive graphical man-machine interface, printed protocols and network transfer of results to other computing systems are provided as well. An optimum combination of measurements and calculations is used to obtain full information about the reactor core and primary circuit.

During the licensing process at the State Office for Nuclear Safety the old VK3 system and SCORPIO-VVER were operated in parallel at Unit 1 of NPP Dukovany, and a large amount of data was collected from both systems. Measured and calculated values from both systems were evaluated and compared as a test of the precision and reliability of SCORPIO-VVER data processing.

Interpretation and validation of signals from measurements

Input signals are transferred to SCORPIO-VVER from two independent data acquisition units: SVRK Hindukus and the Temperature Measurements Back-up System (TMBS). In the first step of data processing all coded signals are converted to corresponding physical units. In the next step values of measured quantities are calculated, e.g. temperatures are obtained from voltage values at RTDs and current values from thermocouples. The validation of measured values is based on stability of signals, redundancy of measurements and independent calculations.

During the tests 346 time points were selected in which the input data and outputs from both SCORPIO-VVER and VK3 were saved over a short time period. Selected data describe various states of the reactor including transient states during the reactor shutdown and start-up. Results from SCORPIO-VVER were compared to corresponding VK3 outputs and to independently calculated values derived from saved input signals. The following general conclusions were arrived at using signals interpretation evaluation:

Results of the SCORPIO-VVER signals interpretation are in all cases identical with results of independent calculations based on the saved input data, and deviations from VK3 outputs are in the range of sensors stability, precision of measurements and numerical precision of the VK3 system.

Some deviations are caused by different methods of signal stabilisation in both systems. For each sensor, VK3 uses weighted average values of signals from some time period – previous measurements have some limited influence on the actual value. In SCORPIO-VVER a redundancy of most measurements is used – unstable signals are corrected in dependency on similar measurements from other sensors. Differences between both systems are visible in transient states, when the old system ignores fast changes and in values from faulty oscillating sensors.

Interpretation of "normalised" signals from SVRK Hindukus

Interpretation of this group of signals is simple. Each signal is transferred from SVRK Hindukus in a normalised form and a linear formula is used for transformation to the appropriate physical unit. Identical formulas are implemented in both systems so that all deviations are caused by fast changes in measured values (when data from both systems were not saved in the exactly same moment) and by differences in signal stabilisation methods, e.g. for signals from ionisation chambers the mean deviation between SCORPIO-VVER and VK3 outputs is less than 0.05% of the full power.

Interpretation of temperature measurements

Two types of temperature sensors are installed in VVER reactors – RTDs and thermocouples. RTD measurements are more accurate because the voltage value (with fixed current) may be directly transformed to temperature. Formulas for thermocouple interpretation are more complicated and results are dependent on determination of temperatures at the cold ends of thermocouples (which are measured by RTDs). The numerical precision of the VK3 system is relatively low, and thus the accuracy of thermocouple measurements interpretation in the old system is lower than the accuracy of RTDs.

In the RTDs interpretation the deviations between outputs of both systems are very small. The mean deviation of all measurements (55 sensors in 346 time points) is approximately 0.05°C. Higher statistical deviations were found in the interpretations of the thermocouple measurements. There, the mean deviation is approximately 0.14°C, and some individual measurements show significantly higher differences. In comparison to the precision of measurements the deviations are acceptable. For example from the total number of 72 660 processed measurements of the fuel assembly outlet temperatures, only eight individual measurements of extremely unstable sensors had a differences higher than 0.6°C.

Interpretation of SPND signals

Signals from SPNDs are interpreted using transformation coefficients calculated in dependence on power distribution, local fuel burn-up, etc. The method is different from the VK3 system implementation, and results (local linear power values) show higher deviations between outputs of both systems than other compared quantities. Methods and examples of results are described in the independent section later in this paper.

Determination of basic primary circuit parameters

Measured values are used for determination of basic parameters which are needed for subsequent calculations – mean temperatures, temperature rises, reactor thermal power, etc. The accuracy of results depends on the precision of measurements and validation of measured values. As an example a comparison of calculated reactor thermal power is presented. Power is calculated using reactor inlet and outlet temperatures at all six primary circuit loops. Both RTD and thermocouple measurements are available for the cold and hot leg of each loop. The VK3 system uses RTD measurements, and thermocouples are taken into account only in the case of RTD malfunction. SCORPIO-VVER uses the most credible measurements (RTD in most cases), and if the credibility is low the temperature value is corrected using data from other sensors as well. Some systematic deviations are caused by differences in determination of other parameters, e.g. coolant flow rate. A comparison of SCORPIO-VVER and VK3 outputs is presented in the following chart, which shows trends of several quantities during the transient process (down to approximately 50% and back to full power). Values were compared in 69 time points during a period of 13 hours.

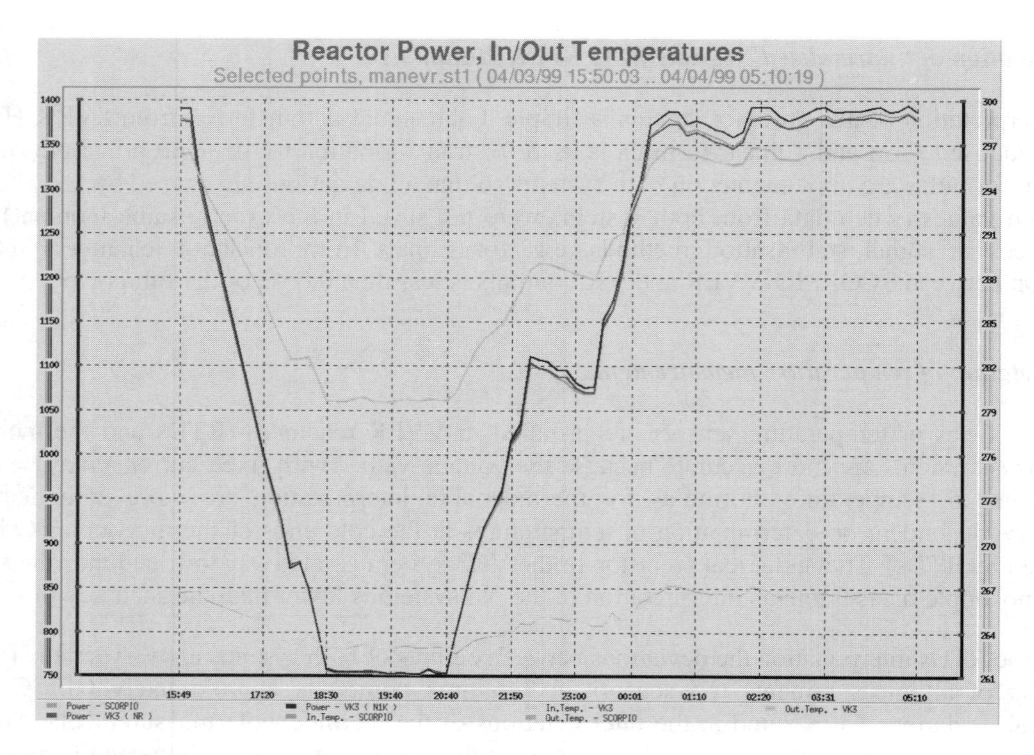

The chart shows trends of the following quantities:

- Mean reactor inlet temperature (°C) by SCORPIO-VVER and VK3.
- Mean reactor outlet temperature (°C) by SCORPIO-VVER and VK3.
- Reactor thermal power by SCORPIO-VVER (MW).
- Reactor thermal power calculated by VK3 using temperature measurements (MW).
- Corrected "representative" reactor thermal power by VK3 (MW).

Mean inlet and outlet temperatures calculated by both systems are very close (overlapping lines). Power values calculated by VK3 directly from temperature measurements show a systematic error at full power and the stability of output is poor. Thermal power values calculated by SCORPIO-VVER are better stabilised and are close to the "corrected" values from VK3, calculated using various calibration factors and other measurements. The mean difference between reactor thermal power calculated by SCORPIO-VVER and "representative" values from VK3 is less than 0.2%.

Power distribution determination

The SCORPIO-VVER module known as "3-D Power Distribution Determination" calculates an array of local relative power values using data from measurements and results of independent calculations provided by the core-follow simulator. This process is called "power distribution reconstruction". Validation of signals in SCORPIO-VVER assigns a credibility factor to each measurement. The interpretation of any measurement in the reconstruction process depends on this factor. VK3 provides a similar function but reconstructed arrays are mostly based on measurements; there is no real simulator calculation and no signals validation (only faulty detectors are excluded).

Two methods of reconstruction are implemented in SCORPIO-VVER. The "traditional" method uses local measurements with higher priority, and results are more similar to the VK3 output. The "advanced" method is more based on global interpretation of measurements and simulator results, and it is able to reach acceptable results with a fewer number of valid sensors.

Reconstruction in SCORPIO-VVER is performed in two steps. First, a 2-D power distribution array is prepared using simulator calculation and 210 validated thermocouple measurements of coolant outlet temperatures at fuel assemblies. Any tilt or radial systematic deviations of measurements and calculation are detected and simulator results are recalculated. Local anomalies in measurements are taken into account depending on the credibility of corresponding sensors. In the next step an axial power distribution in each assembly is calculated from simulator results and linear power values which are obtained from SPND measurements.

In VVER-440 reactors the height of the core is only 2.42 m and there are no problems with axial xenon oscillations. Limiting factors for reactor operation are mostly temperatures at assembly outlets and radial power peaking factors. This means that an accurate determination of the 2-D power distribution is more important than the axial distribution in assemblies. The accuracy of the axial distribution reconstruction is usually lower due to insufficient stability of SPND signals.

Comparison of 2-D power distribution reconstruction results

Two hundred eighty-eight time points in which the reconstruction outputs from both VK3 and SCORPIO-VVER were saved over a short time period were selected. Parts of the data describe transient states including reactor shutdown and start-up. Values of the local power in assemblies obtained from both methods of reconstruction in SCORPIO-VVER were compared to the reconstruction output of VK3 (for all 349 fuel assemblies), to results of simulator calculations and to results of independent calculations based only on temperature measurements (only for 210 fuel assemblies with thermocouples). There are only a few deviations higher than 0.1 MW (i.e. approximately 2.5% of the mean fuel assembly power) and the behaviour of these deviations is statistical. Mean deviations between VK3 output and both types of SCORPIO-VVER reconstruction are in all analysed time points less than 0.055 MW.

The 2-D power distribution reconstruction provided by the VK3 system is close to a "pure" interpretation of temperature measurements. For most fuel assemblies the value from the SCORPIO-VVER reconstruction is between the result of simulator calculation and output from VK3. Results of the "advanced" method are closer to the simulator results and more different from VK3 output. Both methods of reconstruction partially eliminate any systematic errors in simulator calculations and statistical deviations in temperature measurements as well.

Systematic deviations of SCORPIO-VVER and VK3 reconstruction results were studied as well. The following chart represents a core map with mean deviations between results of the "traditional" method provided by SCORPIO-VVER and results of VK3 reconstruction in each fuel assembly (control assemblies were excluded). Mean values were obtained from 232 analysed time points representing a stable reactor operation at full power.

The statistical behaviour of higher deviations distribution is evident. No systematic radial deviation is visible. A very similar picture was obtained from comparison between SCORPIO-VVER reconstruction and distribution calculated directly from temperature measurements.

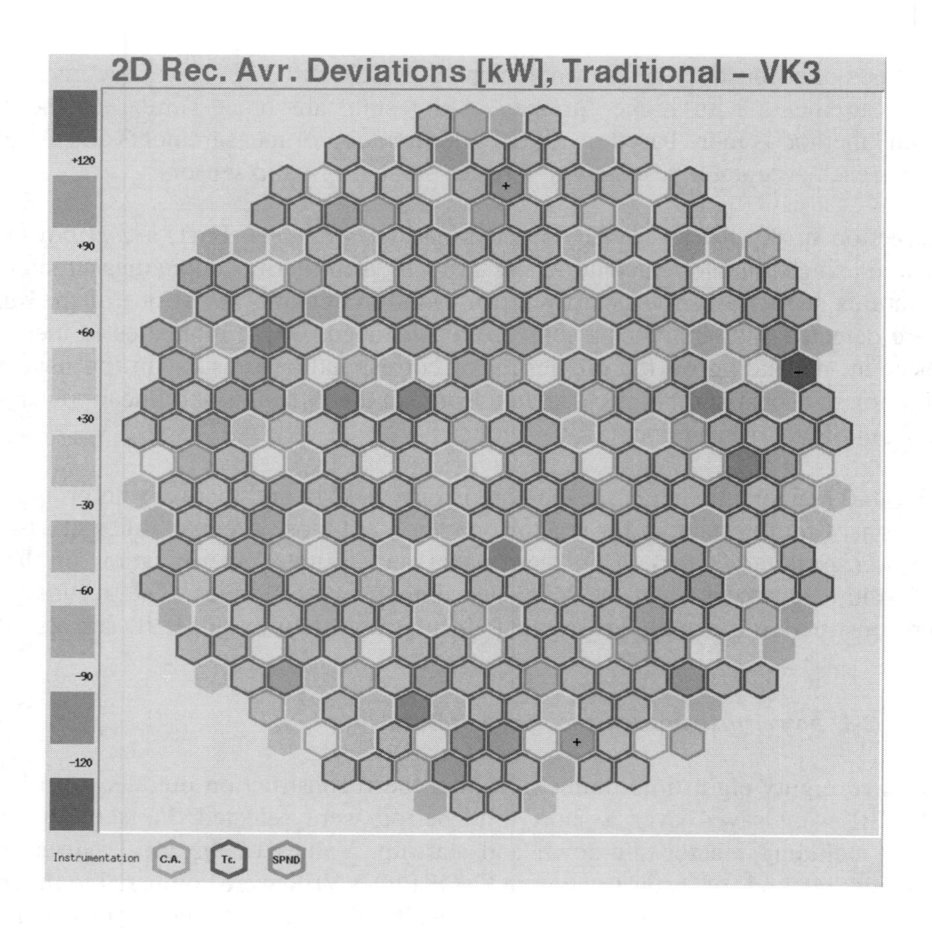

Limit checking and thermal margin calculation

This SCORPIO-VVER sub-system performs calculations of various parameters such as nodal power peaking limits and fuel assembly (FA) temperature rise on the basis of the 3-D coarse mesh power distribution. Parameters are checked in accordance with the present NPP Dukovany specification for operation.

A detailed 3-D pin-wise power distribution is produced for determination of F_Q and $F_{\Delta H}$ power peaking factors and assessment of all margins to the safety (DNBR, LOCA and saturation temperature) on the basis of sub-channel analysis. Pin-wise power distribution is processed to provide information for PCI-margin calculation and off-line reactor vessel fluence calculation. This module also provides the calculation of transformation coefficients for SPND detectors.

SCORPIO-VVER FA sub-channel characteristics methodology

The 3-D pin-wise power distribution calculation is based on the so-called modulation method in which local homogeneous solutions (in each FA) are multiplied by heterogeneous form functions. Heterogeneous fission form functions $f_G^f(\bar{r})$ for fission of group G are obtained off-line by comparison of 3-D pin-wise power distribution calculation with a homogeneous solution normalised to meet the following modulation equation for pin-wise core power distribution $q(\bar{r})$:

$$q(\bar{r}) = A \sum_{G=1}^{2} \phi_G^{\text{hom}} \sum\nolimits_f^G (\bar{r}) \; f_G^f(\bar{r})$$

where: $\phi_G^{\text{hom}}(\bar{r})$ is local homogeneous flux of group G determined

$\sum_f^G(\bar{r})$ is local homogeneous macroscopic fission cross-section

A relates fission reactor rates to power

These local homogeneous values are determined by the 2-D interpolation on the basis of a coarse mesh finite-difference calculation produced in the 3-D Power Distribution Determination sub-system.

Fuel assembly (FA) temperature rises

FA coolant temperature rises •TKAZ is calculated for core individual FAs on the background of assembly-wise power distribution and coolant flow distribution in the reactor. This flow distribution takes into account pressure loss coefficients, which are different for different type of FA (standard old FA, profiled new FA and control FA), bypass for barrel cooling and coolant flow reduction caused by the bypass through the labyrinth sealing of control fuel assemblies and by the bypass through the perforation of fuel assemblies shroud.

Sub-channel output coolant temperature

The most loaded sub-channel output coolant temperature is the maximum of sub-channel output temperatures in the core. All FAs in-core are checked, because of differences in pin-wise power distribution and coolant flow rates in individual FAs. It was taken into consideration that the most loaded sub-channel need not be in the most loaded assembly.

Hot sub-channel output coolant temperature is determined by the use of statistically determined enthalpy rise hot channel factor.

Departure from nucleate boiling ratio (DNBR)

$DNBR_{min}$ is the minimum ratio of critical heat flux to real heat flux in the core, where all full rods and all axial positions are taken into account. Critical heat flux is calculated with the use of correlation PI-3f, which shows good statistical characteristics in an extension range of rod bundle geometric parameters and over a wide range of thermohydraulic conditions. A conservative estimation of minimum advisable value $(DNBR_{min})^{lim}$ is arrived at, which takes into account the above-mentioned requirements (evaluation of DNB correlation uncertainties, power, coolant flow, temperature and pressure uncertainties, code inaccuracy, influence of manufactory tolerances and technological deviations).

Checking of safety margin to the maximum linear power peaking (LOCA)

The maximum linear power peaking of any fuel rod shall not exceed LOCA limiting value including calculation and measurement uncertainties, manufacturing tolerances and additional conservatism, but excluding densification effects. All these factors are statistically processed in one factor and the options for taking into account burn-up dependent LOCA limit are included in the SCORPIO-VVER algorithm.

Validation process and tests

Data collection from reactor operation was carried out at periodic 12-hour intervals. Tests of assembly-wise and pin-wise power distribution and check-up of margins to safety limits were systematically performed from the very beginning of the transient reloads. The aim of the tests was to prove both the functionality of the new monitoring system SCORIO-VVER, meaning the verification of safety limits margins satisfaction on the operation, and design level using comparisons of reconstructed assembly-wise and pin-wise power distribution with off-line design calculations and to define uncertainty coefficients more precisely. The tests also included periodic independent off-line comparison of measured and calculated assembly and pin-wise parameters. The compared fine mesh parameters were: enthalpy rise hot channel factor F_{dh}, heat flux hot channel factor F_Q, LOCA margin of the pin linear power Q_{lin}, coolant saturation temperature margin and departure from nucleate boiling margin DNBR.

As a result of the tests the following values of statistical uncertainties were included into licensing documentation of SCPRPIO-VVER this year:

- Uncertainty of radial FA power reconstruction methodology: 4%.
- Uncertainty of axial FA power reconstruction methodology: 3%.
- Uncertainty of pin-wise FA form factors prediction: 2%.

Models for calculating prompt response of self-powered neutron detector (SPND)

Three models (options) for calculating prompt response SPND have been taken into account. The first model (ACTUAL) is based on standard Russian (at present valid in VK3) Cymbalov methodology. The second one (NORMAL) also has Cymbalov form, but with different polynomial dependencies. The third one (ADVANC) involves explicit Rh wire burn-up calculation.

The preparation of formulas for coefficients was developed over following phases:

- *A/ Basic calculation of the states by the WIMS7 code (transport calculations of neutrons)*
  Transport of neutrons in fuel assembly and in the SPND structure has been calculated by the WIMS7 code in a 2-D hexagonal structure.
- *B/ Transport calculations of electrons by the code MCNP4B and calibration process*
  Transport of electrons from Rh wire to the steel collector is represented in our formulas by the Beta escape probability (β_{esc}). A precise calculation of this parameter is more complex and also requires assessment of effects such as the potential barrier between Rh wire and collector. From our analysis it was seen that β_{esc} strongly depends not only on electron energy but also on Rh wire diameter. This fact lead us to adopt the following procedure with regard to β_{esc}: relative values (burn-up dependence) of β_{esc} were calculated by the Monte Carlo MCNP4B code, reference SPND with defined dimension was chosen to assess the absolute value of β_{esc} from the long time operation and impacts of the Rh diameter and insulator changes on β_{esc} were calibrated (measured). In our SPND library preparation for NORMAL formula relative β_{esc} dependence on SPND burn-up calculated by MCNP4B was used. For ADVANC formula two-group representation of β_{esc} has been used.

At present, standard SPNDs of Russian provenience (POSIT SPND) are used, as well as new SPNDs of American provenience (IST SPND) at NPP Dukovany. Six strings each assembled from seven IST SPND are also used in the core of Unit 1, on which SCORPIO-VVER was rested. The derivation of interpretation formulas for these new SPNDs was based on the methodology mentioned above. POSIT SPND was chosen as a reference SPND, and it was used also as a reference SPND in the calibration measurements provided on the reactor TRIGA in Cornell University. The impact of deviations of dimensions and insulator isotopic composition between POSIT and IST SPNDs on the relative values of β_{esc} was derived from the calibration process. Absolute values of β_{esc} for IST SPND were calculated from these relative values and absolute value of β_{esc} POSIT SPND as reference.

- *C/ Parametrisation of data from A/ and B/ into three different formulas by the code APRO*
  The versatile code APRO, which is a standard used for preparation of the macrocode MOBY-DICK's MAGDA libraries, is also used for processing WIMS7 and MCNP4B calculations into three different formulas (ACTUAL, NORMAL and ADVANC). In general this code calculates cumulative charge in iteration procedure and finally optimised polynomial formulas (of degree up to 25).

- *D/ Validation and corrected coefficients determination*
  Validation of SPND interpretation formulas was realised in several stages during the implementation phase of SCORPIO-VVER (factory acceptance test, site acceptance test, commissioning phase). The last validation test has been realised on the measured data gathered during one cycle operation. This global test was aimed to the comparisons between old monitoring system VK3 and SCORPIO-VVER.

  Very instructive comparisons are FA powers comparisons calculated from SPND signal interpretation (by cubic spline interpolation and integration) against FA powers from radial power reconstruction. These comparisons are used for derivation of additional calibration coefficients and in fact represent differences between SPND and thermocouple (TC) measurements (FA radial power distribution from reconstruction is constructed only on the basis of TC measurements and deviations between TC and reconstruction are not greater than 2%). For the NORMAL formula we obtained deviations of less than 4%, except for two symmetrical SPND strings (KNI). The KNI with faulty SPNDs were excluded. NORMAL and ADVANC have nearly the same interpretations, while the interpretation of ACTUAL is slightly worse. Other systematic comparisons have shown good compliance between TC FA power and SPND FA power also for IST SPNDs, which endorses credibility of our β_{esc} prediction for IST SPND.

Predictive functions of SCORPIO-VVER

The system includes four modules for fast predictive calculations of most important operational parameters: reactivity controllers values (boron, control assemblies positions), peaking factors (both nodal and pin-wise), safety parameters margins, and some other limited parameters as functions of time for a specified power (load follow) transient, up to a maximum of 72 hours ahead. The resulted curves are presented on the operator display.

One of the modules also enables critical parameter (boron or control assembly positions) calculations for reactor hot zero power for any future instant. The resulting time dependence of critical parameter is also presented graphically on the operator display. Cold shutdown boron concentration value can be calculated and presented at any reactor condition as well.

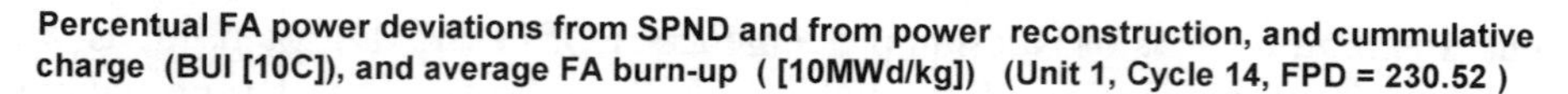

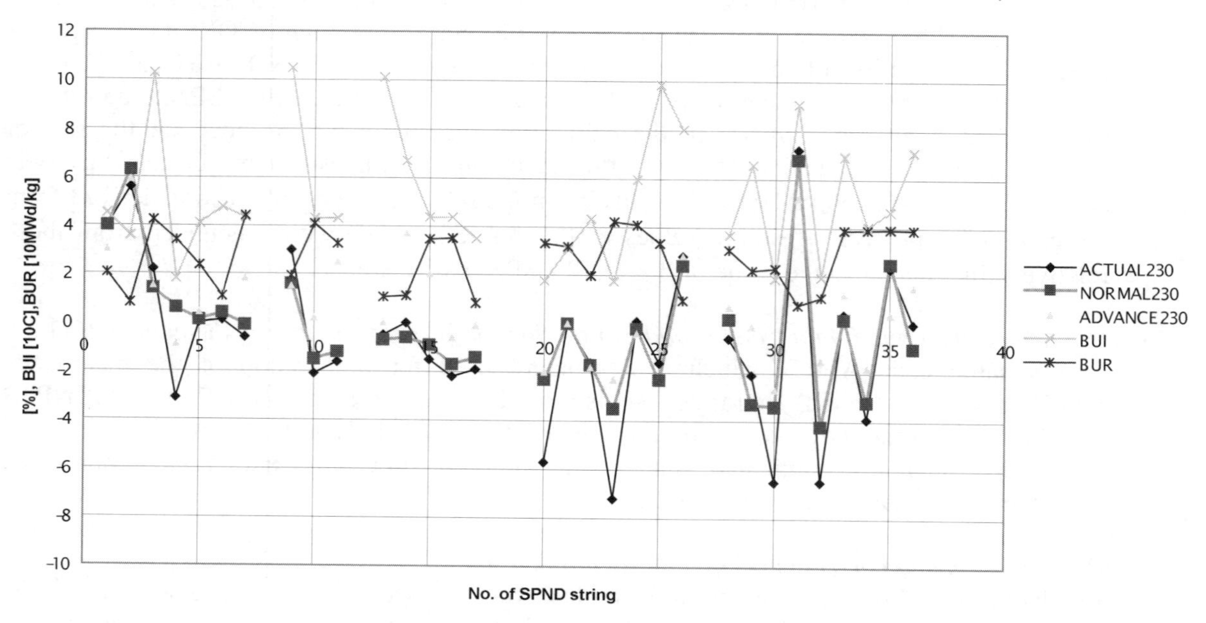

Strategy generator

This module (SG) was originally created within HRP for PWR applications. It proved to be significantly useful at some units (e.g. Ringhals, Sizewell B). The implementation at VVER-440 was suggested by IFE Halden. However, the original module had to be adapted according to NPP Dukovany specific needs and requirements.

The aim of SG is to provide fast assistance for reactor control during a planned power transient. The main requirements are minimising boron changes and maintaining the control assemblies band (Band 6) height position within its permitted, power dependent, range. The power and xenon-iodine densities are calculated for 2 points, without solving the neutronics and hydraulics equations. Instead, pre-calculated coefficients are used to predict responses to changes in power, inlet temperature, control assembly position, boron and xenon concentration. These coefficients are generated by performing a series of simulations with the off-line version of the simulator code.

The only operator input is a graphical specification of total reactor power as a function of time. The rest of the necessary data is taken from the latest outputs of the core-follow modules. SG proceeds to calculate deviation of reactivity at the starting point from its value for a known reference state, assuming a linear correlation between reactivity and the above parameters except of total reactor power (a quadratic dependence is used). This deviation is to be maintained during the whole transient.

A regular step length of ten minutes may be somewhat modified depending on power course specification. The xenon and iodine concentrations at the ends of steps are calculated by assuming that the power in each half of the core changes linearly during time steps. The new values of boron and Band 6 position which make the reactor critical are found by calculating the deviation of reactivity from the reference value, and adjusting controllers until the deviation is the same as at the starting point.

Predictive simulator (SIMPred)

This module is nearly the same code as the core-follow simulator module. However, some functions are omitted, and some are added. To increase the speed of predictive calculations significantly, 24 mesh points per assembly cross-section calculated in the core-follow simulator are reduced to only six. The calculation is enabled only for 60-core symmetry.

The module can be applied to solve one of five types of predictive tasks:

- *Task 1.* Solution of a power transient without recalculation of critical parameters (SG output values are used); 3-D coarse mesh power distributions, etc., are calculated, including nodal peaking factors.
- *Task 2.* Solution of a power transient with recalculation of critical parameters (the same requirements for controllers are applied, only discrete (by node height) values of Band 6 position changes are possible, with "fine" corrections of reactivity "relict"; 3-D distributions are also calculated.
- *Task 3.* Calculation of start-up critical boron concentration for specified Band 6 position and coolant temperature; possible during reactor shutdown only.
- *Task 4.* Calculation of start-up critical Band 6 position for specified boron concentration and coolant temperature; possible during reactor shutdown only.
- *Task 5.* Calculation of cold shutdown boron concentration.

The initial core state conditions are provided by core-follow simulator using the latest of "dump" files generated periodically during operation. If the delay of the used dump file against the actual moment exceeds some limit, or if the actual core parameters differ significantly against the dump ones, an actual state catch-up procedure (i.e. a description of the operational history since the latest dumping) has to be carried out. Core-follow simulator adaptation coefficients are transferred to the predictive simulator to be used for Task 1 or 2 solutions.

Predictive limit checking, thermal margin calculation and PCI margin calculation

Modules are automatically activated during the predictive simulator calculation. For the same transient, specified first for the SG, these modules perform predictive calculations of hot channel factors, safety parameters margins, maximum assembly-wise coolant temperature rise, margin of linear power with respect to a probability of PDI fuel damage, etc. The coarse mesh (nodal) 3-D power distribution and fuel burn-up distribution passed by predictive simulator module for each time step serves as the basis for the calculations.

First experience with predictive functions

The trial operation of the system offered little opportunity for testing and validation of predictive functions after the site acceptance tests at Unit 1 in March 1998. A realisation of the only planned transient has been approved to enable the system testing, including validation of the predictions. Unfortunately, only a few hours of the transient could be utilised for validation for the planned course of reactor power had to be interrupted due to technological troubles (the moment is marked with the vertical line on the figures).

Comparison of predicted and actual values of basic reactivity controllers

Scorpio Prediction

(Unit 1, 3rd April 1999, 4 p.m.)

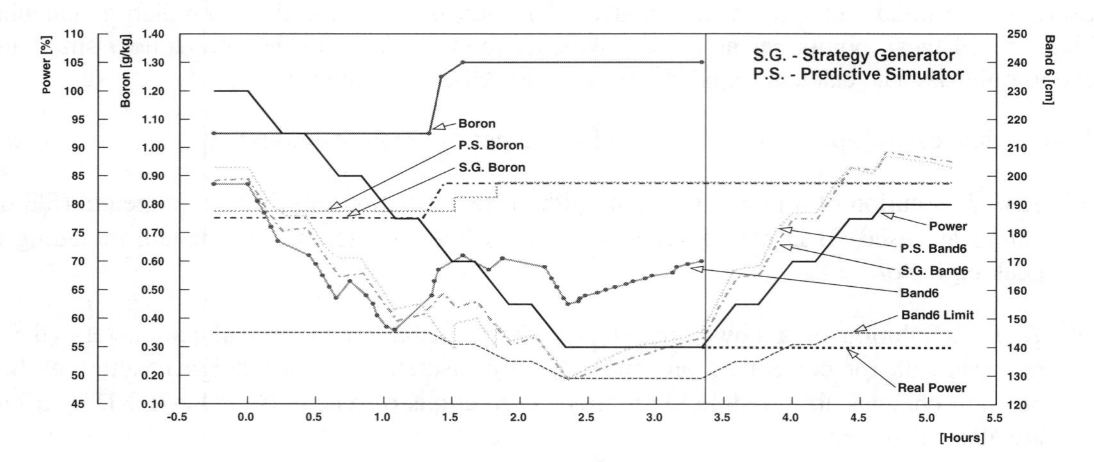

The validation results of SG and predictive simulator (Task 2) predictions of reactivity controllers are presented on the chart. The deviation of the predicted and actual boron is caused by the inaccuracy of the initial concentration value passed from the core-follow simulator. The suggested amount of boration (0.12 g/kg) was due to an operator mistake violated (0.25 g/kg), therefore, since the boration, the curve of actual Band 6 position is shifted but it appears to stay parallel. In general, a good conformity of Band 6 actual position and of both predicted ones can be seen. An exact determination of the starting point of boration (to prevent Band 6 to get below the lower limit) can also be noted.

The following figure presents the comparisons of predicted and actual (monitored with the core-follow part of the system) values of maximum pin-wise linear power and minimum departure to nucleate boiling ratio.

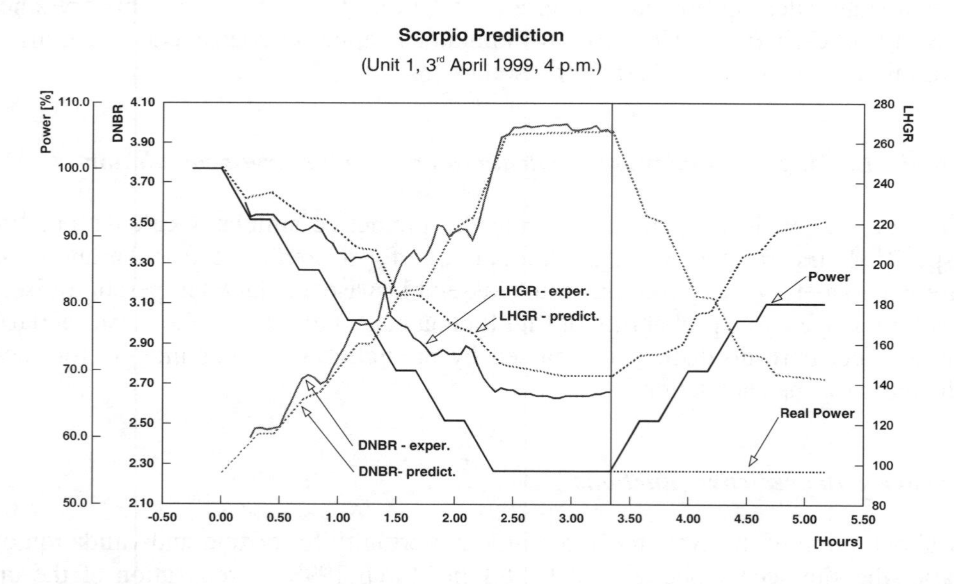

Only two possibilities for the verification of the predicted critical boron for reactor start-up (Task 3 of predictive simulator) occurred so far. The results were good, the inaccuracy was 0.1-0.2 g H_3BO_3/kg H_2O.

Conclusions

The interpretation of input signals in the SCORPIO-VVER system was tested with a large amount of input data in various reactor states. All signals from measurements were processed correctly and results are very close to the output from the old VK3 system. Validation and stabilisation of measured values gives better stabilised results in SCORPIO-VVER. Due to complex signal validation both statistical and systematic errors are partially eliminated and parameters of the primary circuit based on measurements are calculated in SCORPIO-VVER with higher stability and credibility.

The power distribution determination module in SCORPIO-VVER produces better stabilised output, which is closer to the result of the simulator calculation. Inaccurate measurements are eliminated by the SCORPIO-VVER power distribution reconstruction methods as well as systematic errors of simulator calculations. Additional on-line analyses based on power distribution reconstruction and core-follow simulator calculations provide a complex information for operators and reactor physicists including pin-wise characteristics and checking of all limited parameters.

Predictive functions can assist during the load-follow transient after prospective involvement of NPP Dukovany units into weekend main frequency regulation, and can replace the involvement of the reactor physics office in off-line calculations of critical parameters.

RECENT IMPROVEMENTS IN ON-LINE CORE SUPERVISION AT LOVIISA NPP

M. Antila, J. Kuusisto
Fortum Engineering Ltd.
Vantaa, Finland

Abstract

An on-line core supervision system (RESU) based on monitoring of local fuel limits has been in use at the Loviisa VVER-440 reactors for more than twenty years. Minor modifications were made ten years ago to upgrade the computer hardware. In April 1998 Loviisa obtained a licence for 1 500 MW_{th} power. Power uprating and introduction of new fuel types gave rise to the latest improvements in the core supervision system, which is called RESU-98. In August 1999 the Finnish Safety Authority (STUK) officially approved RESU-98, which is now in use at the Loviisa NPP. RESU-98 includes essentially the same computer codes, which are used in reload planning. The extensive in-core instrumentation is utilised to adjust the theoretical 3-D power distribution to get a best-estimate result. In this paper a general review of the RESU-98 system is given including instrumentation, methods, core monitoring, predictive functions and validation. Special attention is paid to recent improvements.

Introduction

An on-line core supervision system (RESU) based on monitoring of local fuel limits has been in use at the Loviisa VVER-440 reactors already for more than twenty years. Minor modifications were made ten years ago when the computer hardware was upgraded. In April 1998 Loviisa obtained a licence for 1 500 MW_{th} power (originally 1 375 MW_{th}). Power uprating and introduction of new fuel types gave rise to the latest improvements in the core supervision software system, which is now called RESU-98. The fast development of computing capacity was utilised in the modernisation. The in-core instrumentation system remains unchanged.

RESU-98 includes essentially the same computer codes that are used in reload planning. These are HEXBU-3D, which is a nodal code and ELSI-1440, which is used for pin power reconstruction. Coolant mixing between sub-channels inside the fuel assemblies is also taken into account when evaluating the hot sub-channel enthalpy rise and DNB margin. The extensive in-core instrumentation including 132 local self-powered neutron detectors of Rh-type and 192 fuel assembly outlet thermocouples are utilised to adjust the theoretical 3-D power distribution to get a best-estimate result. The Finnish Safety Authority (STUK) gave its approval for the new RESU-98 system in August 1999. The system is now in on-line use at the Loviisa NPP.

In this paper a general review of the RESU-98 system is given including core limitations, instrumentation, methods of core performance calculations, core supervision, predictive functions and validation. Special attention is paid to recent improvements.

Core limitations

The on-line core supervision system is used to monitor the core thermal and hydraulic limitations to assure that the core is operated within the limitations and assumptions made in the safety analyses. In the case of the Loviisa reactors the most important limitations with respect to on-line monitoring are those imposed on the linear heat rate and sub-channel outlet temperature.

Linear heat rate

The linear heat rate upper limit is 325 W/cm. A safety factor of 1.12 is included in the calculated values. The upper limit of 325 W/cm was used in the transient and accident analyses including large break LOCA. The linear heat rate limit decreases with burn-up mainly to prevent excess fission gas release from the fuel pellets. Experience has shown that it is possible to design the loading pattern such that the requirement of 325 W/cm set for the maximum linear heat generation rate can be met even with uprated power. In the beginning of cycle there is typically some 5% margin left to this limit, when the control rods are fully out of the core. In the end of cycle there is somewhat more margin left to the burn-up dependent limit.

Sub-channel outlet temperature

The hot sub-channel outlet temperature limit is 325°C (bulk boiling limit). A safety factor of 1.16 for the enthalpy rise is included in the calculated values. No coolant mixing between sub-channels was previously assumed. Coolant mixing between adjacent fuel assemblies is not possible because of the assembly shroud tubes. Taking into account flow redistribution inside the assembly and mixing between sub-channels has an effect of reducing the enthalpy rise of the order of 5%. The mixing credit is based on the application of the FLUENT CFD code package for analysis of mixing in the VVER-440 fuel assembly. At the moment the sub-channel outlet temperature limit is still the most severe limitation at 1 500 MW power especially for Loviisa-1, where the core flow is 4% less than in the case of Loviisa-2. Uprated turbine power also requires somewhat increased steam pressure, which tends to increase the cold leg and thus core inlet temperature.

DNB

The sub-channel boiling limit assures that quite a big DNB margin prevails in nominal operating conditions, where a typical DNB ratio is about 3.3. No direct limit value has been imposed on DNB against transients. An initial state with hot sub-channel on boiling limit and conservative axial power distribution is assumed in the analyses. Wide margins remain in all the analysed cases.

Fuel burn-up

The current burn-up limit for the Loviisa reactors imposed by the Finnish safety authority (STUK) is 40 MWd/kgU for assembly average burn-up. An application has been sent to STUK for the approval of assembly burn-up limit of 45 MWd/kgU corresponding to maximum rod average burn-up of 53 MWd/kgU. STUK has not yet made the final decision. This limitation can also be monitored on-line by the core monitoring system but limit checking is performed by the reactor engineer.

In-core measurements and their interpretation

The core of the Loviisa reactors has been reduced from the original 349 fuel assemblies to 313 assemblies to protect the pressure vessel from neutron irradiation. In the reduced core of the Loviisa reactors there are 192 fuel assembly outlet thermocouples and 33 self powered neutron detector assemblies, each containing four local rhodium detectors, one full-length vanadium detector and one core inlet thermocouple. The Rh-detectors are 25 cm long and are located approximately at elevations of 20, 40, 60 and 80% in the core. The measurement signals are scanned by the plant computer. The in-core sensors are shown in Figure 1.

The interpretation of assembly outlet temperature measurements to assembly power values is possible because there are shroud tubes around the fuel assemblies, which prevent mixing of coolant between adjacent assemblies. The measured enthalpy rise gives information of assembly-wise power distribution. There are, however, some uncertainties associated with the core inlet temperature distribution, local coolant flow rate and incomplete flow mixing at the location of the outlet thermocouple. The inlet thermocouples at the tip of the neutron detector assemblies were originally meant for measuring local core inlet temperature. It has, however, turned out that the neutron flux affects the thermocouple wires passing through the active core to such an extent that the required accuracy is not reached for the measurements. Some attempts were made in the past to model the decalibration but with poor results. This is why loop average value is used for the inlet temperature distribution. Because of the fuel assembly inlet orifices the coolant flow distribution can be assumed to be constant. Our recent studies with computational fluid dynamics (CFD) codes have also confirmed the old suspicion that flow mixing at the elevation of the outlet thermocouple is still incomplete, so that the thermocouple measures the temperature of the water coming mainly from the central region of the fuel assembly. With identical fuel assemblies this has only a minor effect on the measured enthalpy rise distribution (assembly-wise power distribution) but results in a bias on the measured enthalpy rise values. The bias has up to now been interpreted as excess bypass flow. With respect to enthalpy rise distribution the thermocouple measurement accuracy is of the order of 2%.

Figure 1. Core map of the Loviisa VVER-440 reactors showing the locations of in-core instrumentation sensors

(T = outlet thermocouple, X = neutron detector assembly, S = control rod)

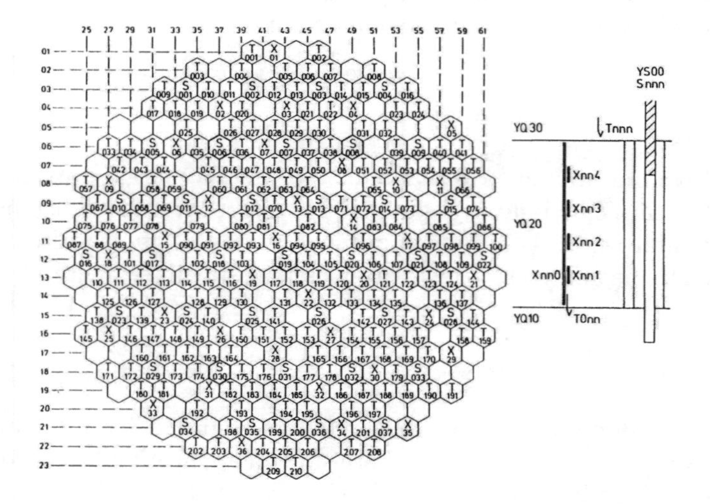

The neutron detector signals are first interpreted into nodal fast flux (E > 0.625 eV) values taking into account the properties of the surrounding fuel (neutron spectrum) and the depletion (delivered charge) of the detector. Nodal power density value is obtained by the conventional two-group cross-sections. The fast flux is chosen as the basic field quantity because of its relatively smooth behaviour within a fuel assembly and its continuity between fuel assemblies [1]. Thus fast flux is suitable for fitting between neutron detector measurements and reactor theory. Fitting is formally carried out as an adjustment of the fast flux distribution. Neutron detector signal interpretation is based purely on theoretical modelling of the sensor response. The full-length vanadium detectors have been used only to verify the depletion of the rhodium detectors. At the moment V-detectors are being changed to background cables in the replacement detector assemblies because the depletion verification is judged to be completed. In spite of that residual deviations up to 7% exist after fitting to the theoretical flux distribution the root mean square deviation being of the order of 2-3% (see Table 1). The big residual deviations are quite few in number and seem to be random in nature with respect to detector age, depletion and properties of the surrounding fuel. Adjusting the interpretation model can not eliminate them. Individual on-line calibration would be the only means to eliminate such deviations. This is not, however, necessary due to the applied fitting method with filtering properties.

Methods of core performance calculations

Determining the 3-D power distribution

Theoretical reference

The theoretical flux and power distributions corresponding to the prevailing reactor core state are calculated on-line with the latest version [2] of the code HEXBU-3D [3], which is a three-dimensional core simulator program for hexagonal fuel assembly geometry. The program was originally tailored to the features of the VVER-440 type reactors. Other hexagonal reactors such as VVER-1000 can also be calculated. HEXBU-3D solves the two-group neutron diffusion equations by a nodal method. The program includes a detailed description for big follower-type control assemblies of the VVER-440 reactor and all the relevant power-dependent feedback effects are considered. The neutronic properties of fuel assemblies are described by homogenised two-group diffusion parameters, which are dependent on fuel burn-up and various feedback effects. The two-group diffusion parameters are calculated by the code CASMO-HEX [4]. CASMO-4 or other codes can also be used. Boundary conditions on the outer boundaries of the reactor and on the control absorbers are determined either by extrapolation distances or by albedo matrices. The solution domain is the whole reactor in the on-line application. Two-dimensional test calculations in a core of the VVER-440 reactor have shown that with given cross-sections HEXBU-3D solves the diffusion equation to an accuracy within 1% in average assembly powers. HEXBU-3D is the standard reload design tool used for the Loviisa reactors. The 3-D xenon dynamics included in the code make it possible to follow the xenon state and predict different xenon transients in detail.

Fitting to the measurements

The purpose of the fitting procedure is to combine theoretical and measurement information in such a way that the best-estimate power distribution is obtained. In the modernised RESU-98 system the fitting methods remain the same as in the original RESU [1]. No assumptions of core symmetry are enforced. The full core is always considered as such.

In the adopted approach [1] the theoretical reference distribution for the neutron flux or power density (now calculated by HEXBU-3D instead of the axial flux synthesis method [1]) is adjusted by a multiplicative correction function in order to bring the result closer to the measurements:

$$\Phi^{fit}(x,y,z) = C(x,y,z;\underline{M})\Phi^{ref}(x,y,z)$$

The vector $\underline{M}$ refers to the measurements. The correction function C is a local amplitude correction imposed on the theoretical distribution. It is mainly required to be smooth and gradually varying in the core. The fitting procedure is described in more detail in [1]. Fitting to the measurements is accomplished by yielding to the measurements while simultaneously imposing restrictions on the correction functions. As a result the fitted flux or power distribution will not necessarily coincide with any single measurement. This acts as a filter for measurement uncertainties and makes the method independent of individual random detector failures.

Fitting to the neutron detector measurements and to the core outlet thermocouple measurements is performed in consecutive stages.

First the fitting of the fast flux distribution to the Rh-detector measurements is performed according to the methods described in [1]. In this case the resulting $C(x,y,z)$ is a smoothly varying three-dimensional correction function. The corresponding power distribution $P^{fit}(x,y,z)$ is then obtained from $\Phi^{fit}(x,y,z)$ by using the original power cross-sections of HEXBU-3D corresponding to the state of the core.

Next an assembly-wise power distribution $P^{ref}(x,y)$ is obtained by axial integration. This is further used as a reference in the fitting to the thermocouple measurements. In this case the resulting $C(x,y)$ is a two-dimensional correction function, by which the three-dimensional distribution $P^{fit}(x,y,z)$ is also multiplied to get the final fitted power distribution, which includes information from both neutron detector measurements and core outlet thermocouple measurements.

After fitting to the in-core measurements the core power distribution is normalised to the best available value of the reactor power. Normally the thermal power calculated from the plant heat balance is used.

The core monitoring system can also be run in different modes with fitting to only neutron detector measurements or thermocouple measurements or without fitting at all. Having available a good theoretical reference distribution calculated by HEXBU-3D this makes it possible to perform direct on-line verification of the theoretical distribution as well as of functioning of individual detectors, if for example detector failure is suspected.

Reconstruction of assembly internal power distribution

The assembly internal pin power distributions are reconstructed by the code ELSI-1440 [5]. ELSI-1440 calculates the two-dimensional, pin wise assembly internal power distribution of a hexagonal VVER fuel assembly (VVER-440 or VVER-1000) using two-group diffusion theory in the form of an inhomogeneous boundary value problem, where the fast flux values at the edge of the calculation area are fixed based on the results of interpolation from the macroscopic, whole core power distribution calculation by HEXBU-3D. The analytical flux problem is discretised using the method of finite differences and a mesh of one point per fuel pin cell. The resulting system of numerical equations is solved iteratively with successive over-relaxation technique. Two-group cross-sections are tabulated as a function of rod burn-up for every pin cell. Rod burn-up is followed for all 126 rods of an assembly.

The assembly internal power distribution is divided into two components in order to speed up the pin power reconstruction. The microscopic component takes into account the effect of water gaps, rod burn-up and so on. It is calculated in the nominal core state. The macroscopic component takes into account the effect of changing instantaneous macroscopic fast flux distribution as compared to the nominal state. The microscopic distribution is updated every two weeks as burn-up proceeds. The effect of macroscopic distribution is taken into account always, when the core state is calculated. For example control rod movements give rise to changes in the macroscopic distribution. The method makes the pin power reconstruction fast and thus suitable for on-line application without noticeable degradation of accuracy.

Calculation of assembly internal flow distribution and flow mixing

According to the detailed CFD calculations and measurements turbulent mixing and coolant flow redistribution inside the assembly shroud tend to smooth out the sub-channel enthalpy rise peaking factor as compared to the isolated sub-channel method. The effect is of the order of 5-7% depending on the location of the assembly in the core. It is taken into account by a reconstruction method whose accuracy corresponds to the detailed CFD calculations performed by using the FLUENT code package [6].

The reconstruction is based on the use of an effective flow distribution, which includes both the characteristic flow distribution of the assembly and the effect of turbulent mixing. The characteristic flow distribution is taken directly from the CFD calculation. It is used together with the pin power distribution corresponding to nominal core state to calculate the effect of turbulent mixing. This is done by using the thermal diffusion coefficient (TDC) to describe the enthalpy exchange between adjacent sub-channels. The TDC is tuned so that the reconstruction method reproduces the result of the original CFD calculation. It has turned out that the effective flow distribution is not sensitive to small changes in the pin power distribution.

The effective flow distribution is analogous with the microscopic part of the pin power distribution and is updated at the same time. It is then always used in the simple isolated sub-channel model, when the instantaneous core state is calculated. Thus the effective flow distribution together with the real pin power distribution produces the sub-channel enthalpy rise distribution. The method is very fast and still accurate.

Evaluation of fuel operating conditions

Evaluation of local fuel operating conditions is based on the fitted macroscopic 3-D power distribution, assembly internal pin power distribution and assembly internal effective flow distribution.

The linear heat rate is calculated taking into account the engineering factor. Minimum margin to the rod burn-up dependent limit is calculated for every fuel assembly. Maximum fuel temperature is calculated from a correlation, which depends on fuel type and linear heat rate. The correlation is based on calculations with the fuel performance code ENIGMA. Maximum sub-channel outlet temperature is calculated based on the outlet enthalpy. Sub-channel outlet enthalpy is calculated for every sub-channel taking into account the pin power distribution and effective flow distribution. Minimum DNB margin is calculated from the Gidropress correlation. Local heat flux and enthalpy values are used.

Fuel rod load changes are monitored by comparing the present power distribution to the previous power distribution. The calculation interval is one hour. A simple mathematical fuel rod power preconditioning model is included. It is implemented in the form of the so-called preconditioned power distribution, which follow the changes in the prevailing power distribution with given time coefficients upwards and downwards, respectively. The preconditioned power is allowed be exceeded by a certain amount, which depends on the power level and fuel burn-up. Up to now the power changes are controlled in the plant technical specifications by limiting only the changes of electric power. The model parameters have been tuned to be consistent with these official limits.

Auxiliary functions

Alarm limits for the local SPND measurements are updated automatically once an hour after calculation of the core state. Each Rh-detector is assumed to respond to the power changes of its nearest neighbouring nodes and their neighbouring nodes. The power amplitude in the response domain is increased until the linear heat rate limit or change rate limit is reached. The prevailing detector signal is multiplied by this amplitude ratio to get the alarm limit for the detector. These limits are valid for reasonable changes from the prevailing core state. The limits are used to realise continual fast core monitoring through direct local measurements.

Direct prediction of the critical boron concentration of the present core state including xenon poisoning is performed automatically once an hour or on demand with the HEXBU-3D code. Prediction of critical boron concentration 24 hours in advance has also been implemented. In this prediction core power level, the position of control rods and coolant inlet temperature are assumed to remain unchanged during the whole prediction period. Possible changes in the xenon state are taken into account.

Implementation of core supervision

The RESU-98 core supervision system is an integral part of the plant process computer system. Scanning of measurements, limit alarming, display of data and reporting is performed by utilising the process computer system software. The core performance calculation programs are run under the system like any other performance calculations, such as plant heat balance. All output of information to the operating personnel and other users occurs via the general sub-systems of the software.

Automatic limit alarming

An alarming sub-system is included in the process computer system software. It is used for limit checking of direct measurements and of calculated quantities as well. Alarm limits may be fixed or they can be updated automatically. Alarm messages of different priority appear on the operators' alarm display screens and on the alarm printer.

Several quantities related to the core thermal and hydraulic operating conditions are evaluated and are monitored to be within permissible limits through the use of pre-set alarm limits. Alarm limit checking is performed every time these quantities are evaluated.

Continual monitoring via calculated alarm limits for the individual direct in-core measurements combines both the monitoring of thermal and hydraulic limits and the monitoring of fuel ramping limits.

Displaying of data

Display formats are available for showing the direct in-core measurements (and their alarm limits) in digital form or on a core map and for showing distributions of calculated quantities on coloured core maps. The results of the latest calculations are displayed. Available display formats are:

- Core map and list of assembly outlet temperatures.
- SPND signals and their alarm limits.
- Core map of hot sub-channel boiling margin.
- Core map of linear heat rate margin.
- Assembly-wise power distribution.
- Assembly-wise burn-up distribution.
- Assembly-wise maximum rod burn-up distribution.

Trend display formats including historical data for user defined calculated quantities or direct measurements can easily be defined by the operator. These formats are quite useful for tracing changes in core operating conditions or in the readings of in-core measurements. Historical data is available up to one year of plant operation. This longest period consists of daily averaged values of measurements and other quantities in the system database.

Reporting

A covering collection of reports is available. These include the summary core performance report, various 2-D and 3-D distribution maps, etc. Available reports include:

- Reactor performance report.
- 2-D power distribution.

- 2-D burn-up distribution.
- 3-D power distribution.
- 3-D fast flux distribution.
- 3-D burn-up distribution.
- Assembly internal rod power distribution.
- Assembly internal rod burn-up distribution.
- Assembly-wise isotopic composition.
- Fuel rod ramping report.
- Report of "reactor measurements" including in-core measurements.
- Summary report of fitting to in-core measurements.
- SPND depletion report.
- Xenon prediction report.
- Diagnostic reports.

Printer reports are obtainable on request, at specified times or in some instances by programmed automatic activation.

Activation of programs

Calculation of the core state, including maintenance of burn-up and xenon states, is repeated automatically once an hour using one hour averaged measurements. The operator may activate core performance calculations at any time using instantaneous, five minute or one hour averaged measurements as input. Typical response time is a few seconds. There are also several other supporting programs available, which are mainly intended for the use of the reactor engineer.

Development of predictive tools for operator support

The latest developments in computer capacity have made it possible to use the accurate core simulator HEXBU-3D as a basic tool also in the prediction calculations. A forward predicting mode including possible change of all criticality parameters is under development. For this purpose the HEXBU-3D code was modified so that, in addition to effective multiplication factor and boron concentration, a criticality search can also be performed using reactor power, control rods and coolant inlet temperature. Priority of the criticality control parameters is given by the user. In this way various transient scenarios starting from the prevailing core state can be calculated and optimised with respect to power production and for example the need of boration or boron dilution. The limitations of control rod position, boron dilution rate (pump capacity) and concentration of dilution water, thermal margins and fuel rod power change rates are to be taken into account as constraints in such predictions.

Validation

The on-line core monitoring system RESU-98 is based on the use of validated codes. Basic cross-section data is produced with the CASMO-HEX cell code. HEXBU-3D is used to calculate the theoretical reference power distribution. ELSI-1440 is used to reconstruct the pin power distributions. The same code system is also used in the core design calculations. A large amount of experience and validation material exists.

On-line validation of the fitted 3-D power distribution is available at any time. A special report exists for this purpose to evaluate the residual deviations and corrections after fitting to the in-core measurements. Various fitting options can also be tried to judge the quality of the power distribution and/or individual measurements. One such example is given in Table 1, where some characteristic parameters of power distribution and residual deviations are given corresponding to different fitting options and assumptions.

Table 1. Results of RESU-98 calculations of the core power distribution using different degrees of fitting to the measurements

(Loviisa-1 beginning of Cycle 23 with 111 functioning Rh-detectors)

Case 1: Normal fitting parameters, base case • Case 2: Flux fitting omitted
Case 3: Temperature fitting omitted • Case 4: Both flux and temperature fitting omitted
Case 5: 30% random failures assumed in flux and temperature measurements
Case 6: One dropped control rod assumed (simulated) in the theoretical reference distribution

Parameter	*Case 1*	*Case 2*	*Case 3*	*Case 4*	*Case 5*	*Case 6*
Indicators of the power distribution						
Peaking factors						
– assembly power	1.30	1.30	1.30	1.29	1.30	1.32
– average axial power	1.24	1.24	1.24	1.24	1.24	1.24
– 3-D nodal power	1.61	1.61	1.60	1.60	1.61	1.64
– linear heat rate[a]	1.99	2.00	1.98	1.98	1.99	2.04
– subch. enthalpy rise[a]	1.65	1.65	1.65	1.64	1.65	1.68
Asymmetries, %						
– axial offset	-1.5	-2.9	-1.5	-2.9	-1.7	-1.3
– sector 1 tilt	-1.2	-1.2	0.3	0.0	-1.3	-2.2
– sector 2 tilt	0.1	0.1	0.2	-0.1	0.3	-0.5
– sector 3 tilt	0.6	0.6	-0.1	0.1	0.6	1.3
– sector 4 tilt	0.7	0.7	0.4	-0.1	0.5	0.7
– sector 5 tilt	0.4	0.3	0.2	-0.2	0.3	0.1
– sector 6 tilt	-0.5	-0.5	-1.0	0.2	-0.4	0.6
Residual rms-deviation of in-core measurements						
– Rh-plane 4, %	2.10	2.59	2.10	2.59	1.65	6.37
– Rh-plane 3, %	2.08	2.47	2.08	2.47	2.17	7.42
– Rh-plane 2, %	2.25	3.38	2.25	3.38	1.20	7.28
– Rh-plane 1, %	2.45	4.31	2.45	4.31	2.31	7.30
– core outlet thermoc., C	0.63	0.62	0.95	0.80	0.69	1.53

[a] Including all engineering safety factors.

Case 2 shows that omitting the flux fitting has some effect on the axial power distribution as compared to the base Case 1. Case 3 without fitting to the temperature measurements is close to the base case. Case 4 without any fitting is also reasonably close to the base case, which means that the pure theoretical power distribution is of adequate accuracy. Case 5 with 30% random failures in the measurements is practically identical with the base case. Case 6 with simulated control rod position measurement error in the theoretical reference distribution demonstrates how fitting to the measurements corrects the power distribution quite close to the base case. Thus fitting to the measurements is of particular value in revealing anomalous core conditions.

Summary

An on-line core supervision system based on monitoring of local fuel limits has already been in use at the Loviisa VVER-440 reactors for more than twenty years. The system is used to monitor the core thermal and hydraulic limitations to assure that the core is operated within the limitations and assumptions made in the safety analyses. Power uprating and introduction of new fuel types gave rise to the latest improvements in the core supervision software system. The new RESU-98 includes essentially the same computer codes that are used in reload planning.

The Loviisa reactors have an extensive in-core instrumentation consisting of fuel assembly outlet thermocouples and SPN detectors. Interpretation of assembly outlet temperature measurements to assembly power values is possible because there are shroud tubes located around the fuel assemblies. The neutron detector signals are interpreted into nodal fast flux values taking into account the properties of the surrounding fuel and the depletion of the detector. The in-core measurements are utilised to adjust the theoretical 3-D power distribution to get a best-estimate result.

The theoretical flux and power distributions corresponding to the prevailing reactor core state are calculated on-line with the latest version of the code HEXBU-3D, which is a three-dimensional nodal core simulator program for hexagonal fuel assembly geometry. HEXBU-3D is the standard reload design tool used for the Loviisa reactors. Fitting procedures are used to combine the theoretical and measurement information to get the best-estimate power distribution. In the modernised RESU-98 system the fitting methods remain the same as in the original RESU.

The assembly internal pin power distributions are reconstructed by the code ELSI-1440. The assembly internal power distribution is divided into two components in order to speed up the pin power reconstruction. The microscopic component takes into account the effect of water gaps, rod burn-up and so on. It is calculated in the nominal core state and is updated every two weeks as burn-up proceeds. The macroscopic component takes into account the effect of changing instantaneous macroscopic fast flux distribution as compared to the nominal state. The method makes the reconstruction fast and thus suitable for on-line application without noticeable degradation of accuracy.

According to the detailed CFD calculations and measurements turbulent mixing and coolant flow redistribution inside the assembly shroud tend to smooth out the sub-channel enthalpy rise peaking factor as compared to the isolated sub-channel method. The effect is of the order of 5-7% depending on the location of the assembly in the core. It is taken into account by a reconstruction method corresponding to the accuracy of the detailed CDF calculations performed by using the FLUENT code package. The method is very fast and still accurate.

Evaluation of local fuel operating conditions is based on the fitted macroscopic 3-D power distribution, assembly internal pin power distribution and assembly internal effective flow distribution. Linear heat rates, maximum fuel temperatures, maximum sub-channel outlet temperatures and minimum DNB margins are evaluated on a local level. Fuel rod load changes are monitored by comparing the present power distribution to the previous power distribution. Alarm limits for the local in-core measurements are updated automatically once an hour after calculation of the core state.

Direct prediction of the critical boron concentration of the present core state including xenon poisoning is performed automatically once an hour or on demand by using the HEXBU-3D code. Prediction of critical boron concentration 24 hours in advance has also been implemented. The latest developments in computer capacity have also made it possible to use the accurate core simulator HEXBU-3D as a basic tool in the prediction calculations. A forward predicting mode is under development. For this purpose the HEXBU-3D code was modified so that in addition to effective multiplication factor and boron concentration criticality search can also be performed using reactor power, control rods and coolant inlet temperature. Priority of the criticality control parameters is given by the user.

RESU-98 is an integral part of the Loviisa plant process computer system. Scanning of measurements, limit alarming, display of data and reporting is performed by utilising the process computer system software. The core performance calculation programs are run under the system like any other performance calculations, such as plant heat balance. Display formats are available for showing the direct in-core measurements (and their alarm limits) in digital form or on a core map and for showing distributions of calculated quantities on coloured core maps. Trend display formats including historical data for user defined calculated quantities or direct measurements can easily be defined by the operator. A covering collection of reports is available. Calculation of the core state is repeated automatically once an hour. The operator may activate core performance calculations at any time. Typical response time is a few seconds.

The on-line core monitoring system RESU-98 is based on the use of validated codes. The same code system is also used in the core design calculations. On-line validation of the fitted 3-D power distribution is available at any time. The Finnish Safety Authority (STUK) has given approval for the new RESU-98 system in August 1999. The system is now in on-line use at the Loviisa NPP.

REFERENCES

[1] P Siltanen, M Antila, "Combining In-Core Measurements with Reactor Theory for On-Line Supervision of Core Power Distribution in the Loviisa Reactors", OECD Nuclear Energy Agency Specialists Meeting on In-Core Instrumentation and the Assessment of Reactor Nuclear and Thermal-Hydraulic Performance, Fredrikstad (Norway), 10-13 October 1983.

[2] E. Kaloinen, "New Version of the HEXBU-3D Code", II Symposium of AER, Paks, 21-26 September 1992.

[3] E. Kaloinen, R. Teräsvirta, P. Siltanen, "HEXBU-3D, a Three-Dimensional PWR Simulator Program for Hexagonal Fuel Assemblies", Technical Research Centre of Finland, Research Report 7/1981, Espoo 1981.

[4] M. Anttila, "The Fuel Assembly Program CASMO-HEX", Input Manual VTT Energy, Technical Report RFD-17/96.

[5] "ELSI-1440, a Two-Dimensional Pin Power Program for Hexagonal Fuel Assemblies", IVO Power Engineering, YDIN/GBR-GT3-30

[6] P. Gango, "Application of CFD Models for 3-D Analysis of Single Phase Thermal-Hydraulics in the VVER-440 Fuel Assembly", 7th AER Symposium on VVER Reactor Physics and Reactor Safety, Hörniz, Germany, 23-26 September 1997.

REFINEMENT OF SIEMENS CORE MONITORING BASED ON AEROBALL AND PDD IN-CORE MEASURING SYSTEMS USING POWERTRAX™

Ivo Endrizzi, Michael Beczkowiak
Siemens AG, Germany

Guido Meier
Kernkraftwerk Gösgen, Switzerland

Abstract

The Siemens/KWU core monitoring concept is consistently based upon two complementary in-core instrumentation systems, Aeroball and PDD. The Aeroball system is a fast, high resolution measuring system used for reference purposes. The PDD system yields continuous local power density data in three dimensions that feed directly into the margin calculators for maximum linear heat generation rate and minimum DNB ratio of the automatic Power Density Limitation System. This standard core monitoring concept can be refined and extended using POWERTRAX™, a state-of-the-art PWR core monitoring software system. Coupling it to the in-core measurement systems enhances the transparency of core behaviour for the reactor operator, simplifies the interpretation of monitoring signals, supports the elaboration of optimised operating strategies, and permits the use of realistic margins for monitoring the limiting conditions for operation (LCO).

Introduction

The PWR core monitoring concept developed by Siemens/KWU is consistently based on in-core instrumentation. This concept combines two complementary systems with different tasks – the Aeroball system as movable flux mapping system, and the Power Density Detector (PDD) system as a monitoring system employing fixed in-core detectors.

The calibration process functionally links the two systems; the monitoring signals are calibrated at regular intervals under reference conditions using the results of the Aeroball system.

This core instrumentation concept has been implemented since 1974 in all Siemens/KWU PWR plants; it is now in field service in 13 plants and cumulative operational experience totals 210 reactor years. This concept is still competitive with modern developments and is also well suited to support advanced signal processing making use of adaptive physical core models, or to feed on-line core monitoring software for analyses considering actual core conditions.

The Aeroball system

The Aeroball system is a flux mapping system based on movable activation probes. The activation probes consist of 1.7 mm diameter steel balls forming stacks guided in tubes (outside diameter 3 mm). The balls are made of steel alloyed with 1.5% vanadium, which acts as neutron sensitive material. The ball stacks normally reside outside the core. To take a measurement, all the stacks are transported into the core at the same time where each extends over the entire active core height. The balls are transported pneumatically by applying carrier gas pressure. Once in the core, they are activated by the neutron flux for three minutes. This short irradiation time makes flux mapping like a snapshot and enables accurate measurement even in semi-transient conditions (xenon-redistribution). During irradiation, a process computer registers all data needed for subsequent evaluation. On completion of irradiation, the carrier gas pressure is applied from the other direction, driving the ball stacks out of the core and into a detector array that is located outside the biological shield. The gamma radiation emitted by the balls is then read by detector arrays arranged in a measuring table. The recorded counts are the primary data that are further evaluated by a computer to yield the 3-D power density distribution and other parameters representative of core conditions. The shortest time interval between two Aeroball measurements is 10 minutes.

This fast and accurate instrumentation system will also be the reference system in the European pressurised water reactor (EPR) as the next generation PWR.

The Power Density Detector system

The Power Density Detector system is a monitoring system which employs self powered neutron detectors with prompt response, i.e. fixed in-core detectors called Power Density Detectors (PDDs). These signals are used for core control and core protection purposes; they are processed together with selected other process variables to yield continuous monitoring signals representative of core conditions.

The *in situ* calibration capability of the PDD system as well as the proximity of the PDDs to the hot spots or hot channels in the core provides a higher accuracy than that achievable by ex-core instrumentation. As a consequence, the allowance for monitoring system imperfections is smaller for the PDD system than for ex-core systems. This provides additional margins for core design and flexibility of operation.

The accuracy of the PDD system was also the most important aspect for the choice of this type of instrumentation for the EPR.

Number and distribution of PDDs

The number and distribution of PDDs over the core is determined by considering the two main requirements related to the application of monitoring signals:

1. Core surveillance.

2. Core protection.

The PDD distribution within the core must support the capability to detect and assess local power increases caused by flux and power redistributions that occur under non-steady-state conditions. The PDD distribution must also make an allowance for proper signal redundancy.

The operating concept in Siemens/KWU PWRs needs only slightly inserted control rods at nominal power. This, in conjunction with general laws of neutron transport in PWR cores, has the consequence that power density perturbations are not locally confined but spread over large regions within the core. Therefore even a relatively small number of detectors strategically distributed over the core is sufficient to register all power density increases of interest. Examples of power shape perturbation modes relative to an unperturbed reference condition together with the arrangement of PDDs are given in Figures 1 and 2. PDDs are placed at locations where power density increases during perturbed conditions.

Signal redundancy is attained by increasing the number of in-core detectors above the theoretically required minimum in order to achieve a high degree of reliability as required for safety grade instrumentation. This is a "functional" redundancy in the sense that power density increases are "seen" by several detectors. If one or more of these detectors fail, the remaining detectors are still able to register the increase. In this way, failure of one or even more in-core detectors does not degrade monitoring system performance and does not significantly impair the accuracy of the derived monitoring signals used for core surveillance and protection.

According to our experience, these design principles equip the PDD system with the required high degree of reliability. In German licensing practice it is recognised that the quality of PDD signals is high enough to allow them to be used for local core protection.

PDD calibration principle

The power-to-signal ratio of any neutron flux detector changes as core burn-up progresses. Likewise, the reference power distribution changes with core burn-up. Calibration of the PDDs must therefore be matched to the reference conditions at regular intervals. Reference values for this calibration are provided by the Aeroball system. They are defined as follows:

1. Each PDD is assigned a section of the core volume as a surveillance zone. The surveillance zones are chosen such that the flux increases at the detector locations are sufficiently representative of the peak power density increases within the zones. The surveillance zones surround the PDDs and together cover the entire core volume.

2. Reference values for calibrating the PDDs are maximum power densities in absolute units (W/cm) occurring under reference conditions within the associated surveillance zones.

By this simple calibration, usually performed every 14 EFPD under unperturbed reference conditions, the accuracy of the Aeroball system is transferred to the PDD system. Under perturbed conditions, PDD signals change in proportion with the neutron flux at the detector location. As a consequence, the calibrated PDD signals are able to follow or track the absolute value of peak power density. Experience shows that with this simple calibration principle and simple conventional signal processing, it is possible to achieve good tracking accuracy for the derived monitoring signals.

PDD signal processing and utilisation

The calibrated PDD signals represent an adequately detailed 3-D image of the highest linear heat generation rates distributed over the core. For core control and core protection purposes, this PDD information is condensed to yield a limited number of monitoring signals.

The most important variables associated with the monitoring signals are peak power density and DNB ratio. For an easier interpretation by operating personnel, it is appropriate to indicate core conditions in terms of margins to various limits. To do this, the PDD signals, together with information on actual coolant conditions at core inlet (pressure, temperature, flow rate), are fed to "margin calculators" which calculate the margins on-line and in real time; the margins are inputs to the Power Density Limitation System. The criteria that define the limits are:

1. The limiting conditions for operation (LCO) that are obtained by analyses of anticipated operational occurrences (e.g. loss of flow event) or hypothetical accidents (e.g. loss of coolant accident). The derived operational limits include conservative allowance to compensate monitoring system imperfections.

2. Protection of fuel rods against damage by pellet/clad interaction (PCI).

3. Termination of fast transient power perturbations that may be caused by spurious control bank movement or sub-cooling transients.

These limits correspond to three different levels in terms of power density. On violation of the above-mentioned limits, staggered automatic countermeasures are initiated by the Power Density Limitation System to restore normal operating conditions. The intensity of these actions increases in proportion with variable departure from established set points and avoids severe protective measures by taking appropriate actions at the right time.

The operating range of practical use is defined by the lowest level given by the LCO and by the magnitude of the incorporated accuracy allowance (uncertainty) of the monitoring system. This range can be used for operational flexibility or can be exploited for advanced in-core fuel management strategies. The economic benefit gained from the strategies that make an optimum use of the thermal-hydraulic potential of the reactor require:

- A core monitoring system that minimises the allowances to cover system imperfections.

- A realistic computation of the margins to the LCO taking into account actual core conditions.

Tracking accuracy

Any core monitoring system achieves only limited accuracy in providing information on margins to the LCOs. The resulting uncertainties must be compensated by appropriate downward adjustments of the set points of the Power Density Limitation System. The magnitude of the allowance depends on the accuracy with which the monitored variable follows or tracks its physical value under normal and upset conditions. This type of accuracy is referred to as “tracking accuracy”. For an ideal monitoring system the measured value derived from the monitoring signal corresponds exactly to its physical value. The correlation between these two values follows the 45° line when plotted. Any deviation from this ideal line is designated as a “tracking error”.

For Siemens/KWU plants, tracking accuracy of the monitoring variables peak power density and DNB ratio derived from the PDD signals can be checked experimentally. This can be done by deliberately inducing a perturbation of power shape that reduces the margins to the LCO. During this perturbation, the “real” values of the monitoring variables are determined by the Aeroball system. This reference system provides the most accurate information about the key safety parameters and their margins to the LCO and is even fast enough to determine these parameters in semi-transient conditions.

Numerous measurements have been performed to verify the system accuracy. Typical results are given in Figures 3 and 4. In Figure 3, peak power density registered by the Aeroball system is plotted as reference on the x-axis, with the peak signal value recorded at the same time by the PDD system on the y-axis. In Figure 4, the margin to the LCO related to DNB ratio is expressed in terms of permissible power increase (PEX). PEX calculated by the process computer on the basis of the actual full core power density distribution and actual coolant conditions assessed by Aeroball measurement is compared to the corresponding margin given by the DNB calculational circuit of the Power Density Limitation System. This circuit uses a linear correlation and appropriate coefficients provided for each cycle by the thermal-hydraulic design to combine calibrated PDD signals and coolant parameters for the formation of power density limits with respect to DNB ratio. Both figures show that the measured data deviate little from the ideal 45° line, thus indicating very good tracking accuracy of the monitoring variables. This is the proof that the monitoring signals derived from the PDDs applying simple conventional signal processing techniques do not require allowances to compensate tracking errors.

Existing plant process computer software

The present plant process computer program uses a linear interpolation method to reconstruct power density in between instrumented fuel assemblies; this requires theoretical information generated by external steady-state core design calculations in the form of sets of analytical factors. These factors are stored in the process computer for each operational control rod configuration at discrete cycle burn-ups. In addition, pre-calculated burn-up dependent activation-to-power conversion factors are needed at the location of the Aeroball probes. Burn-up calculations are performed on demand, usually every 14 EFPD, on the basis of a relative power density distribution assessed by the Aeroball system and considered as representative for the specified time interval. After each Aeroball measurement, the plant process computer calculates the key safety parameters, the calibration factors for the PDDs and the margins to the LCO.

POWERTRAX™ core monitoring software system

The POWERTRAX™ core monitoring software system (CMSS) is a comprehensive, advanced on-line power distribution monitoring and operations support system for PWRs. It uses on-line plant computer data as well as measured activation values obtained with the Aeroball system to update a three-dimensional nodal neutronic model of the core which is used to perform the monitoring and the operation support calculations.

For the three-dimensional modelling of Siemens/KWU plants, POWERTRAX™ utilises PRISM, the reactor simulator code of the CASCADE-3D integrated code system for reactor core design and safety analysis. Using PRISM for on-line core monitoring, the data initialisation at BOC can use design data directly without the need to produce process computer data in a separate step.

The major POWERTRAX™ calculational functions are described below giving a brief review of the implemented CMSS modules.

Core state acquisition module (CSAM)

This module is responsible for generating the POWERTRAX™ calculation points. These are generated by processing the core state data from the process computer. CSAM also updates the core exposure by adding in the integrated power over time for each time step.

The core state data from the process computer is reviewed for the presence of changes in the core conditions significant enough to warrant updating the reactor simulator core follow model. Events that require the running of an update or a surveillance calculation are defined as trigger events. Sample triggering events consist of changes in power level, changes in control bank positions, changes in the RCS soluble boron concentration, and the presence of new measured activation data. In addition to these event triggers there are periodic triggers which are initiated at specific time intervals.

Three-dimensional reactor core calculations

The nodal core simulator PRISM performs reactor core calculations in the POWERTRAX™ system. PRISM is based on a fast running steady-state flux solver. Continuous representation of the microscopic cross-sections covers all possible combinations of thermal-hydraulic parameters in stationary reactor states. For an accurate and efficient determination of pin wise power, exposure, fluxes and detector signals, the full 3-D interpolation and modulation scheme is directly integrated into the code.

INPAX

The INPAX code is responsible for the generation of the "measured" power distribution and the associated power peaking factors by combining the measured activation data with the calculated power distribution from PRISM. The code then performs various edits on the power distribution.

Operating Strategy Generator

The function of the Operating Strategy Generator (OSG) is to model the reactor behaviour through some future anticipated time intervals. The OSG can be run with the optimum strategy algorithm activated, in which case it attempts to determine the best way to operate the plant through the planned manoeuvre. The "best operating strategy" is defined as the one that accomplishes the planned manoeuvre with the maximum capacity factor, minimum coolant treatment and within applicable operating limits. The OSG can also be run with the optimum strategy algorithm turned off. In this case the OSG simply runs through a set of calculations that define the planned manoeuvre. This option can be used to provide an initial starting point for other POWERTRAX™ predictive functions, such as the estimated critical conditions and the shutdown boron concentration calculations.

The output from the OSG (Figures 5 and 6) shows the user's input scenario, the final operating strategy developed by the strategy selection algorithm (if engaged) and the predicted core performance for those operating conditions. These outputs, expressed in "control room terms" (such as steps, amount of boric acid addition, injection rates and total volumes processed) as well as the resultant units, such as boron concentration in the RCS, should allow an engineering user to quickly evaluate the results. These parameters are used by the reactor operators to manoeuvre the plant through the desired scenario.

Benefits of the POWERTRAX™ core monitoring software system

The adaptation of the POWERTRAX™ core monitoring software system to Siemens/KWU PWRs combines several features that result in the following benefits:

- The use of the three-dimensional core simulator obviates the need to generate off-line pre-defined sets of theoretical factors for operational control rod configurations at discrete burn-ups for the power density reconstruction because POWERTRAX™ involves running the three-dimensional core design code PRISM together with the adaptation code INPAX. The measured activation values are combined with the three-dimensional power distribution calculated for actual core conditions considering control rod position, burn-up and xenon dynamics.

- Core design verification, which is based on the comparison between measured and predicted key parameters, is no longer restricted to reference conditions; the comparison can be extended to non-steady-state conditions and can also comprehend the axial power profile of individual FAs. The three-dimensional core simulator calculates the most accurate design value available on-line for each physical parameter at any plant operating condition.

- The Operating Strategy Generator module can fine-tune reactor operation during load follow, thereby reducing soluble boron processing. This module can determine the optimised control rod insertion at part load which permits the return to full power with a required gradient by withdrawing control rods into their full power demand position. It can also elaborate an optimised strategy to perform load-follow operation and planned manoeuvres under conditions of reduced operating flexibility (such as stretch-out operation). This prediction enables the reactor operator to adjust, if necessary, the set points of the bank position control to achieve this optimised mode of operation. The impact of a selected bank movement strategy on axial power shape and burn-up distribution over the cycle can also be determined.

- The three-dimensional core simulator continuously computes the signals of the Aeroball system and the PDD system. This permits the plant engineer to perform extensive comparisons and trend analyses to identify drifts in the calibration of the Aeroball and PDD system or failing Aeroball probes or PDDs. In addition, it is possible to directly check the tracking accuracy of the monitoring signals derived from the PDDs during operation without the necessity of dedicated measurements.

- The prediction of critical boron concentrations for a given start-up window permits an estimate of the amount of demineralised water required. If this amount is not available, it is also possible to predict the power level which can be reached and at least maintained, taking into account the capacity of the coolant treatment system. This information could be important for a start-up after a reactor trip near the end of cycle.

- The information provided by POWERTRAX™ can also be used to perform on-line analyses of anticipated operational occurrences (e.g. loss of flow event) or hypothetical accidents (e.g. LOCA) starting from actual core conditions to provide realistic margins to the LCO. These margins can be directly compared with those generated in the Power Density Limitation System. This comparison shows if a set point adjustment is necessary. The accurate and realistic computation of the LCO constitutes a significant refinement to maximise margins for core design and operation.

Conclusions

As explained in the first part of this paper, the information on core power distribution provided by the PDD system is detailed enough to allow simple signal processing procedures to generate accurate monitoring signals representative of the highest heat generation rate, hot channel power and the corresponding margins to the LCO. At present, this straightforward signal processing used for local core protection will be retained in Siemens/KWU plants even though transition to digital safety grade I&C is in progress. The digital safety system upgrade has already been implemented in two Siemens/KWU PWRs.

Nevertheless, a refinement and an extension of core monitoring in Siemens/KWU plants can be achieved by substituting the present plant process computer program with the powerful on-line three-dimensional reactor core monitoring software system POWERTRAX™. The three-dimensional power distribution calculated on-line considering actual core conditions shall be adapted to the activation values measured by the Aeroball system in reference conditions. Between Aeroball measurements the POWERTRAX™ system supplements the local information given by the PDDs, continuously providing all the relevant parameters required to judge core behaviour, to verify the margins to the LCO and to check in-core instrumentation performance. Furthermore, in the event of PDD failures an accurate and consistent substitute information is still available on-line using POWERTRAX™. This advanced monitoring system enhances the transparency of the core behaviour for the reactor operator, simplifies the interpretation of monitoring signals and permits the plant engineer to define realistic margins to the LCO. In addition its quick predictive capabilities support nuclear operations and allow the reactor operator to determine optimised operating strategies for accomplishing planned manoeuvres.

POWERTRAX™ represents a state-of-the art PWR core monitoring system; it has been fully operational at two Westinghouse PWRs since 1995.

Figure 1. Axial power shape perturbations

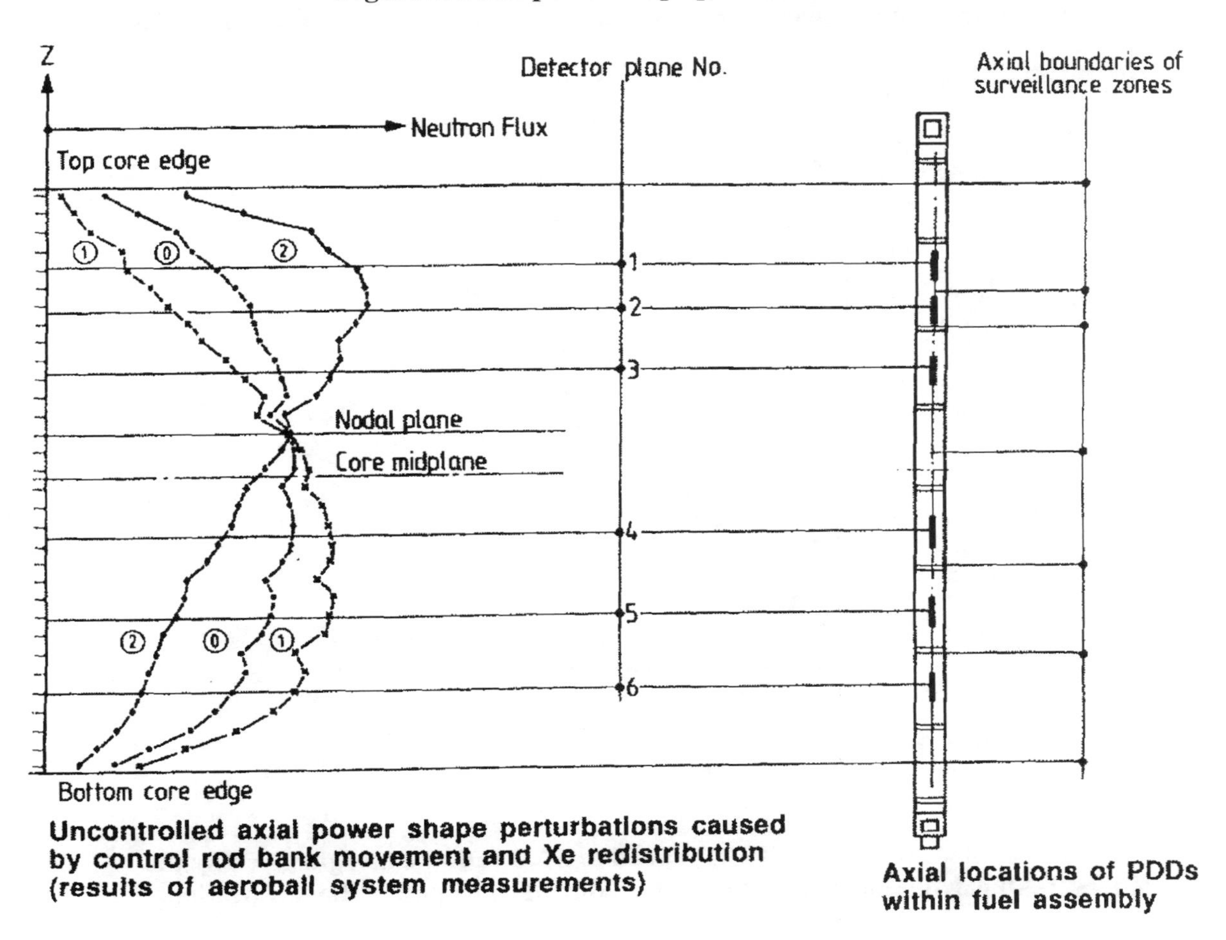

Figure 2. Radial power shape perturbations

• PDD Locations

-20%
-5%
0%
+5%
+7%
+8%

spurious insertion of one control assembly

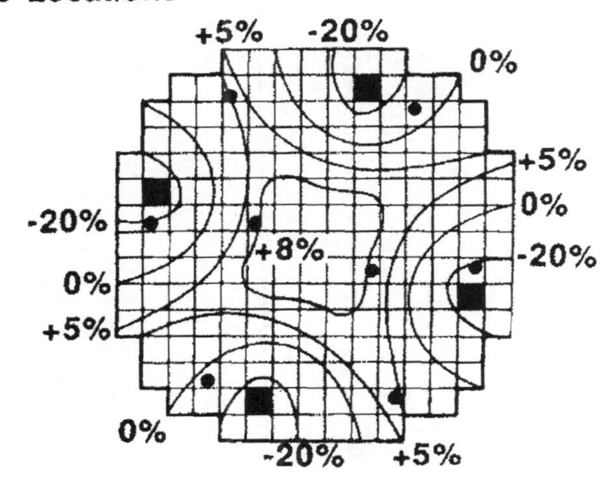

2/3 insertion of one control assembly quadruplet

Figure 3. Peak signal value by the PDD system vs. peak power density by the Aeroball system measured during load change operation and simulated control rod malfunctions

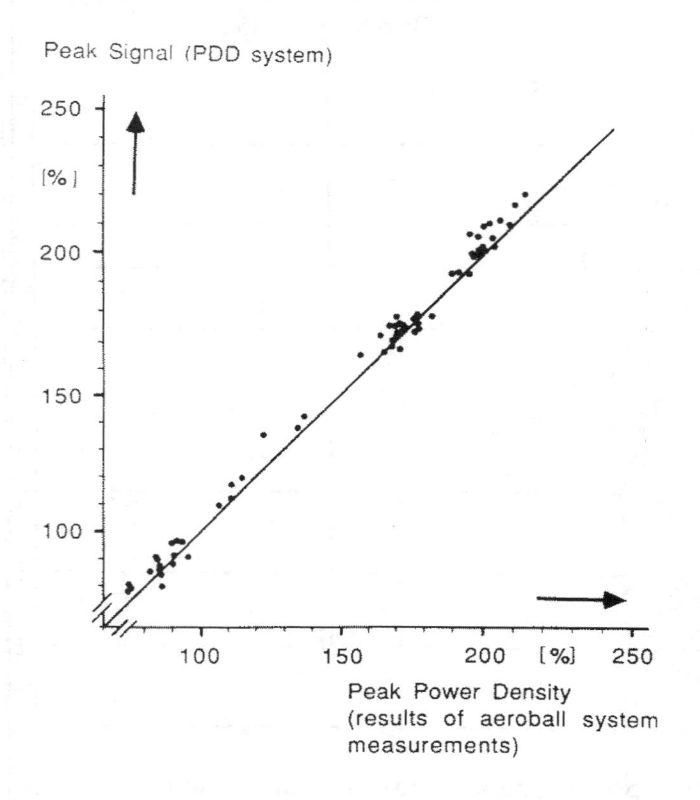

Figure 4. Margin to the LCO related to DNB ratio vs. margin given by the DNB calculational circuit measured during axial power shape variations

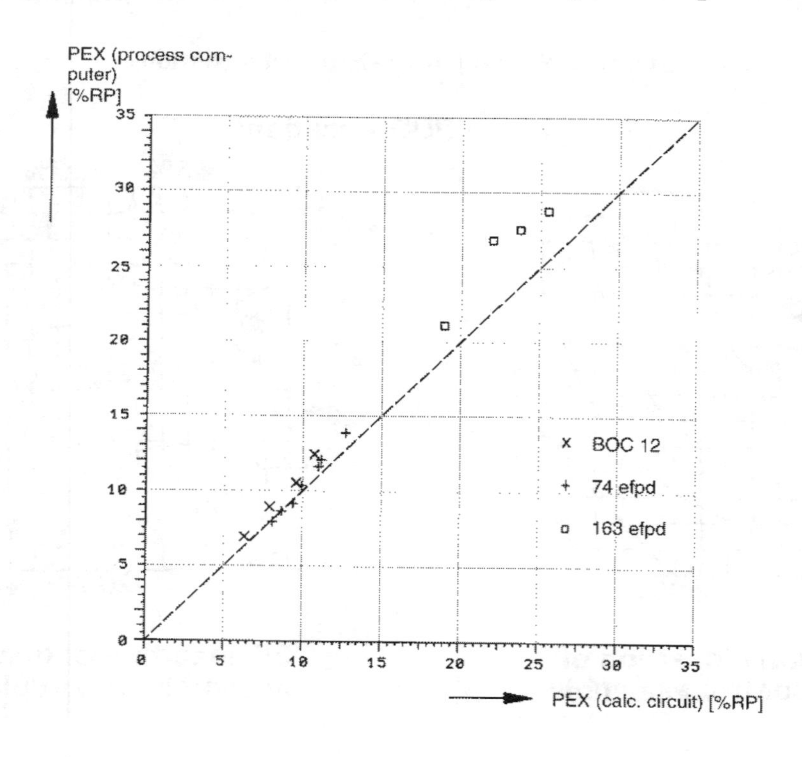

Figure 5. Results of OSG optimisation calculation (Part 1)

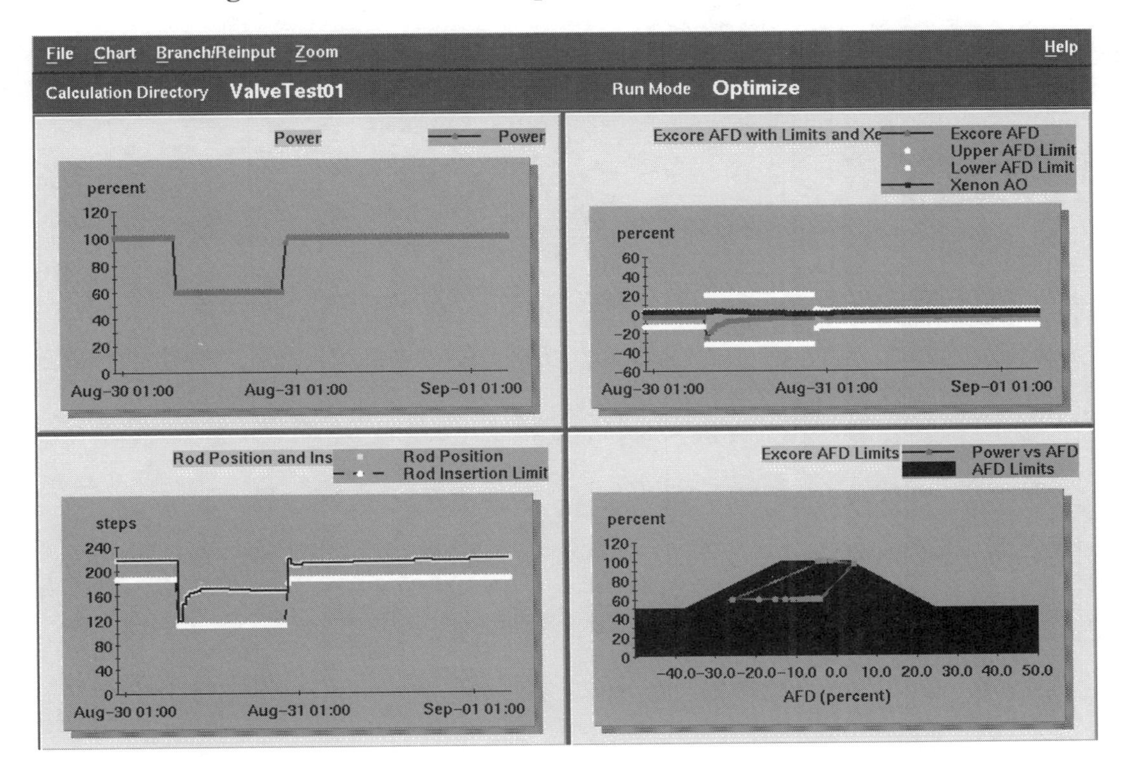

Figure 6. Results of OSG optimisation calculation (Part 2)

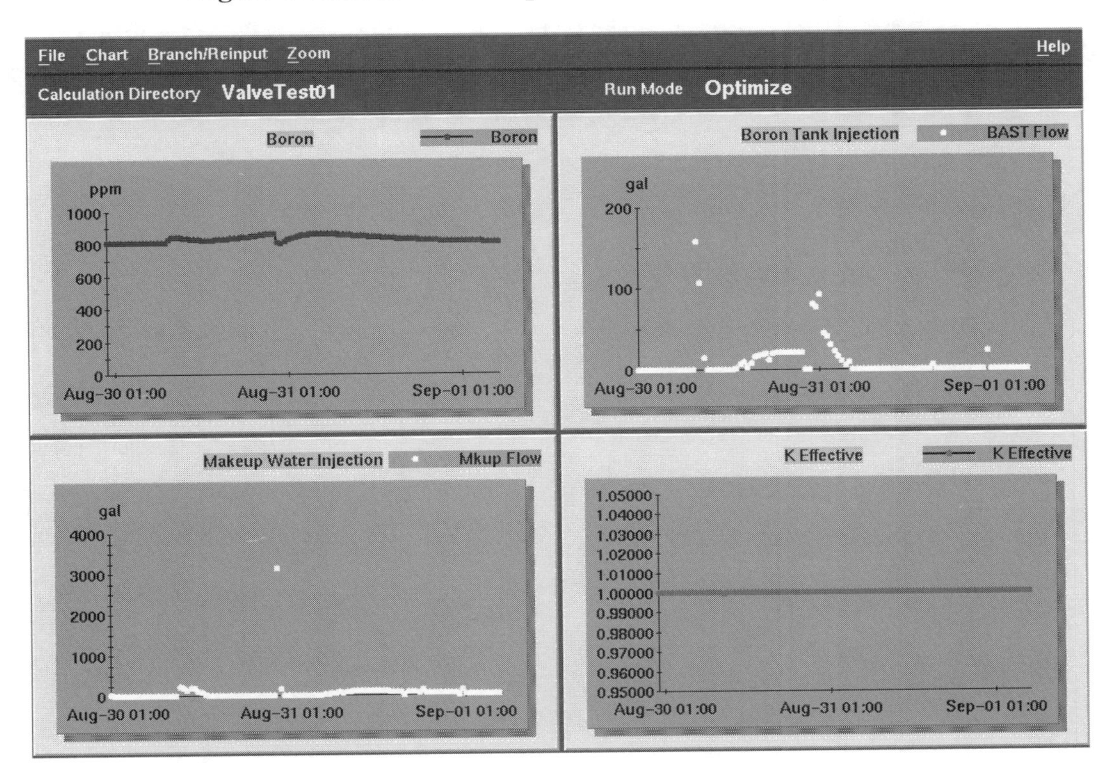

REVIEW OF THE CURRENT STATUS OF CORE MONITORING SYSTEMS AND THE FUTURE TREND IN PWRs IN JAPAN

Yoichiro Shimazu
Graduate School of Engineering, Hokkaido University
E-mail: shimazu@qe.eng.hokudai.ac.jp

Masahiro Masuda
Nuclear Fuel & Core Safety Engineering Department
Kobe Shipyard & Machinery Works, Mitsubishi Heavy Industries, Ltd.
E-mail: msmasuda@ncore.kobe.mhi.co.jp

Abstract

Basically all PWRs (and BWRs) in Japan are operated as base load plants and power uprating is not yet an urgent concern. Thus requirements from the utilities for core monitoring systems are not strongly focused on improvement of the accuracy of local power in the core. However, on-line core monitoring systems have been continuously installed in PWRs in Japan. In this paper we review the installation history and usage of the systems to clarify the incentives for utilities to install on-line core monitoring systems. It is found that not only the function of local power monitoring but also other functions that can provide economic benefits to the utilities are important factors in the decision-making process of the installation. Some examples and the future requirements are discussed based on the trend of the plant operation.

Introduction

On-line core monitoring systems based on modern core simulators have been offered to Japanese utilities for the past 10 years. The utilities have selected the BEACON system for their plants. This system has been licensed from Westinghouse to Mitsubishi Heavy Industries, Ltd. (MHI) and MHI has carried out modifications or customisation of the system in order to satisfy the utilities' particular requirements. The first BEACON system was installed in Ikata Unit No. 3 of Shikoku Electric Power Company during its initial start-up stage in 1994 [1]. The installation process was completed five years after the first presentation of the system to all PWR utilities in Japan. It took several more years for the systems to become widely installed. The objective of this paper is to obtain a certain insight with regard to incentives which encourage the utilities to decide to install such systems based on the review of published papers, and to try to see develop an idea of future trends.

Before going into the discussion, a brief introduction of the BEACON system is provided [1,2]. BEACON is an on-line core monitoring system taking continuous reactor operating signals from plant process computers, which are used to correct 3-D power distribution calculated by the integrated 3-D core simulator. Based on this technology it can continuously evaluate or monitor 3-D in-core power distribution as accurately as that of actual measurement. The 3-D core simulator is used not only to evaluate current power distributions but also to predict core behaviours and reactivity characteristics. It can also process the measured data of in-core flux distribution. Thus the system is a powerful tool not only for operators but also for nuclear engineers at the site, helping to detect anomalies at an early stage, perform core analysis or predict optimal operating strategy after unplanned shutdowns from an economical point of view.

Installation history

At present 23 PWR power plants are operating in Japan, and seven plants among them have installed the BEACON system. It is expected that more plants will soon install the system. A chronological installation history is shown as follows [2,3]:

1994	Ikata Unit No. 3 , Shikoku Electric Power Company, 3 loop plant
1996	Genkai Unit No. 3, Kyushyu Electric Power company, 4 loop plant
1996	Genkai Unit No. 4, Kyushyu Electric Power company, 4 loop plant
1997	Tsuruga Unit No. 2, Japan Atomic Power Company, 4 loop plant
1999	Ohi Unit No. 2, Kansai Electric Power Company, 4 loop plant
1999	Ohi Unit No. 1, Kansai Electric Power Company, 4 loop plant
1999	Mihama Unit No. 3, Kansai Electric power Company, 3 loop plant

Incentives of the utilities for the monitoring systems

Some papers and proceedings presented by the utilities and MHI concerning the core monitoring system have been reviewed. The first paper describes the basic features of the BEACON system and its original functions as introduced by Westinghouse: core monitoring, prediction of core behaviour and core analysis. There are few discussions, however, which focus on the advantages obtained through detailed linear heat rate monitoring or start-up economies after shutdowns. This is because all power plants in Japan are operated as base load plants and at present no urgent need for uprating exists. The unplanned shutdown rate per year is less than 0.2 per plant, which is far less than the world average of 2.5 per plant (except Japan as of 1995). Rather it is clear that Japanese utilities are more interested in gaining economic benefits by reducing the duration of refuelling outage in order to obtain a high capacity factor. The key issues relating to the reduction of the duration of refuelling outage are core physics tests, power distribution measurements before and during power escalation and in-core and ex-core nuclear instrumentation calibration in the area of core monitoring.

The other area in which the monitoring system could be effectively applied is the function of continuous on-line monitoring. Currently, plants are operated for longer cycles such as 13 equivalent full power months, and much longer cycles are to be expected. In such longer cores axial xenon oscillations become divergent in the later cycle. Thus even a small perturbation could initiate sustaining xenon oscillations and further divergent oscillations. It is preferable to control or suppress the oscillation as effectively and quickly as possible when necessary. Xenon oscillation control problems have been solved and effective strategies are already adopted in actual operations. One of the strategies, which is very powerful and widely in use, is the so-called "bang-bang control" method. It is quite difficult, however, to completely eliminate the oscillation by one control rods manoeuvre as the theory expounds. The difficulty comes from the fact that it requires precise timing and strength of the control rod manoeuvre based on the extrapolation of the oscillation behaviour. If the optimal timing and strength are missed, small oscillations still remain and additional control should be necessary. In order to solve the operational inconvenience the monitoring system has been effectively used.

Shortening of the start-up schedule

Power distribution measurement [3]

It has been a standard procedure to measure power distribution at hot zero power, 50% and 75% of rated power, respectively, before reaching the rated power in Japanese PWR utilities. The measured data are processed using design information of flux and power distribution in order to evaluate, by interpolation and/or extrapolation, the assembly power of which assembly is not directly measured. As the power escalation schedule cannot be precisely fixed in advance, the design information is prepared assuming the reactor is in stable condition. Thus it requires that power distribution should be measured with the reactor as stable as possible in order to obtain an accurate power distribution evaluation, which means waiting for a certain period until the reactor transient, for example the xenon distribution transient, dies out before the measurement. This waiting interval must be minimised to reduce the outage time.

The BEACON system follows the actual reactor operation on-line. It has the capability not only of generating the required information but also of processing the data. However, as the power distribution calculated by BEACON is basically normalised to that of the design code at around rated power, the accuracy of the BEACON calculation must be verified at various power levels in order to process the measured data. Some modifications of the data evaluation procedure have been carried out and the accuracy has been checked based on the actual reactor data in Genkai Unit No. 4, Turuga Unit No. 2, Ikata Unit No. 3 and Ohi Unit No. 1. This enables to use the system under any reactor conditions to measure power distribution, effectively eliminating the waiting interval before the power distribution measurement.

In-core and ex-core nuclear instrumentation calibration [3]

The power range ex-core nuclear instrumentation system (NIS) is used not only to measure power range but also to measure axial offset (AO; AO = (Pt - Pb)/(Pt + Pb), Pt and Pb are power in the upper half and the lower half of the core, respectively). It has four channels, each of which consists of upper and lower neutron detectors of long ion chamber. In order to monitor in-core AO as accurately as possible, periodic recalibrations are necessary. The first one is carried out at the start-up stage at around 75% of rated power. The conventional procedure has been to obtain correlation between in-core AO and each detector current during a xenon oscillation, which is purposely introduced by a control rod operation. The xenon oscillation is controlled after the data acquisition. However this procedure takes time for the introduction of the oscillation; the data acquisition requiring about eight hours, followed by the control of the oscillation, in total about 10 to 20 hours.

On the other hand efforts evaluating the accuracy of an analytical procedure have shown its applicability to actual NIS calibration [4]. The analytical procedure evaluates the relative correlation between in-core AO and detector current based on a response matrix of the detector. Then normalising the correlation to one point measurement, the final correlation is evaluated, which is equivalent to that based on the conventional procedure. In the procedure the required analytical information are the response matrix of the detector, various three-dimensional power distributions and the corresponding fast flux distributions. This information can be calculated in an off-line computer but it is much more convenient to use the on-line monitoring system.

The applicability of the BEACON system for the procedure has been proved in actual plants. This procedure eliminates the conventional tedious process of the data acquisition as described above, thus reducing the power escalation time by 10 to 20 hours.

Based on the new application of the system the conventional power escalation schedule, which takes about 60 to 90 hours, can be reduced to 40 hours as shown in Figure 1. This is a reduction of about two days, which can be considered a very good contribution to higher capacity factor.

Axial xenon oscillation control [5]

New control methods

A new method for xenon oscillation control was proposed and has proved to be very effective in domestic PWRs [6,7]. The concept of the method is based on three axial offsets, two of which (AOi and AOx) are employed in addition to the conventional axial offset of power distribution of AOp. AOi is defined as the axial offset of power distribution, which would give the current iodine distribution under equilibrium condition. AOx is similarly defined for the xenon condition. These AOs can be simply calculated in a two-point reactor model. With these three AOs, xenon oscillation control is clearly defined because xenon oscillations are eliminated when these three axial offsets have the same value. Based on this fact a graphical guidance for the operator was proposed which shows the trajectory of the plot of (Aoi-AOx, AOp-AOx) in the X-Y plane. The trajectory obtained in this way shows the following very characteristic behaviour:

- When the xenon oscillation is stable (neither divergent nor convergent) the trajectory is an ellipse with its major axis inclined at a fixed angle to the x-axis. The major axis lies in the first and third quadrants. The centre of the ellipse is located at the origin.

- The direction of the plot is always clockwise. The plot goes completely around the ellipse during one cycle of the xenon oscillation.

- When the oscillation is divergent the trajectory is an elliptic spiral which becomes larger and vice-versa.

When control rods are moved and the power distribution changes the trajectory responds as follows:

- When the power distribution changes to the bottom of the reactor to give a negative change to AOp, the trajectory moves in the negative direction to the y-axis and vice-versa.
- When control rods stop the trajectory begins to draw an ellipse or an elliptic spiral of the same characteristics from the point.

When the plot is at the origin where three AOs are identical, no xenon oscillation exists.

Control rules

As explained above the characteristics of the trajectory are very simple and definitive, and making the plot move to the origin as follows can eliminate the xenon oscillation:

- Identify the position of the current plot.
- If it is in the right-hand side of the y-axis and leaving from the origin then give negative AOp to cross the major axis. If the plot is in the left-hand side of the y-axis then vice-versa.
- Wait until the plot reaches the y-axis.
- When the plot reaches the y-axis then lead the plot to the origin on the y-axis by changing AOp.

These control rules assume that the AOp can be changed as a step. In an actual reactor, however, it may take some time to move control rods. In such cases the rules must be appropriately modified.

The guidance system can easily be installed in the monitoring system because the required information is available continuously. The function has been installed in the system. One of the examples is shown in Figure 2. It has been proved that the new control procedure installed in the monitoring system is very effective and helps to improve the operational reliability.

Future trends

Accurate monitoring of local power

MOX fuel is going to be used in some reactors. MOX fuel has different burn-up characteristics from that of uranium oxide fuel. It may require additional consideration from the viewpoint of fuel integrity. It may also require stricter monitoring of fuel power history to verify the design power history or the assumption in the safety analysis. This is true not only for MOX fuel but also for other fuels of aggressive design. From this point of view it is clear that the capability of estimating local power will be a more urgent requirement.

Integrated fuel management [1]

Fuel management systems are still based on a localised batch process. However it will be favourable to trace or follow all of the information for each fuel assembly beginning from the acceptance at the site to the discharge from the site. The main part of the information is related to the operation history in the reactor, which will be generated in the monitoring system. Thus integration of the existing fuel management system with the monitoring system is quite useful and effective.

Relaxation of technical specifications [1]

The current technical specifications are based on ex-core monitoring systems and periodical in-core power distribution measurements. Thus the monitoring parameters are not direct parameters such as linear power but rather AO or tilting factor. This kind of monitoring requires an additional margin to take into account the uncertainties of monitoring and estimation of direct safety parameters. It results in more strict specifications or narrow operating area, which may cause some economical penalties. Utilising in-core information, which is continuously obtained through the monitoring system, can solve these problems. In order to achieve these objectives a new concept of core monitoring will be required to be established. The monitoring system will become indispensable when it is based on the concept of achieving economical operation with the same level of safety.

Conclusion

On-line core monitoring systems in Japanese PWRs continue to be installed. This trend has been established due to the economical benefits that the monitoring system can provide. It is not directly related to the capability of local power monitoring of the system. However in the future the monitoring system will be used in wider area such as monitoring local power, for example in MOX fuel cores, sophisticated fuel management systems and relaxation of technical specifications. In order to follow this trend it is important to demonstrate clear economic benefits to the utilities, sufficient to cover the investment for the monitoring system.

REFERENCES

[1] N. Fujituka, H. Tanouchi, Y. Imamura, D. Mizobuchi, T. Kanagawa, M. Masuda, "Experience and Evaluation of Advanced On-Line Core Monitoring System BEACON at Ikata Site", INCORE'96, Mito, Japan, 16-17 October 1996.

[2] C.L. Beard, "Experience with the BEACON Core Monitoring System", OECD Specialists Meeting on In-Core Instrumentation and Reactor Core Assessment, Pittsburgh, PA, 1-4 October 1991.

[3] M. Masuda, H. Yasui, Y. Shimazu, S. Ueda, S. Yuge, Y. Imamura, A. Tatematsu, "Reduction of Start-Up Period using BEACON System", Proc. JAES, Fall Meeting, 1998 (in Japanese).

[4] T. Yamaguchi, S. Honda, Y. Shimazu, H. Yasui, "Pre-Calibration of In-Core/Ex-Core Nuclear Instrumentation System Based on Theoretical Calculations for Start-Up at Refueling Outage", Physor'96, Mito, Japan, 16-20 September 1996.

[5] Y. Yasui, M. Masuda, Y. Shimazu, A. Tatematsu, C. Fujita, "Operation Guidance of BEACON During Xe Oscillation", Proc. JAES Fall Meeting, 1998 (in Japanese).

[6] Y. Shimazu, "Continuous Guidance Procedure for Xenon Oscillation Control", *J. Nucl. Sci. and Tecnol.*, Vol. 32, No. 2, pp. 95-100, February 1995.

[7] Y. Shimazu, "Verification of a Continuous Guidance Procedure of Xenon Oscillation Control", *J. Nucl. Sci and Technol.*, Vol. 32, No. 11, pp. 1159-1163, November 1995.

Figure 1. Comparison of power escalation schedule

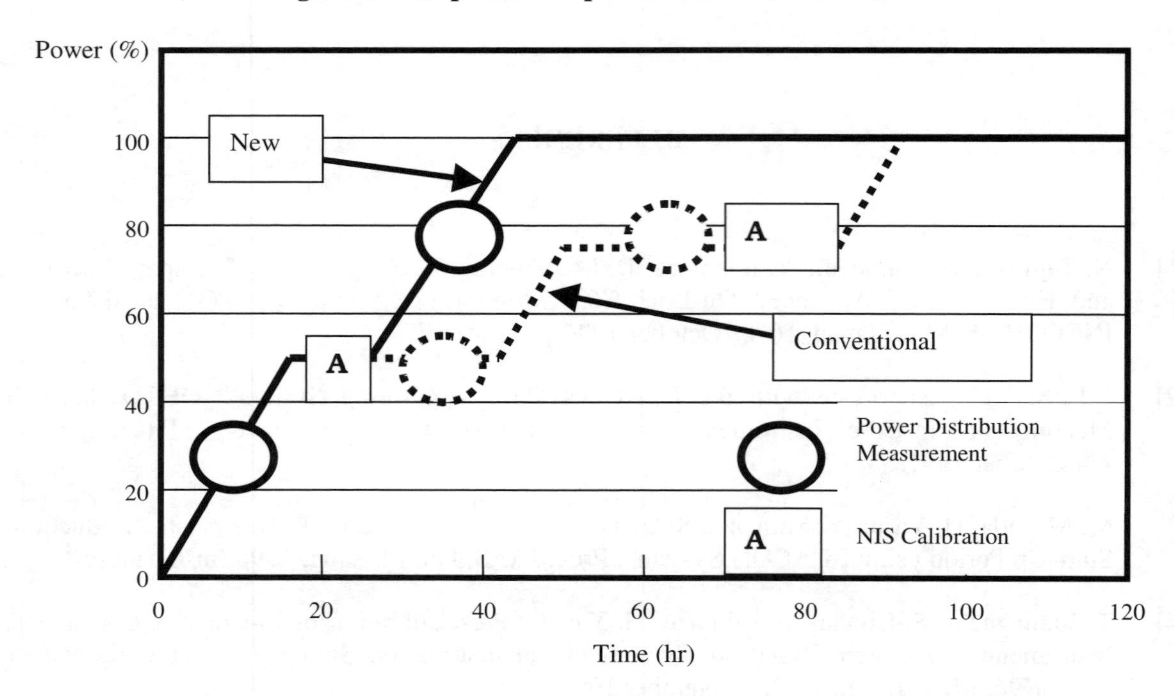

Figure 2. Example of xenon oscillation control

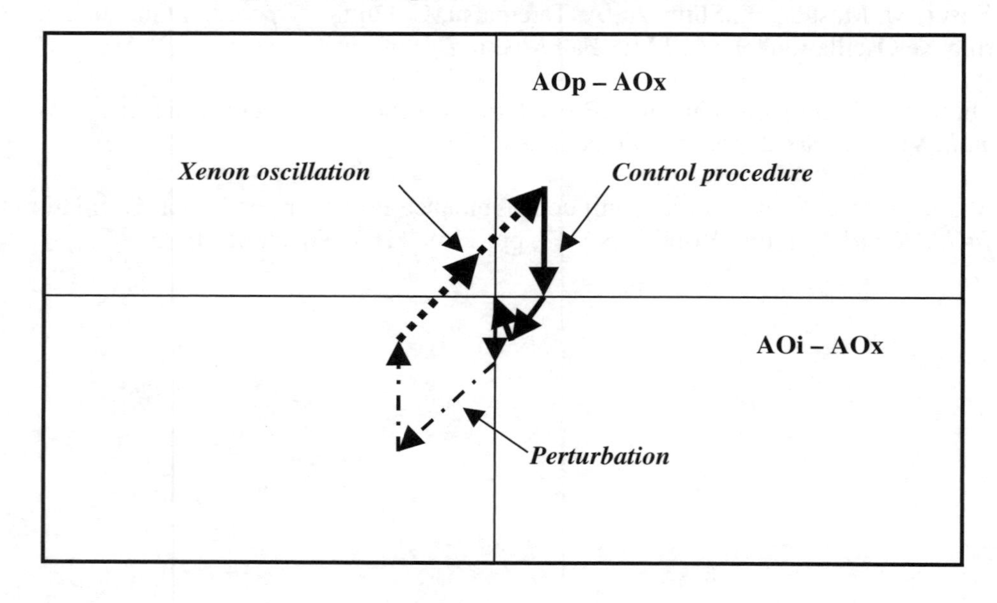

CORE SUPERVISION METHODS AND FUTURE IMPROVEMENTS OF THE CORE MASTER/PRESTO SYSTEM AT KKB

S. Lundberg, J. Wenisch
Hamburgische Electricitäts- Werke AG
Überseering 12, D-22297 Hamburg

W. van Teeffelen
Kernkraftwerk Brunsbüttel GmbH
Otto-Hahn-Straße, D-25541 Brunbüttel

Abstract

Kernkraftwerk Brunsbüttel (KKB) is a KWU 806 MW_e BWR located at the lower river Elbe, in Germany. The reactor has been in operation since 1976 and is now operating in its 14th cycle. The core supervision at KKB is performed with the ABB CORE MASTER system. This system mainly contains the 3-D simulator PRESTO supplied by Studsvik Scandpower A/S. The core supervision is performed by periodic PRESTO 3-D evaluations of the reactor operation state. The power distribution calculated by PRESTO is adapted with the ABB UPDAT program using the on-line LPRM readings. The thermal margins are based on this adapted power distribution. Related to core supervision, the function of the PRESTO/UPDAT codes is presented.

The UPDAT method is working well and is capable of reproducing the true core power distribution. The quality of the 3-D calculation is, however, an important ingredient of the quality of the adapted power distribution. The adaptation method as such is also important for this quality. The data quality of this system during steady state and off-rate states (reactor manoeuvres) are discussed by presenting comparisons between PRESTO and UPDAT thermal margin utilisation from Cycle 13.

Recently analysed asymmetries in the UPDAT evaluated MCPR values are also presented and discussed.

Improvements in the core supervision such as the introduction of advanced modern nodal methods (PRESTO-2) are presented and an alternative core supervision philosophy is discussed.

An ongoing project with the goal to update the data and result presentation interface (GUI) is also presented.

Introduction

Kernkraftwerk Brunsbüttel (KKB) is a KWU 806 MW_e/2 292 MW_{th} BWR located at the lower river Elbe, in Germany. The reactor has been in operation since 1976 and is now operating in its 14th cycle. The core is loaded with two batches of SVEA-96/L fuel. The remaining bundles are of the SVEA-64C/L design. The core supervision at KKB is performed with the ABB CORE MASTER system. This system is mainly composed of a 3-D simulator (PRESTO) supplied by Studsvik Scandpower A/S. The core supervision is performed by periodic PRESTO 3-D evaluations of the reactor operation state. The power distribution calculated by PRESTO is adapted with the ABB UPDAT program using the on-line LPRM readings. The thermal margins are based on this adapted power distribution.

PRESTO/UPDAT system

Description

The core supervision at KKB is performed with the 3-D PRESTO code within the ABB Atom CORE MASTER frame. The PRESTO version used in the KKB core supervision system is an older version from 1989. This version is fed with lattice data from the 2-D code RECORD. It also contains a simplified hydraulics model. The hydraulics correlation data for this model is provided either by a more modern PRESTO version, containing a refined hydraulics model or calculations with the ABB stand-alone hydraulics code CONDOR. The main reason for still using this older PRESTO version is due to the licensing situation in northern Germany.

The core power distribution is monitored by in-core detectors, LPRM, located in 30 detector strings with four detectors each. The layout of the detector system is shown in Figure 1. The detectors are neutron detectors. The calibration of the detectors is performed periodically by running gamma sensitive TIP detectors trough each detector string. After a TIP run the gains of the LPRM detectors are adjusted to yield the same "power" as the TIP detectors.

Figure 1. Axial and radial layout of the KKB detector strings

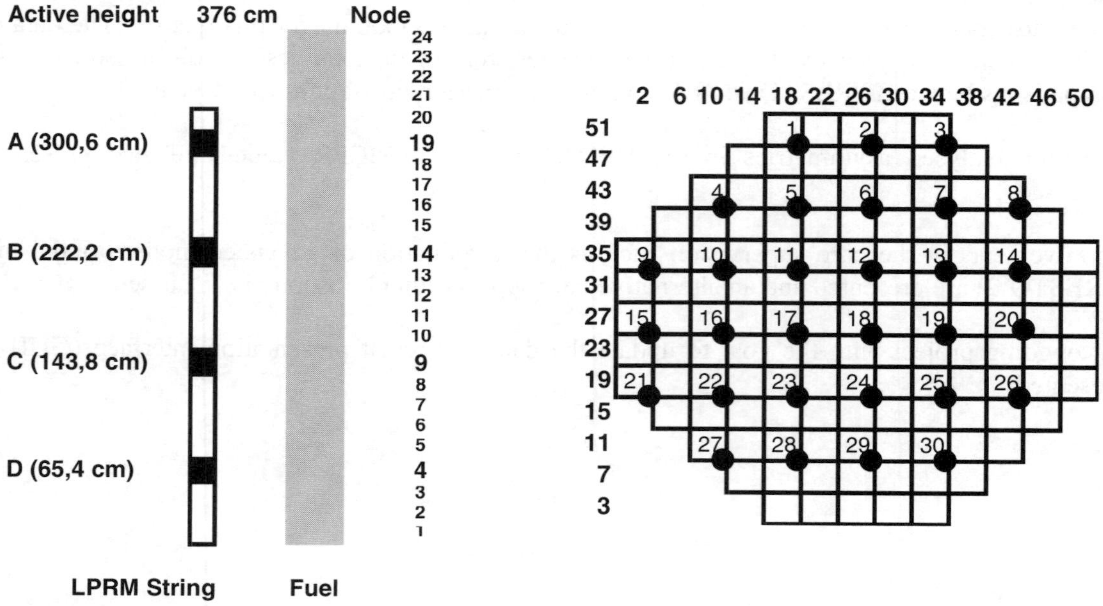

As seen from Figure 1 the radial locations of the LPRM strings are non-symmetric.

A PRESTO calculation is performed periodically, every four hours, or is triggered by changes in the major operation parameters (core power, core flow and control rod position). The neutronic model of PRESTO is a 1.5 group nodal model. This means that the thermal flux calculation is performed in the fast neutron group using the spectrum index, relating the fast flux to the thermal flux and a group gradient correction factor.

After the 3-D power distribution and a set of predicted LPRM readings have been calculated the UPDAT code starts. This code collects the actual LPRM readings from the 120 LPRM neutron sensitive detectors. The measured LPRM readings are compared to the readings predicted by PRESTO and the calculated power distribution is adjusted based on this comparison. The information from the last TIP run is also utilised in the adaptation. The adaptation is described by the equation:

$$P_{UPD} = P_{PTO} * \Sigma_i \; W_i * (SLPRM/PLPRM) * TIPCOR$$

where: P_{UPD} is the adapted nodal power
P_{PTO} is the calculated nodal power
W_i is geometrical weighting factors
SLPRM is the measured LPRM readings
PLPRM is the PRESTO predicted LPRM readings
TIPCOR contains the correction factor at the last TIP measurement

The UPDAT model functions fairly well in steady state situations. If a control rod has been moved, however, causing changes in the power shape a new PRESTO calculation of the LPRM signals is required. The TIPCOR distribution, though, is no longer valid. In order for UPDAT to function properly a new calibration will be required.

When the adapted power distribution is available the thermal margins are calculated, based on this power distribution.

The adaptation is momentary, i.e. the burn-up updating performed in the core supervision system is done with the calculated power distribution and not with the adapted distribution.

Quality of PRESTO/UPDAT power distribution

The quality of the adaptation is dependent on the quality of the detector signal, their axial location related to their description in the 3-D code used and how well the predicted LPRM readings and power distribution has been calculated.

The quality of the calculated to measured power distributions are quantified by calculating the nodal and radial standard deviations for the TIP comparisons. Typical standard deviation, with the system used at KKB, is in the range of 4-8% nodal and between 2.5-3.5% radial. An example of the nodal and radial standard deviations from Cycle 13 is shown in Figure 2.

PRESTO/UPDAT comparison in KKB Cycle 13

The thermal margins are expressed as the LHGR and the MCPR relative utilisations (MFLPD and MFLCPR). They are calculated by UPDAT, based on the adjusted power distribution and presented to

Figure 2. KKB Cycle 13 TIP standard deviations

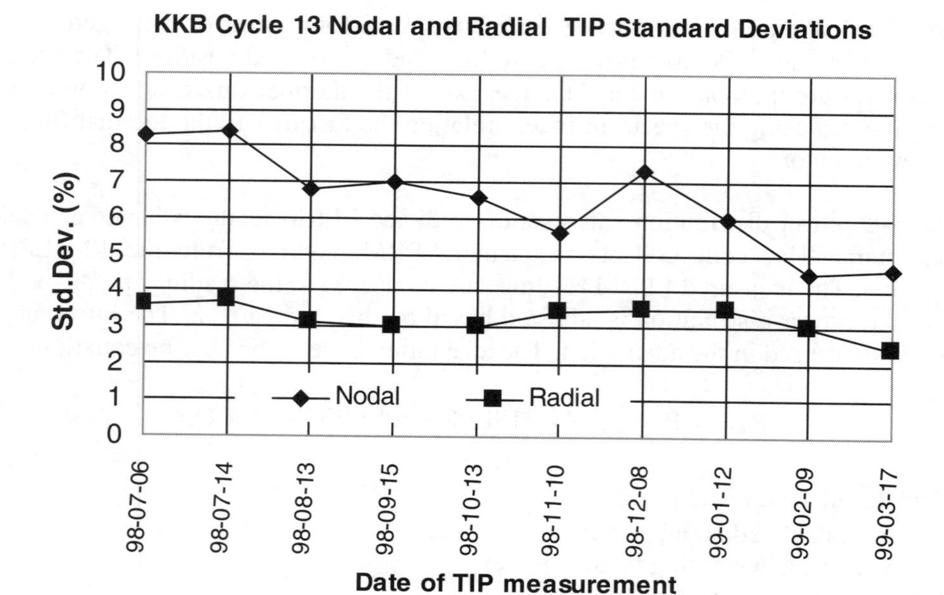

the reactor operator. These values normally differ from the MFLPD and MFLCPR values calculated directly by PRESTO. Examples of these differences from KKB Cycle 13, after the rated power was reached in July 1998, are shown below for the parameters MFLPD and MFLCPR.

A good core supervision system should show small differences between the measured and the calculated thermal margins. This is also essential for an optimal off-line core operation support. Such a support can help improve the cycle utilisation. Small measured to calculated differences also provide for reduced design margins in the reload work, which in turn increases the utilisation and the economic revenues from a core loading.

A typical difference for the MFLPD parameter is about 7% and about 3% for the MFLCPR parameter (average for Cycle 13 after start-up).

Figure 3 shows the reactor thermal power during Cycle 13. A period with extensive load following down to about 80% was performed in January 1999. The power reductions represent periodic testing and load following to deeper power levels.

Figure 4 shows the relative difference between UPDAT and PRESTO for the MFLPD parameter during Cycle 13. The difference is fairly constant between 2 and 10% during the rated operation. The higher values occur when the axial power shapes are top peaked due to an intermediately located control rod (c.f. below). The differences towards the end of the cycle are small although the axial power shape is top peaked. This is due to the absence of control rods in the core.

During load following and periodic tests (off-rated conditions) the differences are clearly larger. The main reason for this is the changed power distributions at off-rated conditions, caused by rod insertions and xenon transients. This leads to inaccurate TIPCOR distributions. The relatively large deviations (up to 45%) are due to the fact that the older PRESTO version used in the core supervision at KKB does not account for the power dependant LHGR limits. If these were taken into account the differences would be reduced to a band of ±25%. The MFLPD values are proportional to the reactor power and the larger deviations at off-rated power are thus not safety related.

Figure 3. KKB Cycle 13 thermal reactor power

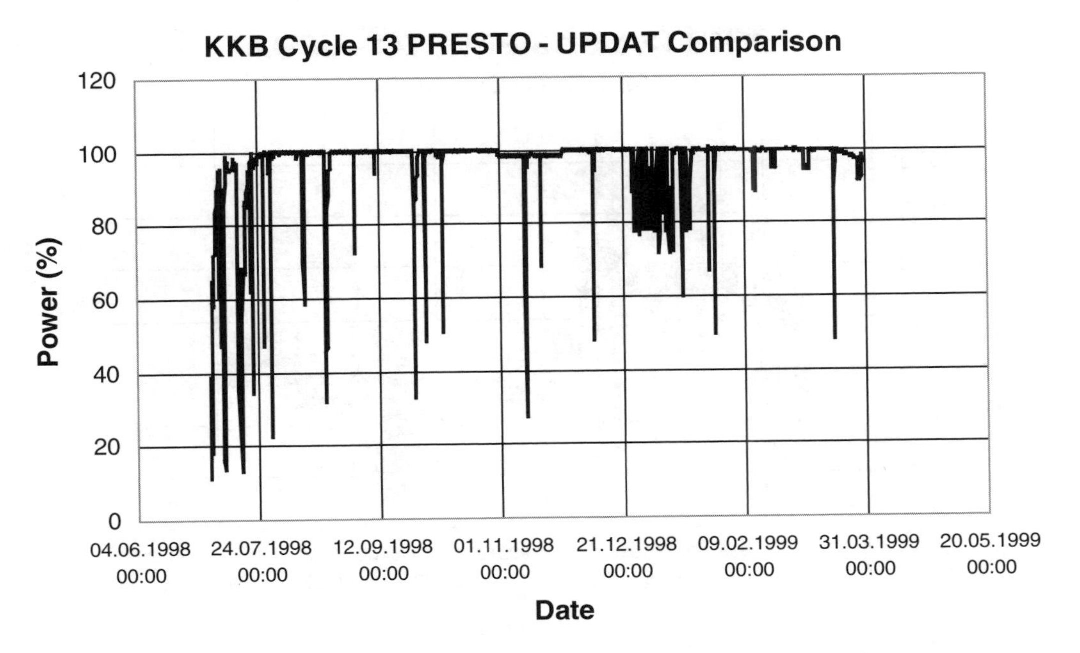

Figure 4. MFLPD difference between UPDAT and PRESTO in KKB Cycle 13

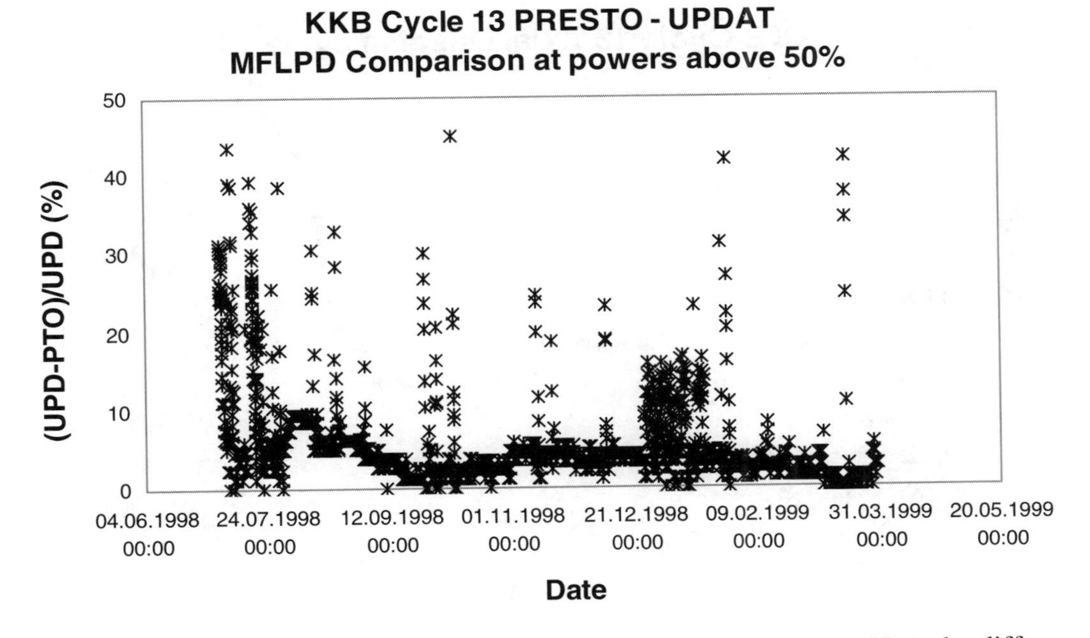

Figure 5 shows the same relative difference for the MFLCPR parameter. Here the differences are in the range 0-5% during the rated operation. During off-rated conditions the differences increase, however less than for the MFLPD.

It is evident that the differences in MFLPD and MFLCPR do not only follow the power level. The differences are also related to the axial power shape, as can be seen in Figure 6. This figure shows the axial offset (AO) during KKB Cycle 13.

Figure 5. MFLCPR difference between UPDAT and PRESTO in KKB Cycle 13

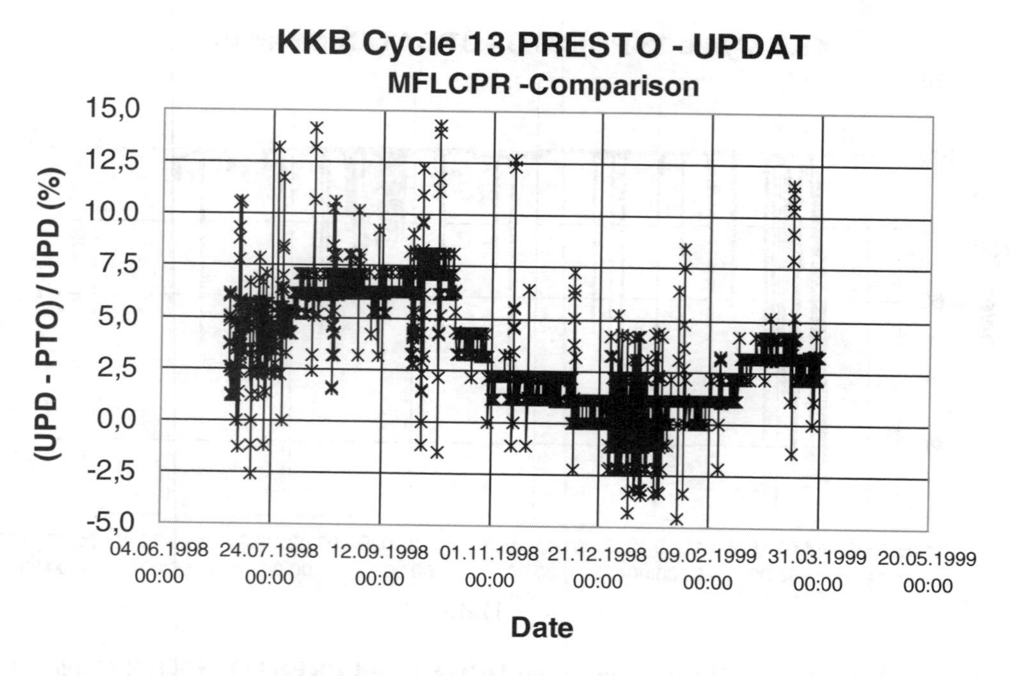

Figure 6. KKB Cycle 13 axial offset

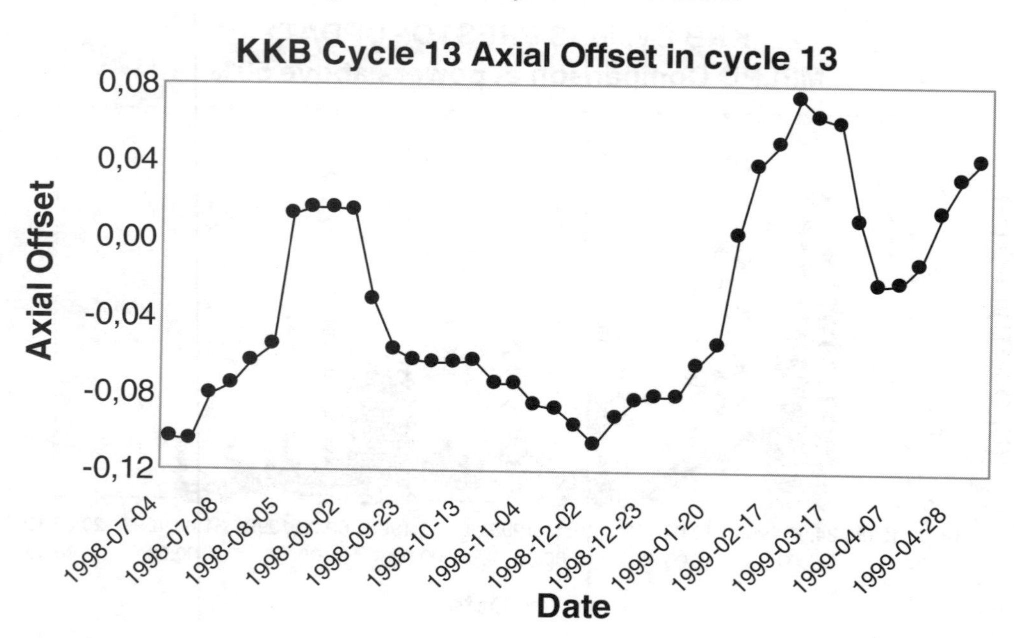

The axial offset is a measure of the axial power shape. It is defined as:

$$AO = \left(P_{high} - P_{low}\right) / \left(P_{high} + P_{low}\right)$$

where: P_{low} = the sum of the nodal power for nodes 1-12
P_{high} = the sum of the nodal power for nodes 13-24

A negative AO represents an axial power shape peaked towards the bottom of the core while a positive AO represents a power shape peaked towards the top of the core.

The smaller differences between UPDAT and PRESTO are seen when the axial power shape is more peaked towards the bottom. This is more pronounced for the MFLCPR parameter than for the MFLPD parameter.

The mid and top peaked power shapes are induced by control rods in intermediate axial locations which increase the deviations. These larger deviations are believed to be related to the treatment of the control rod presence in the lattice calculations. The control rod treatment in the 2-D code RECORD is based on a semi-empirical blackness theory model. In this model the current to flux ratio at the absorber surface is expressed in terms of neutron transmissions and reflection characteristics of the control rod. That is not, however, sufficient to produce accurate rodded cross-sections. The future use of the transport code HELIOS is expected to eliminate this problem.

Therefore one way to improve the core supervision with the present KKB system, i.e. reducing the differences between PRESTO and UPDAT, is to avoid the use of axially intermediate control rod positions. Sufficiently bottom peaked power shapes can be accommodated by using the control rods in deep and shallow positions. The use of improved methods will also eliminate this problem.

UPDAT/PRESTO comparison during reactor manoeuvres

As seen in the figures above the PRESTO/UPDAT differences increase during reactor manoeuvres. A detailed example of this is shown in Figure 7 (MFLPD) and Figure 8 (MFLCPR) for a power reduction in September-October 1998. This power reduction was performed with both a core flow reduction and insertion of control rods. The Delta (parameter) in the figures are calculated as (UPDAT-PRESTO)/UPDAT in %.

The increased differences occur when the power has been reduced and during the return to rated power. One reason for this is that the LPRM detectors are not calibrated for the off-rated situation. In such a situation the axial and radial power shape changes, as compared to the initial (calibrated) shape. This leads to an extra uncertainty in the measured LPRM readings. In the example the control rods is inserted, meaning that the TIPCOR values are inaccurate.

In situations like this gamma sensitive LPRM detectors could help mitigate the increased differences. Gamma sensitive detectors are less sensitive to their geometrical location and need not be re-calibrated at reduced power in order to yield a correct power distribution. Also, in manoeuvres without control rod motion the PRESTO/UPDAT differences increase relative to the rated situation.

The disadvantage with gamma sensitive detectors is however related to their function in the reactor protection system (RPS), as they also respond to delayed neutrons.

UPDAT asymmetries

The LPRM detectors are not symmetrically located in the core. This should not, however, pose a problem for UPDAT. The algorithm used to calculate the adjusted nodal powers only introduces a slight asymmetry. This can be seen by comparing the relative bundle powers (expressed in per cent) calculated with PRESTO and UPDAT, as shown in Figures 9(a) and 9(b).

Figure 7. UPDAT/PRESTO difference for MFLPD during a reactor manoeuvre

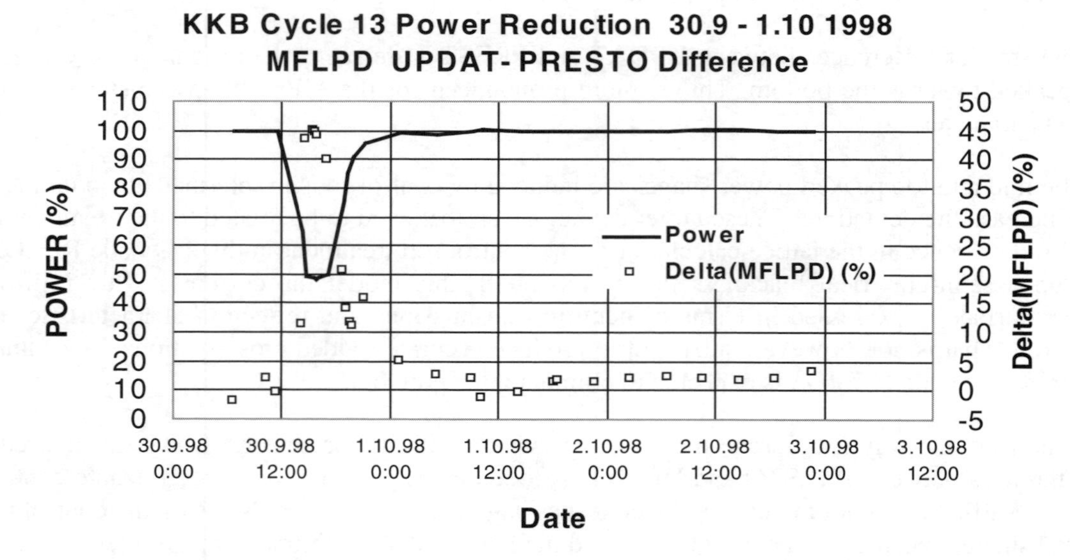

Figure 8. UPDAT/PRESTO difference for MFLCPR during a reactor manoeuvre

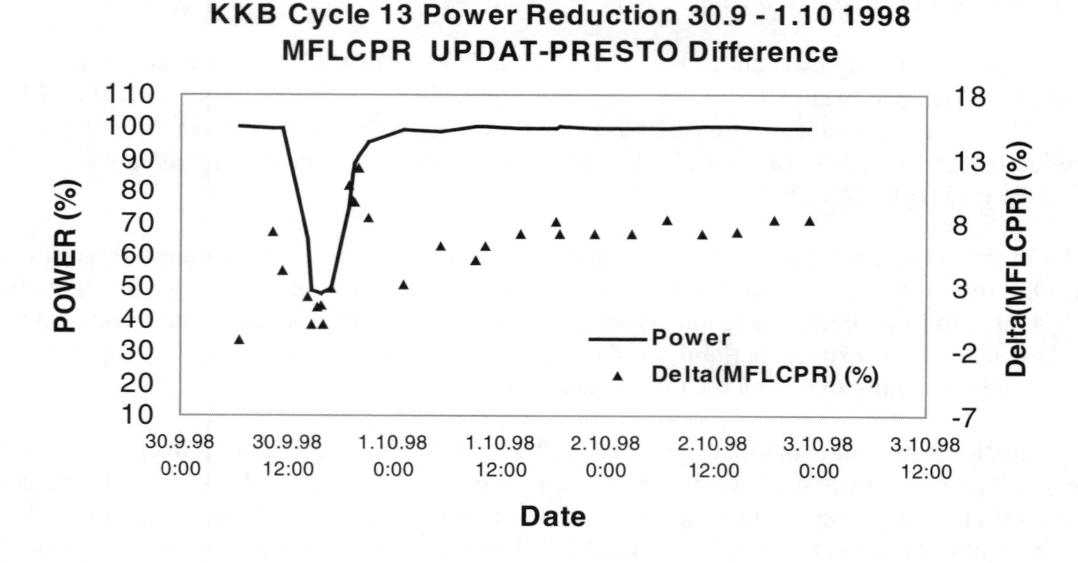

Figure 9(a) shows the relative channel powers in per cent in a 10 × 10 region around the central control rods. The black circles show the positions of the detector strings. The shadowed cells show the location of the control rods. The quarter core rotation symmetrical cells, containing the limiting MCPR are highlighted. The asymmetry in these cells is 0.8%. Figure 9(b) shows the same parameter calculated by UPDAT.

The asymmetry is now increased to 2.2%, i.e. a factor of 2.7 larger than with PRESTO, which is deemed acceptable.

If the MCPR utilisation, important for an optimised operation, is studied the situation is changed. Figure 10(a) shows the MFLCPR * 1 000 for the same 10 × 10 region as that calculated above with PRESTO.

Figure 9(a). PRESTO bundle power at TIP 13.10.98 (KKB Cycle 13)

	17	*19*	*21*	*23*	*25*	*27*	*29*	*31*	*33*	*35*
36	89	96	97	125	104	96	127	96	94	89
34	95	97	128	101	101	122	100	128	97	96
32	96	128	126	101	**129**	103	129	126	127	97
30	127	101	129	125	107	107	125	101	101	125
28	96	122	103	107	97	97	107	**129**	102	104
26	104	102	**129**	107	97	97	107	103	122	96
24	125	101	101	125	108	108	125	129	101	127
22	97	128	126	129	103	**130**	101	126	129	96
20	96	97	128	101	122	102	101	128	97	96
18	89	95	96	127	96	104	126	98	97	90

Figure 9(b). UPDAT bundle power at TIP 13.10.98 (KKB Cycle 13)

	17	*19*	*21*	*23*	*25*	*27*	*29*	*31*	*33*	*35*
36	88	96	97	125	105	99	131	99	98	92
34	94	96	126	101	103	126	104	133	101	100
32	95	127	125	101	**131**	106	133	130	131	100
30	128	101	129	126	110	111	128	104	103	127
28	98	124	105	110	100	100	110	**132**	103	105
26	108	105	**133**	111	101	100	110	104	123	97
24	130	105	105	129	112	111	127	131	102	127
22	101	133	131	134	107	**134**	104	128	130	95
20	101	102	133	105	127	106	105	130	98	95
18	93	99	100	132	100	109	130	100	98	89

Figure 10(a). PRESTO MFLCPR values (*1 000) at TIP 13.10.98 (KKB Cycle 13)

	17	*19*	*21*	*23*	*25*	*27*	*29*	*31*	*33*	*35*
36	687	744	656	840	702	653	849	644	725	690
34	723	751	863	680	689	834	680	865	752	745
32	646	869	854	683	**884**	698	875	852	859	656
30	851	681	876	848	732	732	847	683	678	838
28	652	836	698	732	773	774	732	**885**	690	705
26	705	690	**883**	732	775	775	734	700	837	654
24	836	679	684	849	736	737	851	879	686	856
22	655	860	856	877	699	**886**	687	858	874	647
20	743	751	870	682	837	693	682	862	755	731
18	687	723	647	853	652	704	838	658	749	693

The asymmetry is 0.35%, i.e. the core MFLCPR is very symmetric.

After the UPDAT adjustment of the power distribution and the subsequent CPR calculation in UPDAT the asymmetry increases to 3.3%, i.e. 10 times as high than with PRESTO. This is shown in Figure 10(b).

Figure 10(b). UPDAT MFLCPR values (*1 000) at TIP 13.10.98 (KKB Cycle 13)

	17	*19*	*21*	*23*	*25*	*27*	*29*	*31*	*33*	*35*
36	681	737	655	866	719	676	904	670	756	718
34	719	740	902	685	705	871	708	950	784	778
32	645	905	852	687	**940**	722	915	887	948	682
30	881	686	890	864	751	758	878	704	700	879
28	668	862	714	753	796	799	755	**956**	707	719
26	734	719	**958**	758	802	800	753	714	856	661
24	893	712	712	885	765	761	873	899	694	879
22	687	947	895	919	727	**971**	706	877	930	646
20	778	784	956	711	880	725	709	933	760	727
18	717	754	674	910	682	741	894	677	754	683

For the TIP measurement studied (10.13.98) the power shape in the TIP strings contributing to each of the four symmetric cells is slightly top peaked, with the peak value at node level 15.

Examination of other situations has shown, however, that the asymmetry problem is not related to the axial location of the active control rods. Asymmetries also occur for bottom peaked situations.

The asymmetry introduced by the UPDAT adjustment and the subsequent CPR calculation by UPDAT is no problem as long as sufficient CPR margin is available. If the margin is approaching 0-2%, however, then a 3% asymmetry in a limiting position could cause an additional conservatism, as the highest of the asymmetric values is setting the operation limit. If one of the lower values is the "true value" an extra penalty, negatively impacting the fuel cycle cost, occurs.

Improvements in future core supervision

In the present UPDAT methodology the nodal power is calculated by including the contribution from eight LPRM detectors surrounding the node in question. This is done by weighting the distance from the LPRM detectors to the node. One improvement here could be to include the contributions from a larger set of detectors. This option is available in the ABB Atom POLCA-7 based system. It has not yet been tested, however.

Another improvement would be the use of advanced methods. Examples of such methods are true two-group models using discontinuity factors when calculating the nodal coupling. The use of discontinuity factors greatly improves the core analyses in regions with steep neutron flux gradients, such as control rod presence or large radial burn-up differences. Examples of codes based on this so-called modern nodal theory are POLCA-7, SIMULATE-3 and PRESTO-2. For HEW, operating the two BWRs Krümmel (KKK) and Brunsbüttel (KKB), the PRESTO-2 code is anticipated for future use. Qualification of this code, which uses cross-sections generated with the 2-D code HELIOS, is presently underway for KKB and is almost finished for KKK.

The analyses so far performed for the KKK plant show improvements in the TIP comparisons. The measured detector signals are the same as those used with the older version of PRESTO. This indicates that upgrading the nuclear methods has a positive implication on the results and will thus improve the core supervision.

Examples of the improved nodal and radial TIP standard deviations are shown in Figure 11.

Figure 11. KKK TIP standard deviations with the old and the new PRESTO codes

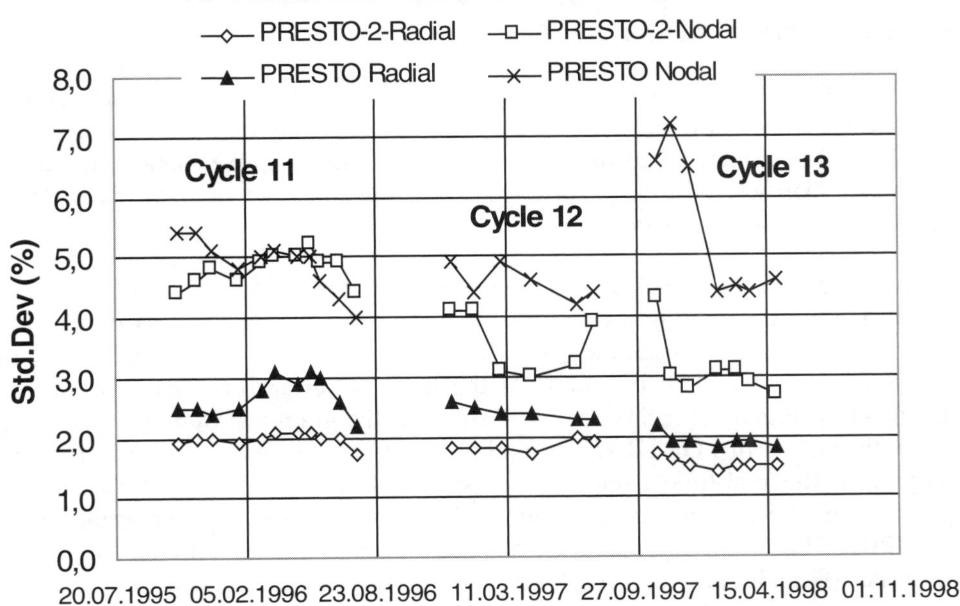

The main reason for these improvements, apart from a more accurate power calculation, is the improved determination of the predicted TIP readings. This is accomplished by an improved detector model, with correction factors for detector string voiding, tilted channel, internal channel water film and channel bowing.

One way to avoid the type of asymmetries described above is to use a methodology with quarter core symmetrically reflected detector readings, producing pseudo-string detector readings. This requires, however, that the calculated asymmetry of the core be small. Such an option is available in the PRESTO-2 based core supervision system installed at the Swiss KKL reactor. It is also used in the GE 3-D MONICORE system.

The improved 3-D methods in the modern codes also provide the option to base the thermal margins evaluation on the calculated power distribution. The measured LPRM signals will in such an application be used for comparisons to calculated distributions in order to validate calculated power distribution against a measurement. Typical acceptance values are of the order of 2% radial and 4% nodal. This strategy has been used at the Swiss KK Leibstadt plant for a number of years.

When using this method it is very important to continue the comparisons to measured data. The measurements do have their weaknesses, however the calculations can, e.g. not reveal mechanical errors such as worn channel corners or errors in fuel enrichment/Gd design in the fuel. An example of a worn of channel corner, due to detector string vibrations, occurred in a reactor a few years ago. The measured signal deviated from the calculated signal due to changed hydraulic conditions. The signal was thus believed to be in error and the calculated power was used to determine the CPR at this location. During the outage the channel failure was discovered and it was concluded that the measurement had been correct. It is thus important to impose limits on the global and local deviations between the measured and the calculated signals before the calculated power distribution is accepted as the basis for evaluation of the thermal margins. In case the limits are violated an adaptive method can be used as backup.

Development of a new graphical user interface

The CORE MASTER PRESTO core supervision system (CMPS) has been operated at KKB since 1988/89. The CMPS is typically installed on a VAX 8300 computer, called the central analysis computer (CAC), and is connected to the plant process computer (VAX 8600) via an ETHERNET cable as a DIGITAL hardware/software cluster, sharing the same set of hard disks (HSC concept). The disks are grouped in form of shadow sets.

In case an outage of the CAC occurs, the plant process computer will take over the CAC functions – and vice-versa – continuing the operation and precluding the loss of data without data inconsistencies. The interface between the CMPS and the user is the User Interactive Module (UIM). The UIM is a screen-oriented shell program. At the lower level are programs for formatted ASCII reports of calculated results by PRESTO and UPDAT and the graphical output is performed with the program RIT, both written in FORTRAN. In addition a display of a few key parameters and status signals directly from the real-time database, e.g. status of relevant flags associated with different CMPS programs and alarm flags, is performed. Scalar trend plots of important process values, PRESTO and UPDAT based thermal margins, XE and I average concentrations, etc., are provided on different time scales (8h, 24h, 2d, 7d, ...). The philosophy of providing access to only a few important pieces of information to the shift personal leads to the implementation of two different UIMs, which increases the maintenance effort.

In order to improve the performance of the CMPS system a project was initiated in 1997/98 to port the CMPS on a COMPAQ (former DIGITAL) ALPHA-processor based computer system still running the OpenVMS operating system. The operating system was chosen due to the very good experience made in the past years. The software itself was ported without changing its functionality and models due to the licensing environment.

The elapsed time of the core supervision modules PRESTO/XENUPD and UPDAT are nowadays reduced to about one to two minutes in full core applications.

The plant process computer is linked to the new system, made up of two computers for fallback purposes, via a network. Only the presentation layer is modified to a system based on modern windows techniques. The new graphical user interface (GUI) is highly modularised, easy to configure (one system for shift and physics department), and based on CORE MASTER 2 features provided by ABB Atom. The basic software for this GUI is the public domain Tcl/Tk (Tool command language/ Tool kit) and C++.

Special emphasis has been made on the display of scalar trend informations. About 60 parameters are predefined and the user may define his own scalar variables, e.g. k_{eff} in pcm over 1. The trend data are automatically updated every minute in case of process data or after the completion of the core supervision modules. In order to reanalyse past manoeuvres, the time interval of interest may be freely defined by the user without any loss of information due to condensation of data. Other interesting features are:

- Periodical update of the "global plant data" and power flow map.
- Radial informations of up to three entities, axial plots.

- Comparison of results from PRESTO and UPDAT, arithmetic operation on distributions.
- Summary of important results for 36 neighbouring bundles.
- Presentation of LPRM data.
- Automatic generation of protocols and support of Tip-measurements.
- Status and trigger function for core supervision modules.

Examples of some of the features are shown in the Figures 12 and 13 (based on test data).

Figure 12. Example of a trend plot for a past manoeuvre and "global reactor data" display

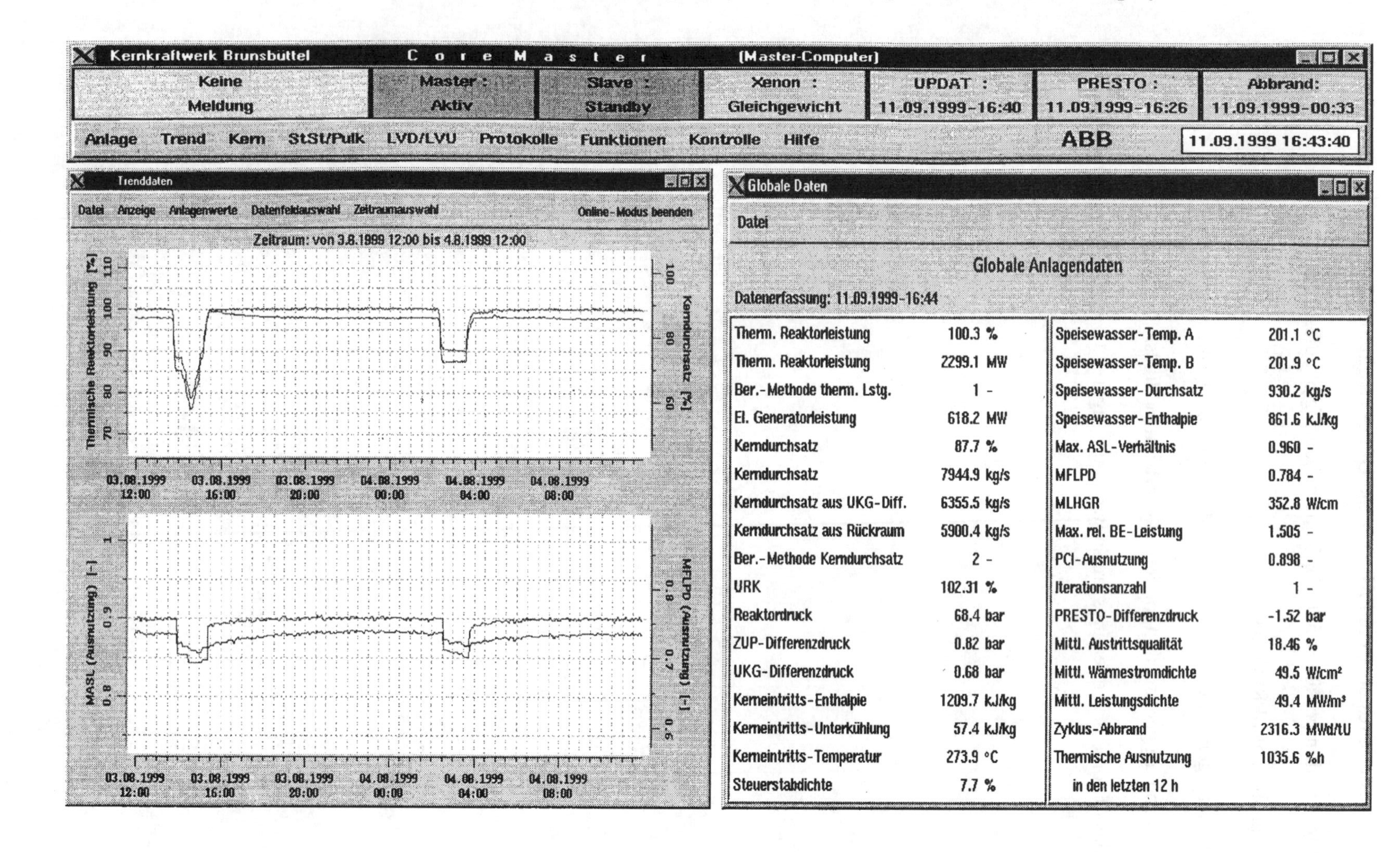

Figure 13. Example of the display of a radial LHGR distribution (maximum values) and one axial plot of two bundles

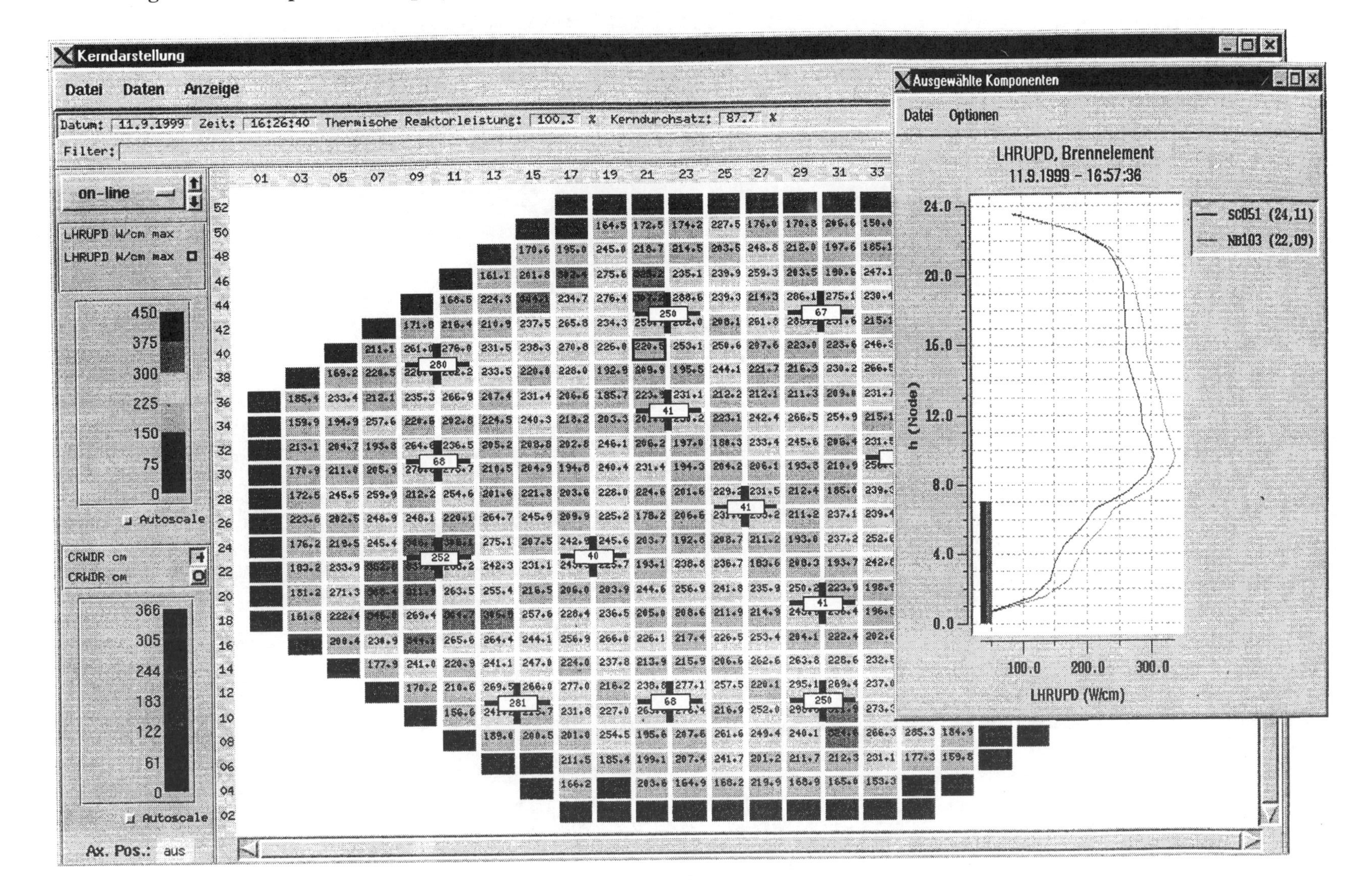

THE SIEMENS ADVANCED CORE MONITORING SYSTEM FNR-K IN KKI-1, KKP-1 AND KKK

H. Potstada, M. Beczkowiak
Siemens AG KWU Erlangen, Germany

M. Frank
Kernkraftwerk Isar 1, Germany

K. Linnenfelser
Kernkraftwerk Philippsburg 1, Germany

Abstract

On-line monitoring of the reactor core based on the FNR-K system provides detailed information on power distribution, significant physical reactor parameters and margins with respect to safety-related limits. Especially the fast operator support function with instantaneous graphical output of essential calculation results relevant for reactor safety has proven to be very useful. The comprehensive possibilities of archiving and reconstruction of reactor state points by the FNR-K system is also a feature of practical interest. Fields of continuous development are the introduction of expert system related techniques for power ramp prediction and integration of the state-of-the-art core simulator MICROBURN B2.

Introduction

Today operation of modern boiling water reactors (BWR) is supported by a process information system which furnishes the operating personnel with measured and computed data to provide an overview of the current plant condition which is as complete as possible. The on-line core monitoring system is an additional component which provides safety relevant information on local physical processes occurring inside the reactor core which is derived on the basis of three-dimensional simulator calculations.

The Siemens/KWU on-line core monitoring system FNR-K (German abbreviation for advanced nuclear calculations for core monitoring) has already been in use on various hardware platforms for a number of years at the Isar 1, Philippsburg 1 and Krümmel BWR plants in Germany.

General FNR-K software functions

As for nearly all core monitoring systems, for BWRs the main components are a data acquisition module, a module for triggering online core-follow calculations, a control module for starting predictive calculations, archiving and file maintenance modules and output modules for colour graphic displays and paper logs. The heart of the system is a core simulator program that delivers the calculation results needed for core monitoring such as thermal limits, linear heat generation rate and exposure update.

The simulator module presently used in the FNR-K is the 1.5 group 3-D core simulator RS3D, which has been service-tested over many years and combines the properties of neutronic and thermal-hydraulic three-dimensional core calculation with three-dimensional gamma and neutron flux measurement performed during operation. An essential advantage of this module is its fast calculation speed even on relatively old workstations (e.g. HP-735), where a full core calculation with significant change of state can be performed in less than 10 seconds.

The fast operator support module

The very fast calculation speed of the core simulator is needed in the fast operator support module. With instantaneous self-updating graphical output (see Figure 1) of essential calculation results relevant for reactor operation, this module offers a nearly real-time view on the quantities MCPR, MFLPD and the maximum excess of the PCI envelope in the core to the operating personnel. Additionally the actual operating point in the power/flow map and information about xenon/iodine concentration history are displayed.

The module is well accepted in all of the three BWRs with FNR-K core monitoring. In particular, KKP-1 took advantage of the prompt information about limiting core parameters by loosening the strict operational rules for increasing power after temporary power reduction for predefined testing purposes. Accelerated return to full power is made possible by utilising the reserves of local conditioning of the fuel and staying close to the limits during power increase. The detailed local treatment of conditioning around inserted control rods is very useful for rapid development of the rated power rod pattern. The possible reduction of delay intervals in comparison with conventional power up procedures results in an overall acceleration of reaching rated power.

Extensive archiving and reconstruction of state points

The FNR-K system has been furnished with extensive possibilities of archiving and semiautomatic reconstruction of past core state points. Up to 1 250 data sets consisting of the essential data needed for graphical representation of 2-D core maps and axial profiles are stored on disk, triggered by event. Thus it is possible to generate history plots of local physical parameters for any core position over a whole cycle. The complete on-line calculation databases of the core simulator are also stored event triggered and in reasonable time distances. With the help of the heat balance data and control rod positions stored with high time resolution, intermediate state points can be reconstructed by simulator calculations. The input data for the calculations are automatically combined after the user has chosen a starting point database and a time range to be reconstructed.

Developments in progress

To enhance the predictive capability of FNR-K, a module for time optimisation of reaching full power after temporary power reduction has been specified and will be implemented and tested as a prototype in the near future. The module has to determine a fast path to a user defined target power automatically by combining core simulator trial calculations of path sections with heuristic rules (Figure 2). The safety relevant core parameters such as MFLPD, MCPR, PCI exceeding tolerance and position in the power/flow map as result of the trial calculations, are taken into account by an inference module, to generate other suitable path sections iteratively. The inference module is furnished with rules of thumb and a decision logic derived from experience.

Another task currently worked on is the integration of the state-of-the-art core simulator MICROBURN B2 with advanced nodal expansion method, two-group neutronics and pin power reconstruction. Integration of this module would be a step towards more calculation accuracy for core states deviating from normal conditions, but it would not be for free. Considerable optimisation efforts must be taken to guarantee a calculation speed suitable for the fast operator support module and the use of faster workstations may be necessary.

Figure 1. Sample output of the fast operator support self-updating display

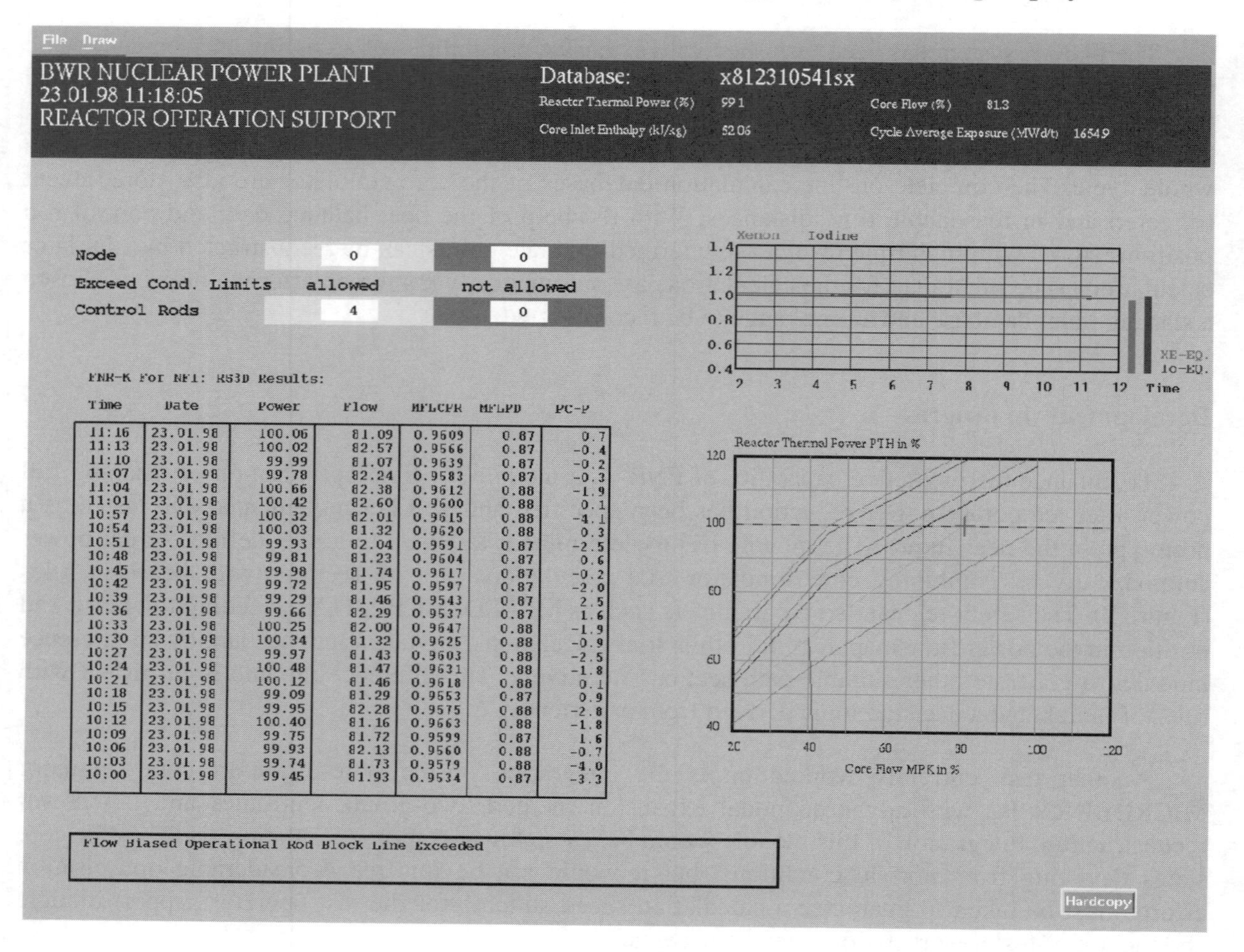

Time	Date	Power	Flow	MFLCPR	MFLPD	PC-P
11:16	23.01.98	100.06	81.09	0.9609	0.87	0.7
11:13	23.01.98	100.02	82.57	0.9566	0.87	-0.4
11:10	23.01.98	99.99	81.07	0.9639	0.87	-0.2
11:07	23.01.98	99.78	82.24	0.9583	0.87	-0.3
11:04	23.01.98	100.66	82.38	0.9612	0.88	-1.9
11:01	23.01.98	100.42	82.60	0.9600	0.88	0.1
10:57	23.01.98	100.32	82.01	0.9615	0.88	-4.1
10:54	23.01.98	100.03	81.32	0.9620	0.88	0.3
10:51	23.01.98	99.93	82.04	0.9591	0.87	-2.5
10:48	23.01.98	99.81	81.23	0.9604	0.87	0.6
10:45	23.01.98	99.98	81.74	0.9617	0.87	-0.2
10:42	23.01.98	99.72	81.95	0.9597	0.87	-2.0
10:39	23.01.98	99.29	81.46	0.9543	0.87	2.5
10:36	23.01.98	99.66	82.29	0.9537	0.87	1.6
10:33	23.01.98	100.25	82.00	0.9647	0.88	-1.9
10:30	23.01.98	100.34	81.32	0.9625	0.88	-0.9
10:27	23.01.98	99.97	81.43	0.9603	0.88	-2.5
10:24	23.01.98	100.48	81.47	0.9631	0.88	-1.8
10:21	23.01.98	100.12	81.46	0.9618	0.88	0.1
10:18	23.01.98	99.09	81.29	0.9553	0.87	0.9
10:15	23.01.98	99.95	82.28	0.9575	0.88	-2.8
10:12	23.01.98	100.40	81.16	0.9663	0.88	-1.8
10:09	23.01.98	99.75	81.72	0.9599	0.87	1.6
10:06	23.01.98	99.93	82.13	0.9560	0.88	-0.7
10:03	23.01.98	99.74	81.73	0.9579	0.88	-4.0
10:00	23.01.98	99.45	81.93	0.9534	0.87	-3.3

Figure 2. Block diagram of the FNR-K rule and calculation based optimisation module for rapid power increase

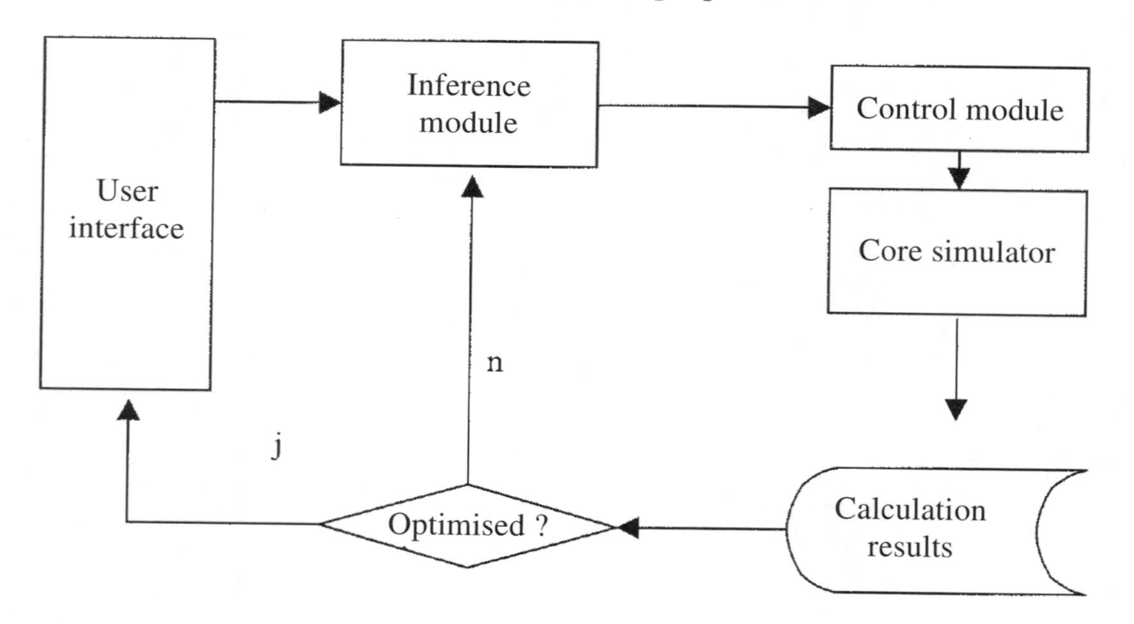

DESIGN AND VALIDATION OF THE NEW ABB CORE MONITORING SYSTEM

Jesper Eriksson
Forsmarks Kraftgrupp AB, SE-742 03 Östhammar, Sweden

Per Kelfve, Carl-Åke Jonsson and Stig Andersson
Nuclear Fuel Division
ABB Atom AB, SE-721 63 Västerås, Sweden

Abstract

New ABB core monitoring systems, based on the state-of-art ABB 3-D core simulator POLCA7, are now being introduced. POLCA7 has been installed in all three BWR units at Forsmark and began operation during 1999. A system design has also been developed for GE-type BWRs and a first delivery made to the Hope Creek BWR in 1999. The system operates in parallel with the current monitoring system, which will be phased out in 2000. The Hope Creek system uses the CORE MASTER 2 system architecture. CORE MASTER 2 is the new engineering environment for LWR core design, developed by ABB and Vattenfall in co-operation.

Background

Core monitoring systems based on the ABB CORE MASTER system design and using the 3-D core simulator POLCA have been supplied to all ABB designed BWRs. POLCA (Power On-Line CAlculation) has been used for monitoring all ABB BWR cores since the early 70s. The method has gradually been refined; for example, automatic tracking of xenon variations was introduced in the 80s. The CORE MASTER system design is also used in the Leibstadt (GE type BWR) and Brunsbüttel (Siemens type BWR) monitoring systems.

Recently, a large development and validation effort has been made to upgrade the system to the most recent version of the 3-D simulator, POLCA7 [1], which is capable of accurately monitoring high burn-up cores with modern advanced 10 × 10 fuels. POLCA7 simulates the 3-D static behaviour of the reactor core, both for BWR and PWR. For PWR core design, POLCA7 has been in use since the early 90s. Today the same code version is in use for both PWR and BWR. POLCA7 is approved by the Swiss HSK and is being reviewed by the US NRC, with BWR fuel and core design as a first goal. PWR core monitoring application, in conjunction with in-core detector upgrades, is another goal.

Integral validation of POLCA7 has been performed for a number of different plants and cycles of operation and reported in [2]. The validation database rapidly increases when POLCA7 replaces the previous tool for BWR core design.

A new core engineering system, CORE MASTER 2, has been jointly developed by ABB Atom and Vattenfall [3], and is built around POLCA7. CORE MASTER 2 is now operating at ABB and Vattenfall for all BWR core design work. The main features are the 3-D core simulator POLCA7, the integrated database and a task-oriented graphical user interface.

POLCA7

POLCA7 [1] simulates the 3-D steady-state behaviour of a reactor core, both for BWR and PWR. POLCA7 solves the coupled thermal-hydraulic and neutronic equations. The two-group diffusion equation is evaluated with an accuracy of 1% or better. Based on the converged neutron flux solution, the power of each pellet of the core may be computed. This detailed solution forms the basis for detector reading simulations and supervision of the local power densities in the core.

Validations of the neutronics models of POLCA7 have been performed for a number of different plants and cycles of operation. The results of the validation confirm improved calculation methods and are reported elsewhere, such as in [2].

One of the most important enhancements made in POLCA7 in relation to the previous simulator used in ABB's on-line core monitoring system(s) is the increase in the accuracy with which thermal margins can be estimated. The increased accuracy is achieved through improvements in models as well as the computation of detailed information that was previously unattainable. For example, the pin power reconstruction capability provides powers and exposures down to the pellet level for all fuel pins in the core. Consequently, thermal load parameters such as linear heat generation rate (LHGR) can be computed using so-called pin-stub powers and the margins to operating limits can be established based on these pin-stub values in combination with pin-stub burn-ups, nodal burn-ups or both. Likewise, monitoring of pellet clad interaction (PCI) phenomena is based on pin-stub LHGR for any number of fuel pins in all fuel assemblies in the core.

Another important improvement that resulted from the pin power reconstruction capability is that the prediction of detector responses can be done in a more realistic (physical) and accurate manner. LPRM responses are computed based on the neutron fluxes reconstructed in the detector location, while gamma-TIP responses are determined from the pin powers of all nodes around the detector. Previous models used only nodal average powers and highly approximate gamma-to-neutron conversion factors to predict the detector responses.

The accuracy of the pin power reconstruction method is closely connected to the accuracy of the nodal solution methodology. In POLCA7, a standard analytic nodal method with a quadratic transverse leakage approximation is used to solve the two-group diffusion equations. The homogenised nodal equivalent parameters include macroscopic and microscopic (isotopic) cross-sections, side and corner flux discontinuity factors, pin power form functions and detector constants for both neutron- and gamma-sensitive detectors. The explicit tracking of the most important actinides and a large number of fission products, and a model to account for intra-nodal depletion and power feedback effects, yields a cross-section model, which implicitly accounts for almost all depletion history effects. In addition, an axial homogenisation model facilitates the modelling of axially heterogeneous fuel designs and the computation of the detailed axial power profiles used in thermal load parameter and detector response predictions.

The BWR thermal-hydraulics (T-H) models in POLCA7 treat the lower plenum, each fuel assembly, a lumped inter-assembly bypass channel, the upper plenum, steam separators, steam dome, downcomer and main recirculating pumps. POLCA7 provides the capability to describe the hydraulic characteristics of the core region in detail. Separate loss coefficients can be provided for important assembly components such as inlet orifices, bottom nozzles, tie plates and spacers. Geometric axial variations in the fuel assemblies and specific leakage paths to bypasses can be described. The T-H module can handle reactor pressures from 1 to 200 bar and a temperature range from room temperature to hot full power.

CORE MASTER 2

CORE MASTER 2 (CM2) replaces the earlier generation of CORE MASTER, developed by ABB Atom and used by a several European utilities for core design and on-line BWR core surveillance. CORE MASTER 2 is primarily intended for core design work. However, the general system architecture and database make it suitable for core monitoring as well. CORE MASTER 2 is presently available for BWR and also for PWR in a prototype version.

CORE MASTER 2 has been jointly developed by ABB Atom and Vattenfall as a calculational tool for fuel engineering analysis. The main features of CM2 are the 3-D core simulator POLCA7, the integrated database and the graphical user interface; each of them is briefly described below.

The CM2 system is built around an integrated database with object oriented definition and structure. Major features of the database manager are data security, traceability and hence quality assurance of input and results. Several groups of data are defined. Engineering data language (EDL) data includes plant information, component information, cycle information, control parameters as well as core input and output data. Cell data includes all nuclear data generated by lattice calculations. Distributions, generated by the core simulator, are multi-dimensional data structures describing the state of the core components as a function of space and time.

Through the advanced graphical user interface (GUI) the CM2 end user may access all functions of CM2 as well as viewing stored and calculated parameters in graphical form, c.f. Figure 1.

Figure 1. Part of CM2 graphical user interface

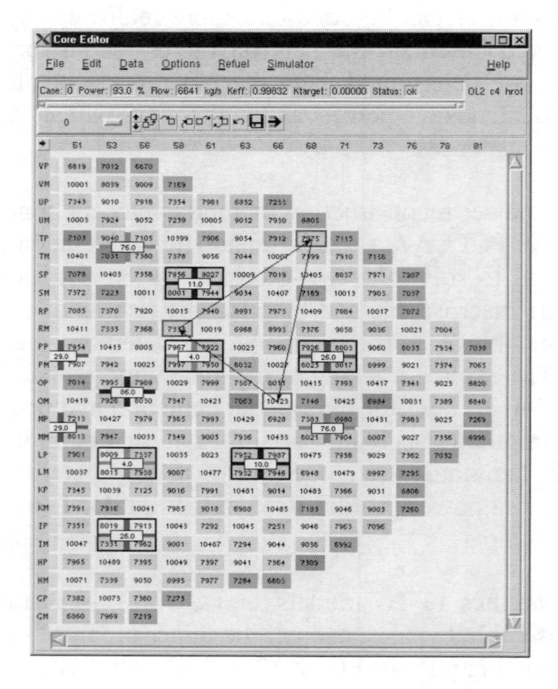

New user interface functions may be created using the command language Tcl/Tk. Several widgets have been defined for standard presentations, such as power distribution displays, fuel bundle data displays, etc. Graphical displays may be printed on hard copy devices, together with automatically generated reports. CM2 may be implemented on UNIX workstations (SUN, DEC, HP, etc.) or PCs.

Core monitoring system configuration

A typical computer configuration for the ABB core monitoring system (ABB CMS) is presented in Figure 2. Signal data from the process are collected and stored in one or more plant process computers (PPC). The plant reactor operator can access the data in the signal database and major CMS functions from one or more operator stations (OS), which might be the same as the PPC.

Figure 2. Typical ABB CMS configuration

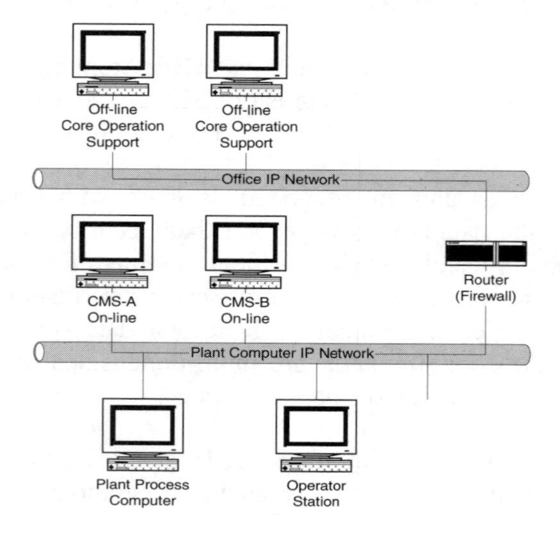

The core monitoring system itself is implemented on two redundant workstations (CMS-A and CMS-B in the figure). One of them is "active" at a given time, while the second is kept on "hot standby", with data synchronised with the active CMS computer, thus prepared to take over in case any error occurs or when requested.

Data communication between the computers of Figure 2 is based on Ethernet and TCP/IP. CMS-A and CMS-B are primarily dedicated for on-line core surveillance and are connected to the plant computer IP network. Additional (off-line) core operation support functions, primarily for the need of the reactor engineer, are implemented on separate CMS computers connected to the office IP network.

Restricted exchange of data between the off-line and on-line CMS systems may be performed through a "firewall" router, denying unauthorised access to the on-line computers.

The communication with the PPC is requested from the CMS computer at system start-up. Only one CMS computer (the "active" one) may establish communication with the PPC at a time. The reactor operator may, when the second CMS computer is on "hot standby", manually switch to this computer to become "active" instead.

In order to test a true, independent CMS system in parallel with the actual on-line CMS system, a new efficient method for parallel CMS operation was developed and put into operation. Process data for the core simulator are transferred to a mailbox area on the on-line CMS system at frequent intervals (once a minute) and are fetched from there by the parallel CMS system without disturbing the ongoing core monitoring. The method is appropriate while validating a new version of the core simulator or new features of the CMS itself.

ABB CMS basic modules and functions

The ABB core monitoring system is structured as a modular software package with distinct interfaces, as indicated in Figure 3. This makes it easy to adjust to different customer specific requirements and computer environments. The ABB CMS basically includes the modules of the right part of Figure 3. The thermal balance calculation module is, due to the high frequency data exchange with the signal database, usually implemented on the PPC, but might be implemented on the CMS computers instead.

The process computer data exchange module of the CMS utilises an ABB developed client/server protocol based on TCP/IP and using sockets. All data to be read from or written in the signal database are defined in a parameter list on the CMS computer side. This makes it easy to extend the data to be exchanged between the PPC and CMS by only adding or subtracting a signal data identifier in the parameter list.

The thermal balance calculation is performed at regular intervals, such as once or twice every minute, giving an updated global reactor power value. The CMS side of the data exchange module requests, also at regular intervals, the power, the total core flow and the control rod positions and some other process signal parameters. In case any of these parameters has changed above user defined limits an automatic evaluation with the 3-D core simulator will be triggered. The automatic triggering is further based on elapsed time, with time intervals that are different at stable operation and during major transients.

Through the CMS graphical user interface the user may interact with the system and look at the results. Several user groups are defined. The reactor operator needs only part of the full information

Figure 3. ABB CMS software package structure

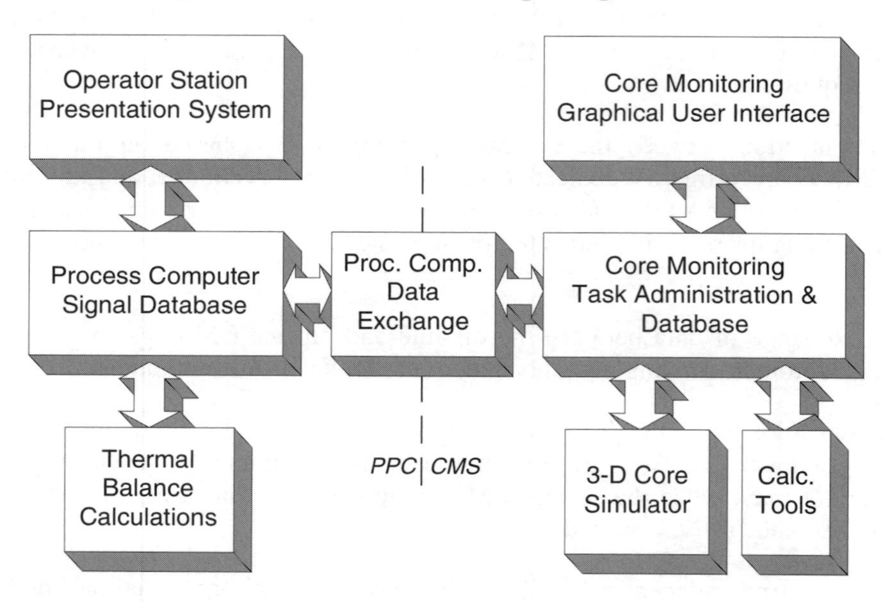

and functionality available. The reactor engineer needs more detailed information and extra tools and should be able to update the data bank at the beginning of the cycle and after detector calibration, etc. The system administrator needs full access in order to set parameters for start-up of the system, block and unblock the system, etc.

The reactor engineer may, after a TIP data collection campaign, initiate a calculation of parameters for LPRM detector calibration. The calculated parameters are either used as "software correction factors" in order to correct the raw detector signals read by the system or as multiplying factors for adjusting the detector amplifiers. The reactor engineer may also start an evaluation with the 3-D core simulator to compare measured against calculated TIP data.

The LPRM detector signals may be used for adjustment of the calculated 3-D power distribution due to the actual detector readings. Based on the modified power distribution new thermal margins are calculated and returned as result from the full 3-D core evaluation, this being the adaptive method "UPDAT". Non-adaptive methods combined with on-line evaluation of uncertainties for the calculated margins are also possible.

The graphical user interface gives the reactor operator information about the last 3-D core evaluation performed. The reactor operator may set the xenon mode to equilibrium, preventing automatic 3-D core evaluations to be triggered as well as reset to xenon in transient mode, allowing such triggering again. He/she may also manually start extra 3-D core evaluations or predictive calculations, look at the results and send reports to the printer. Some standard reports are printed automatically, e.g. if negative PCI or thermal margins have been detected.

The central module contains the database needed for the GUI and a number of programs administrating the tasks requested from the PPC communication or from the GUI. The task administrator starts calculations with the 3-D core simulator and other requested calculations and stores the result in the database, making them available to the GUI. Some resulting data, such as PCI and thermal margins, are returned to the PPC. After each 3-D calculation the data are synchronised from the "active" CMS computer to the other. If requested, specific data are also transferred to the off-line CMS system for off-line studies or evaluations.

All major events are registered as log messages written into a common log file, thus providing the system administrator with a tool for detecting any errors that might occur in the system. Major errors that severely prevent the basic functionality of the core monitoring system are sent to the PPC alarm system.

The system was designed with POLCA7 as the 3-D core simulator. The system, however, is built in such a way that another 3-D core simulator, possibly using a different data storage structure, can be integrated with the rest of the system, demanding only minor efforts and changes limited to the interface specific for exchanging data with the other core simulator.

Experience from Forsmark

Parallel operation with POLCA7 has been performed since 1998 for all three units at Forsmark and on the new CMS computers since June 1999. The new CMS system will begin operating for monitoring during September 1999 for all three plants at Forsmark. Before that, careful validation of the code has been made for all operating cycles in all three units.

POLCA7 models for Forsmark have been set up and validated with both ABB generated lattice data (using PHOENIX4 and the ENDF-B/VI library) and CASMO4, used by Vattenfall. Results for Forsmark 3 were included in [2] and are generally very satisfactory. Here we include some examples from Unit 2, and using CASMO4 lattice data, Figures 4-7. Similar accuracies have been observed in a variety of plant types, with different core and operating conditions, indicating the versatility and strength of the models and data.

The hot eigenvalue in Figure 4 is very stable, with a variation band of 200 pcm in the latter cycles, and almost no variation from cycle to cycle.

Figure 4. Forsmark 2 CASMO4/POLCA7 validation. k_{eff} at TIP measurements, Cycles 4-17

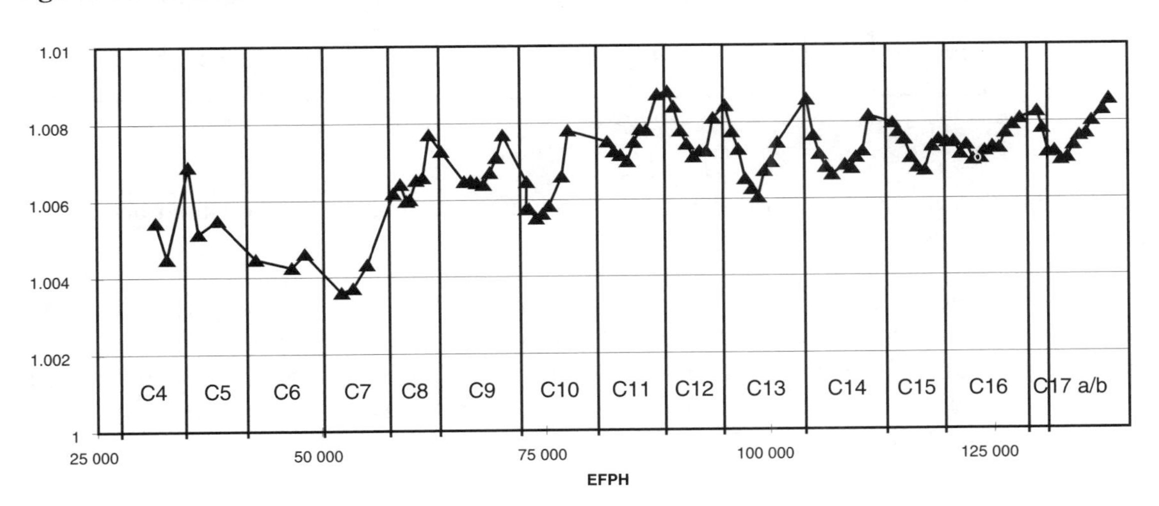

The deviation in predicted local power, as seen in the nodal TIP comparisons (Figure 5) is on the average 4% rms, with values as low as 3% during the later cycles. A larger deviation occurs for Cycle 14 and reflects a somewhat increased deviation in predicted axial power variation during the cycle. At the end of that cycle, however, the deviation is again very low. For most cycles the nodal error is smallest at the beginning and end of the cycle, and largest about mid-cycle, which may point to the difficulty in accurately modelling the gadolinium depletion.

Figure 5. Forsmark 2 CASMO4/POLCA7 validation. Nodal TIP deviation, Cycles 4-17.

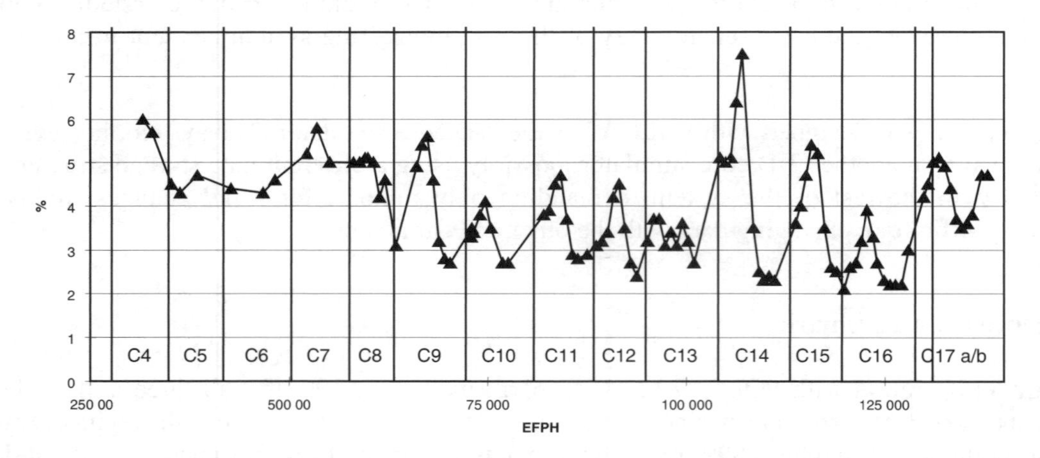

The radial TIP deviation, Figure 6, is also very stable and low, below 2%.

Figure 6. Forsmark 2 CASMO4/POLCA7 validation. Radial TIP deviation, Cycles 4-17.

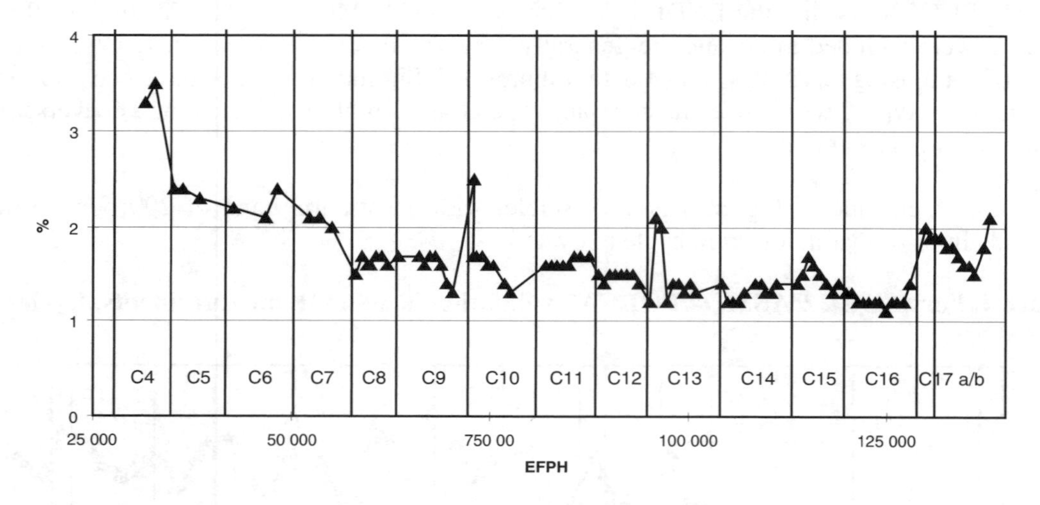

The validation for the Forsmark BWRs also includes a comparison of measured and calculated fuel channel inlet flow in eight core positions (see Figure 7).

This is a unique feature of ABB internal pump BWRs, and particularly interesting for qualification of the thermal-hydraulic model in the simulator. The core in Forsmark 2 experiences a transition from original 8×8, via SVEA-64 to 10×10 lattice SVEA-100 and SVEA-96S in the latter cycles. The calculated and measured inlet flow agrees within 0.2 kg/s or 1.2% rms in flow error for individual channels in all the latter cycles.

Experience from Hope Creek

The first ABB CMS which takes advantage of CORE MASTER 2, was developed and tested for installation at Hope Creek. In addition to the basic CMS functionality, this system includes engineering support functionality specific for GE type BWRs. ABB Atom delivers the CMS as part of a process computer replacement, for which ABB CENP has the main responsibility.

Figure 7. Forsmark 2 CASMO4/POLCA7 validation. Channel-flow rms error, Cycles 4-17.

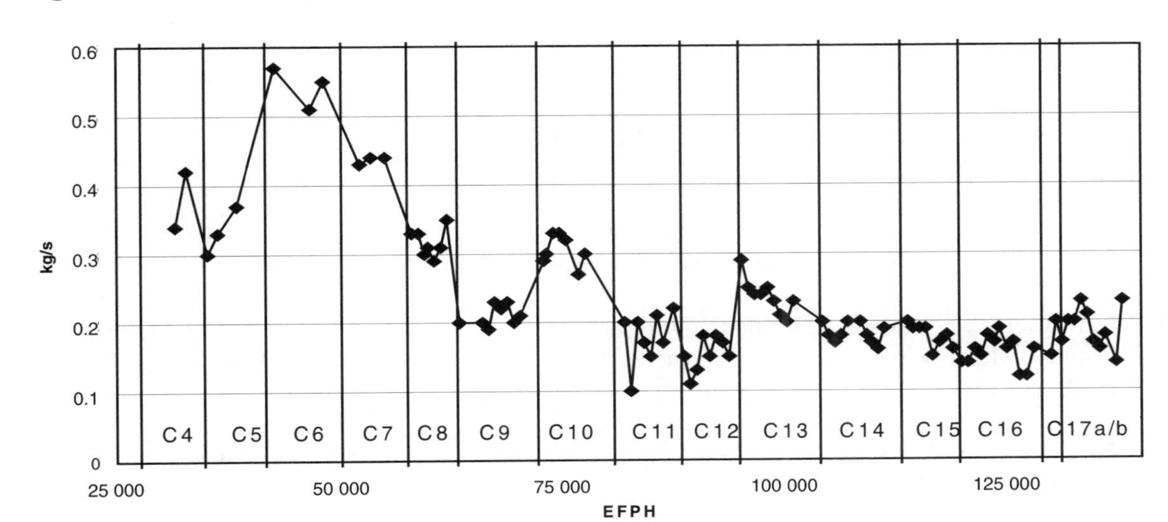

In conjunction with the Hope Creek system implementation, POLCA7 is being licensed at the US NRC for used in core design. The system will begin parallel operation in September 1999 and sharp core monitoring operation during the year 2000.

Conclusions

New ABB core monitoring systems will be going into operation, using state-of-the-art techniques and methods in all areas:

- Effective data exchange method with the plant process computers.
- User friendly graphical user interface.
- Database handlers that secures data.
- Comprehensive task and data administrator.
- Last, but not least, the advanced POLCA7 core simulator.

The system and simulator validation results provide assurance that the new system can replace existing core monitoring systems and comply with all operational and licensing requirements.

REFERENCES

[1] S-O. Lindahl, "POLCA7 – Overview", ABB Atom Report BR 95-1033, 1995.

[2] E. Müller, P. Forslund, Y. Parmar, "Validation of the ABB Atom Core Physics Methods", ANS Topical Meeting on Advances in Nuclear Fuel Management, Myrtle Beach, South Carolina, USA, 1997.

[3] C-Å. Jonsson, A. Ljungblad, "Core Master 2 – A New Reactor Physics Calculational System", PHYSOR 96, Mito, Japan, 1996.

INTRODUCTION OF VIRTUAL DETECTORS FOR CORE MONITORING SYSTEM OF KOREAN STANDARD NUCLEAR POWER PLANT*

Eun Ki Lee, Yong Hee Kim, Kune Ho Cha and Moon Ghu Park
Korea Electric Power Research Institute, Korea Electric Power Corporation
103-16, Munji-Dong, Yusung-Gu, Taejon, 305-380, Korea

Abstract

A novel algorithm known as the virtual detector method (VDM) is introduced to reconstruct the axial power shape (APS) for the on-line core monitoring system of the Korean Standard Nuclear Power Plant (KSNP). A pure statistical method (SM) is also introduced and the results are compared with the currently implemented five-mode Fourier fitting method (FFM). VDM adopts nine virtual detector informations coupled with a regression model based on the Alternating Conditional Expectation (ACE) algorithm. VDM uses Fourier fitting with the information of nine virtual detectors expanded from the currently implemented FFM, which uses five-level detector information. By introducing virtual detectors, we can increase the number of axial detectors, and thus expect the computational errors of APS to be reduced. The two methods (SM and VDM) are applied to in-core mapping data from six cycles of Yong Gwang nuclear power plant Units 3 and 4. For ~3 500 cases of APSs extracted from a cycle of operation which is simulated by a three-dimensional nodal code, the accuracy of the three methods (SM, VDM, FFM) is compared. The average root mean square (RMS) error and average of axial peaking error of SM and VDM resulted in reduction of more than 50% and 70%, respectively, relative to FFM. VDM and SM also show more realistic axial profiles and predict more accurate axial peaking than FFM. These improvements can contribute to a larger thermal margin. SM shows the most accurate results for all cases. VDM can almost obtain the same results as SM, and using far fewer computation steps. VDM can be a useful tool for precisely reconstructing axial power shapes in a core monitoring system.

* This paper was not presented orally at the workshop.

Introduction

The digital on-line core monitoring system of Korean Standard Nuclear Plant (KSNP), the Core Operating Limit Supervisory System (COLSS) [1] provides some key parameters for operators to evaluate the core operating condition. These key parameters include departure from nucleate boiling ratio (DNBR), peak linear heat rate (PLHR), azimuthal tilt, axial shape index (ASI) and licensing power, all of which are based on the various measured plant data. The importance of DNBR and PHLR are particularly emphasised because of their strong relationship with thermal margin. In KSNP, a total of 45 detector assemblies are loaded into the core, and each detector assembly has five-level fixed rhodium detectors. COLSS actually calculates the core average axial power shape (APS) using the normalised five-level detector information collapsed from 225 in-core fixed detectors and the five-mode Fourier fitting method (5-FFM) [1] consisting of a trigonometric function. Although this deterministic method has definite applications, the accuracy of 5-FFM tends to decrease for the power shapes of saddle or highly shifted to top/bottom of the core. This is due to limited information of the measurement. In addition, 5-FFM can show unrealistic axial profiles when estimating the saddle type APSs. Generally, the accuracy of the deterministic method (5-FFM or spline function synthesis) closely depends on the amount of detector information. Recently, we developed two methods to overcome the barrier of limited detector segments. These are a pure statistical method (SM) [2] and a mixed statistical method coupled with a virtual detector concept (VDM). In SM, the 20-node (same with nodal design calculation) APSs are reconstructed using the optimal correlation between each axial node power and five-level detector powers. The ACE algorithm [3] is used to get the discrete optimal transfer functions between each of 20 plane powers and five-level detector signals. On the other hand, VDM uses nine virtual detector informations to reconstruct axial power shapes by using the nine-mode Fourier fitting method (9-FFM). Each virtual detector signal is extracted from the corresponding optimal correlation calculated by the ACE algorithm.

The ACE algorithm is a generalised regression method that yields an optimal relationship between a dependent variable, y, and independent variables, $\{x_i,\ i = 1, \ddot{y}, p\}$. It has some advantages over traditional non-linear regression techniques. The convergence of the optimal transformations is guaranteed and an initial trial or basis function is not required. In addition, iterative modifications of the relationships are not required.

To show the usefulness of the VDM, a total of 21 000 cases of APSs are reconstructed from six cycles of Yong Gwang Nuclear (YGN) Units 3 and 4. These axial profiles are prepared for an overall uncertainty analysis (OUA) from COLSS, which uses the Reactor Operating and Control Simulator (ROCS) code [4]. The reconstructed APSs of the two methods are compared with the reference shapes of ROCS and current five-level FFM.

In the following sections, the ACE algorithm and data sets used in regression are described in detail. Some numerical results and conclusion will be also discussed.

ACE algorithm and VDM

Normally, one can reconstruct the continuous spatial variable by synthesising the discrete measurements from spatially distributed detectors by Fourier, spline and/or non-linear fitting method like neural networks. However, it is not easy to obtain satisfactory accuracy with a limited number of detectors in spatial domain. The objective of this paper is to provide a constructive method of applying virtual detectors. The key factor of VDM is to justify the signal from the virtual detectors. In this work, the signal transferred to the virtual detectors are extracted from the neighbouring real detectors via the correlation specifying the accurate and robust relationships between them constructed by the

ACE algorithm. Then the desired continuous form of the measurement can be synthesised via the fitting methods with enhanced accuracy and an increased number of detectors.

A problem to reconstruct the spatial variation of APS (as depicted in Figure 1) is considered below.

Figure 1. One-dimensional spatial detector system with five real detectors and four virtual detectors

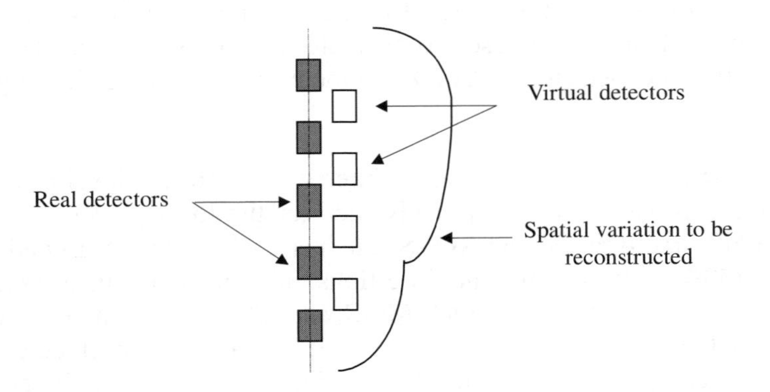

The ACE algorithm is applied to ascertain the optimal correlation between the virtual detector signal (dependent variable), P_v, and the real detector signal (independent variables), $\{D_i,\ i = 1,\_,5\}$. For the systems which have difficulties in extracting the optimal relationship between output y and input variables x_i by conventional statistical methods, the transformation techniques can be generally applicable. For multivariate regression problems with a set of data $\{P_{vi}, D_{1i}, D_{2i}, \_, D_{5i},\ i = 1, \_, N\}$, the optimal transformations of multivariate problems are readily derived by the ACE algorithm as follows:

$$\phi_n(D_n) = S\left[\theta(P_v) - \sum_{d \neq n}^{5} \phi_d(D_d) \middle| D_{ni}\right] (n = 1,\ldots,5) \quad (1)$$

and:

$$\theta(P_v) = \frac{S\left[\sum_{d=1}^{5} \phi_d(D_d) \middle| p_{vi}\right]}{\left\| S\left[\sum_{d=1}^{5} \phi_d^2(D_d) \middle| p_{vi}\right] \right\|} = \frac{S\left[\sum_{d=1}^{5} \phi_d(D_d) \middle| p_{vi}\right]}{S\left[\sum_{d=1}^{5} \phi_d^2(D_d) \middle| p_{vi}\right]} \quad (2)$$

where $S[T|q_i]$ means a conditional expectation at q_i and is determined by evaluating a weighted expectation about T around the neighbouring values in the interval $[i - M, i + M]$ for a given user defined value, M. All transformations should be mean zero functions. The transformations of each variable are coupled with each other and solved by an alternative-iteration procedure so as to minimise the square error of regression:

$$e^2 = \frac{1}{N} \sum_j \left[\theta(P_v) - \sum_{d=1}^{5} \phi_d(D_d)\right]^2 \quad (3)$$

There are several weighting or smooth techniques such as histogram, nearest neighbour, kernel, regression and super-smoother. In this paper, the kernel method is used with a weighting function defined on the real line with a maximum of $z = 0$.

Basically the ACE algorithm produces the discrete optimal transformations for given discrete data sets. Therefore, after the optimal transformations are calculated, we construct a simple regression model based on the least squares method to obtain the continuous polynomial functions describing each transformation. To consider the extrapolation of data, each transformation is divided into three regions. The first and third region of each transformation is approximated by a linear function, and an up to 9th order polynomial function is used to estimate the data sets in the second region. The Data Analyser for Virtual In-Core Detector (DAVID) computer code has been developed to automate all of these procedures.

In YGN Units 3 and 4, the 45 detector assemblies with fixed five-level in-core rhodium detectors are distributed in the core so as to appropriately monitor the local powers. To perform the COLSS OUA, ~3 600 cases of three-dimensional ROCS runs per each cycle are needed to reflect the various core power levels (50% ~ 100%) and the insertions of control element assembly (0 ~ 20 steps, 1 step = 19.5 cm) for burn-up steps of BOC, MOC and EOC giving a data set consisting of core averaged 20-node axial powers and five-level detector information. At each detector level, the detector power information is computed as a sum of 45 radial detector assembly powers, not the whole 177 assemblies and all five informations are normalised to 100. We used these 3 600 data sets per cycle to obtain optimal relationships between the five-level detector powers and each 20-node axial powers or each virtual detector powers. For a reference data set, ~21 000 data sets of six cycles are collected from Cycles 1 to 4 of YGN Unit 3 and Cycles 3 and 4 of YGN Unit 4. To verify the accuracy of the VDM, we also prepared some data sets not included in the calculation of optimal transformation, which are ~250 cases of APSs retrieved from the xenon oscillation simulation and ~20 cases of CECOR results of YGN Unit 3 Cycle 2. CECOR code is a three-dimensional off-line core mapping tool of ABB-CE type reactors and can reconstruct more accurate APSs than COLSS.

Figure 2(a) shows an example of relationships between the eighth virtual detector power and each normalised five-level detector powers of YGN Unit 3 Cycle 2. Figure 2 shows that it is very difficult to find the initial trial functions due to the diffuse and complex nature of the relationship between real and virtual detectors. Figure 2(b) represents the results of regression by the ACE algorithm for the data sets of the eighth virtual detector. It can be seen that the developed algorithm can give smooth discrete optimal transformations.

Figure 2(a). Diagram of relationship between 8th virtual detector power and 3rd (or 4th) real detector power

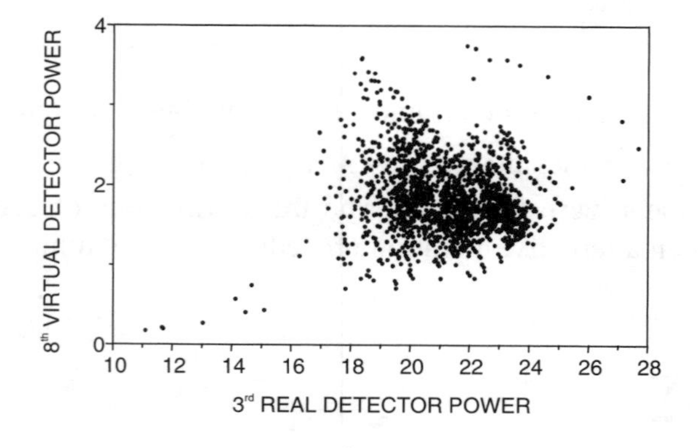

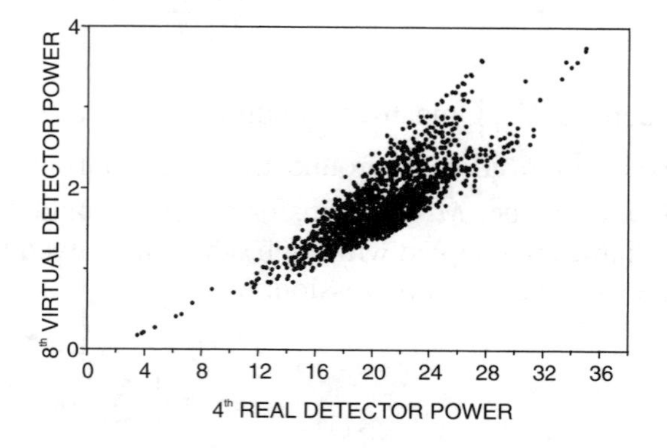

Figure 2(b). Decimal transformations for 3rd and 4th real detector power (only used for 8th virtual detector power)

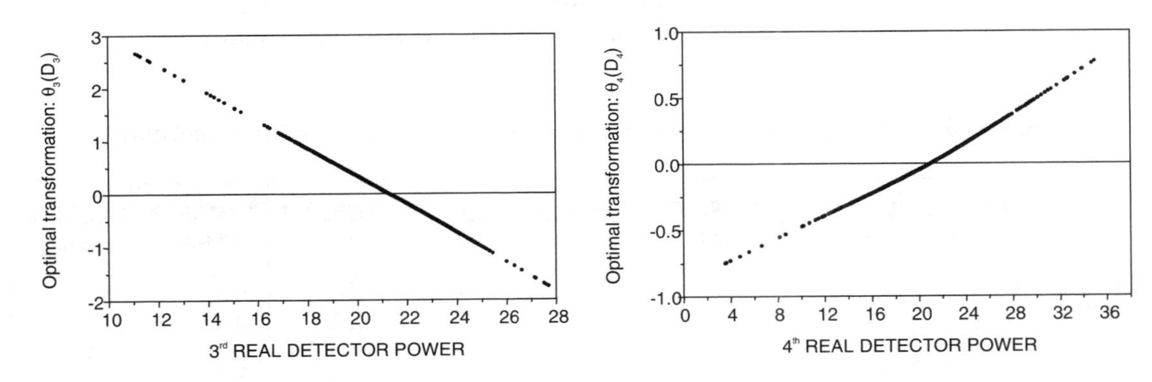

Synthesis of the measurements

The detailed methodology of the statistical method (SM) is described in Ref. [2]. SM has a total of six continuous polynomial functions per axial plane and can axially calculate 20-node powers directly using the optimal correlation functions. The performance of the SM is remarkable, but it requires too many polynomial coefficients.

For VDM, as the number of detector segments increases from five to nine, we can expect to obtain more accurate APSs without excessive increases in transformation parameters. The key feature of VDM is how to get the optimal correlation between real five-level detector powers and each of four virtual detector informations. In VDM with ACE transformation, a total of 54 transformations are needed, while 120 transformations are required for the SM. Once the nine-level detector information is determined by optimal correlation, APS can be computed using the same procedure as the current 5-FFM with an increased number of modes. Thus we can say that VDM is a "semi-statistical method".

For the Fourier fitting method, the following equation is used to generate axial powers:

$$P(z) = \sum_{n=1}^{ND} \left(a_n \cos n\pi B_c (z - 0.5) + b_n \sin n\pi B_c (z - 0.5) \right) \quad \begin{cases} ND = 5 \text{ for current } 5-\text{FFM} \\ ND = 9 \text{ for VDM} \end{cases} \tag{4}$$

where n = number of mode (n = 1,…,ND), a_n and b_n = Fourier coefficients to be determined with ND detector powers, and B_c = fitting parameters minimising the average root-mean-square (RMS) error for 20 node powers and axial peak power error, ΔFz, compared with ROCS power shapes. B_c has an important role in the accuracy of APS. For 5-FFM, due to the burn-up dependency of data sets, three boundary conditions are required to implement COLSS. But for VDM, just one boundary condition is needed because all of data sets for each cycle are used in the regression process.

The Fourier coefficients are determined by the following equation:

$$PD_m = \int_{z_l^m}^{z_h^m} P(z)dz, \ (m = 1, \ldots, ND) \tag{5}$$

where PD_m, Z_l and Z_h mean the detector power, relative lower and upper height at the m-th detector level, respectively.

Numerical results

Table 1 is the summary of the results comparing three methods (SM, VDM, 5-FFM) for all six cycles and two verification cases where the reconstructed 20-node APSs are compared with the ROCS results.

Table 1. The comparison of the APS reconstruction errors for three methods

Cases (No. of data sets)	YGN3C1 (3472)	YGN3C2 (3472)	CASE I (129)	CASE II (129)	YGN3C3 (3468)	YGN3C4 (3468)	YGN4C3 (3468)	YGN4C4 (3468)
Avg. Abs. RMS (%)	0.71[a] 0.94[b] 2.35[c]	0.56 0.94 2.84	0.79 1.11 2.10	0.52 0.94 2.32	0.49 0.72 2.44	0.54 0.82 2.50	0.54 0.68 2.35	0.48 0.78 2.40
Max. Abs. RMS (%)	3.68 3.63 8.97	3.92 3.97 10.00	1.30 1.45 3.18	1.20 1.36 3.95	2.05 3.68 10.14	2.77 3.91 10.84	1.87 3.35 10.11	2.91 3.66 10.29
Avg. Rel. ΔFz (%)	0.53 0.69 2.54	0.40 0.76 3.01	0.44 0.51 1.82	0.25 0.51 2.47	0.42 0.64 2.65	0.41 0.79 2.66	0.34 0.69 2.52	0.35 0.70 2.38
Max. Rel. ΔFz (%)	2.65 2.94 6.16	2.94 3.23 6.37	0.97 0.87 3.47	0.47 0.86 4.52	2.12 2.85 5.84	2.56 2.97 6.58	2.47 2.68 5.67	2.84 2.90 6.10
Avg. Rel. ΔASI	0.00150 0.00160 0.00365	0.00168 0.00179 0.00300	0.00213 0.00213 0.00375	0.00194 0.00201 0.00335	0.00137 0.00148 0.00246	0.00167 0.00172 0.00293	0.00133 0.00147 0.00234	0.00148 0.00155 0.00271
Max. Rel. ΔASI	0.00785 0.00802 0.01567	0.00913 0.01031 0.01463	0.01035 0.01026 0.01018	0.00939 0.00946 0.01025	0.00824 0.00842 0.01057	0.00874 0.00926 0.01114	0.00772 0.00801 0.01053	0.00761 0.00782 0.01055

[a] SM Abs. = absolute
[b] VDM Rel. = Relative
[c] 5-FFM

The average RMS error and the average ΔFz error of SM and VDM are decreased to ~1/3 and ~1/5 of those of FFM, and the maximum RMS error and the maximum ΔFz error is also reduced more than 50%. The results for Case I and Case II should be focused on which are generated from the xenon oscillation simulation at 50% and 80% power level at the BOC of YGN 3 Cycle 2, respectively. These data are not included in the regression of optimal transformation for YGN 3 Cycle 2. Therefore, Case I and Case II are important examples to demonstrate the validity and robustness of the developed optimal correlation. All of the SM parameters are very accurate, but VDM also has very good results. Table 1 explicitly shows that SM and VDM can calculate much more accurate axial power profiles than the current 5-FFM does.

Due to its strong relationship with PHLR and thermal margin, axial peak power is also an important parameter. To emphasise the advantage of VDM, the relative axial peak power differences between VDM and 5-FFM are plotted in Figure 3 for YGN Unit 4 Cycle 3, where the peak power difference of i-th data set, $\Delta F_{59,i}$, is defined by:

$$\Delta F_{59,i} = \frac{F_{Z,i}^{5-FFM} - F_{Z,i}^{VDM}}{F_{Z,i}^{5-FFM}} \tag{6}$$

The positive value of ΔF_{59} means that 5-FFM overestimates the peak power more than does VDM. Figure 3 shows that the positive values are dominant over 75% of the whole domain.

Figure 3. Distribution of F_z Difference (ΔF_{59}) between 5-FFM and VDM for the Case of YGN Unit 3 Cycle 3

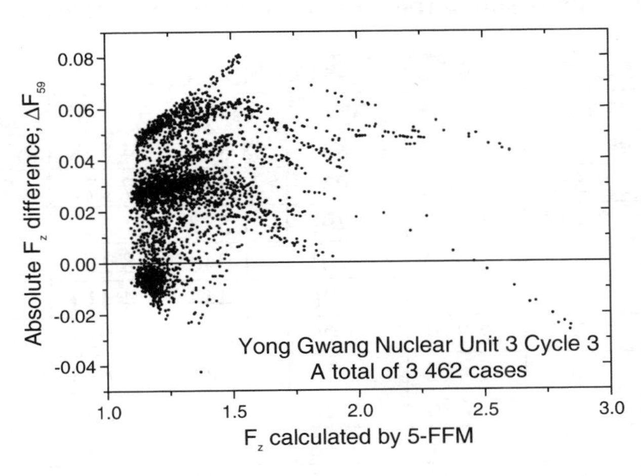

Figure 4 shows the average absolute RMS error for 3 462 data sets of YGN Unit 3 Cycle 3. The error profile of current 5-FFM represents its burn-up dependency. The average error of 5-FFM is ~0.03, while that of SM and VDM is ~0.01. In the case of SM, the average absolute RMS errors are nearly same values over the entire cycle. Although VDM shows a little dependency on the burn-up rate, it is small enough to be neglected compared with the 5-FFM.

Figure 4. Distribution of absolute RMS error for each data sets for the case of YGN Unit 3 Cycle 3

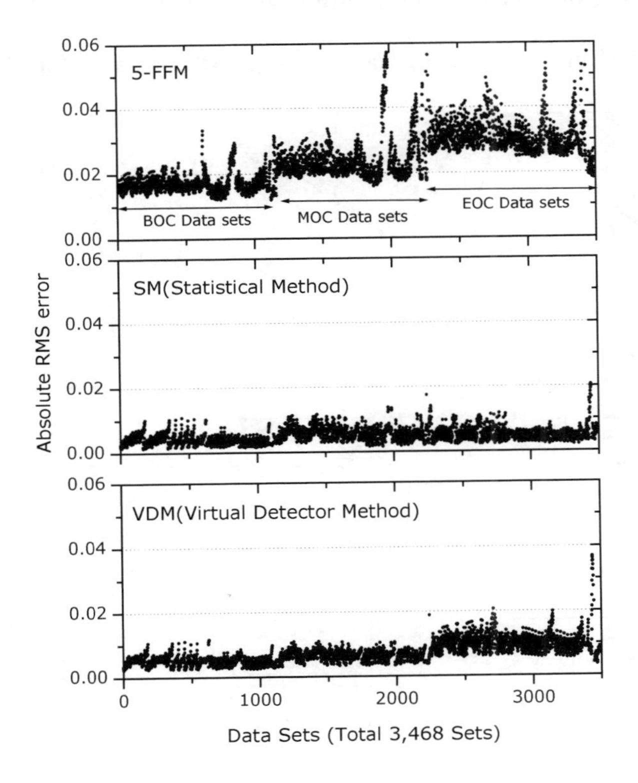

Node-wise relative error distributions for two cycles are in Table 2 where we can ensure the accuracy of the developed methods. SM and VDM give the calculated error less than 1% for ~60 000 nodes, while the 5-FFM gives ~40 000 nodes with 2% error. The number of nodes over 10% error is remarkably reduced with SM and VDM.

Table 2. The comparison of the number of nodes within given error bounds

Error (ε)	YGN Unit 3 Cycle 2			YGN Unit 3 Cycle 3			YGN Unit 4 Cycle 4		
	SM	VDM	5-FFM	SM	VDM	5-FFM	SM	VDM	5-FFM
< 1%	60 732	54 465	21 371	63 841	58 888	24 838	63 577	57 557	24 247
$1\% < \varepsilon \leq 2\%$	7 072	9 552	17 281	4 933	7 739	17 673	5 343	8 913	18 350
$2\% < \varepsilon \leq 3\%$	1 319	2 438	11 576	799	1 527	11 160	676	1 721	11 769
$3\% < \varepsilon \leq 4\%$	349	1 016	8 400	101	1 048	8 167	83	693	7 942
$4\% < \varepsilon \leq 5\%$	121	826	5 232	26	444	4 414	19	598	4 057
$5\% < \varepsilon \leq 6\%$	51	923	2 779	20	62	1 933	14	219	1 896
$6\% < \varepsilon \leq 7\%$	26	404	1 417	14	21	823	7	20	823
$7\% < \varepsilon \leq 8\%$	24	83	686	16	11	331	15	7	296
$8\% < \varepsilon \leq 9\%$	19	25	371	8	9	141	2	4	154
> 9%	127	108	727	22	31	300	44	48	246
Total no. of nodes	69 840	69 840	69 840	69 780	69 780	69 780	69 780	69 780	69 780

Another inevitable phenomena of 5-FFM is that its axial profiles are sometimes unphysical especially when the core axial power shape has mild saddle or flat type. These phenomena should occur due to limited axial detector information. In Figure 5, we can see that VDM prevents these phenomena effectively, and each APS of VDM shows smoother and more realistic behaviour compared with the APSs of off-line core monitoring systems [5].

Figure 5. Comparison of axial power profiles of VDM and CECOR

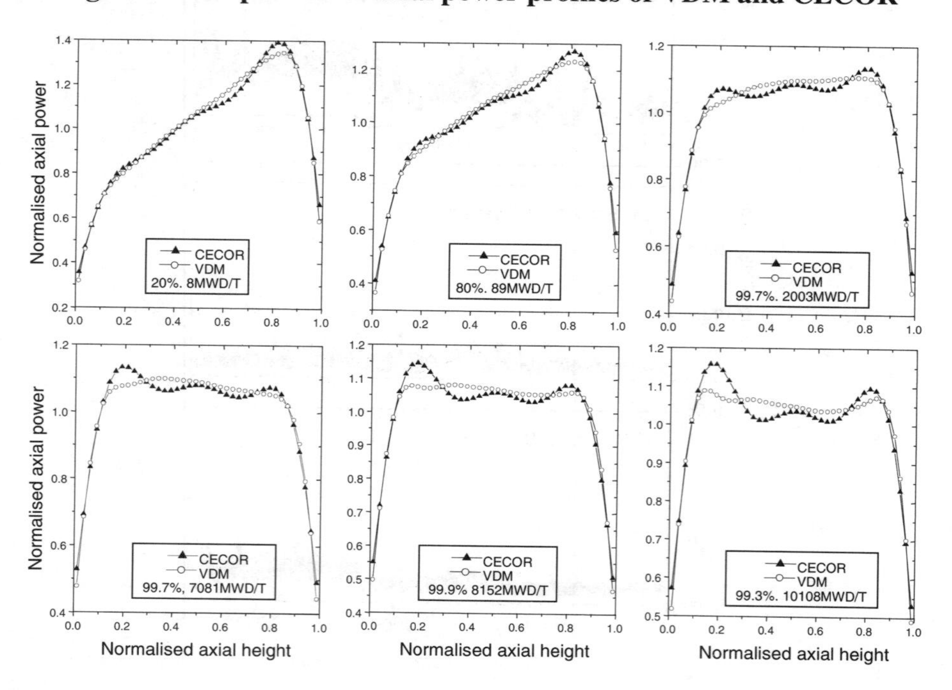

Conclusions

The on-line core monitoring system of KSNP supervises many parameters representing core operating conditions, including peak linear heat rate (PLHR) and departure from nucleate boiling ratio (DNBR). Because PHLR and DNBR are strongly related to thermal margin and depend on the core axial power shape, power profiles should be calculated as accurately as possible.

In this paper, the virtual detector method (VDM) is introduced to enhance the axial power shape reconstruction accuracy by giving the effect of increasing the number of axial detectors. The accuracy and robustness of the method is demonstrated by comparing statistical method (SM) and the currently implemented five-mode FFM for ~21 000 data sets retrieved from six cycles of Yong Gwang Nuclear Units. The VDM with nine-mode FFM corresponding to nine detectors and a total of 54 optimal continuous functions achieved nearly same results with SM. To obtain optimal transformations for SM and VDM, the alternating conditional expectation (ACE) algorithm is used.

The numerical results show that SM and VDM can give more precise and realistic axial power distributions and much better statistical performance than the current five-mode FFM. Especially, VDM can be a viable approach due to its relatively small memory requirement for fitting parameters than SM, and it has a similar implementation scheme to the existing method.

REFERENCES

[1] J.P. Pasquenza, *et al.*, "COLSS: Assessment of the Accuracy of PWR Operating Limit Supervisory System," CENPD-169-P, Combustion Engineering, Inc. (1975).

[2] E.K. Lee, *et al.*, "Reconstruction of Core Axial Power Shape Using the Alternating Conditional Expectation Algorithm," *Ann. of Nuclear Energy*, 26, pp. 983-1002 (1999).

[3] L. Breiman and J.H. Friedman, "Estimating Optimal Transformations for Multiple Regression and Correlation," *Journal of the American Statistical Association, Theory and Method*, 80:391, 580-619 (1985).

[4] R. Loretz, *et al.*, "ROCS User's Manual – Coarse Mesh Diffusion Theory Neutronics Code," *CE-CES-4 Rev. 4-P*, December (1989).

[5] J.L. Biffer, *et al.*, "CECOR 2.0: General Description, Methods and Algorithms," CENPSD-103-P, Rev.1-P, August (1984).

LIST OF PARTICIPANTS

BELGIUM

FRAIKIN, Roger
Tractebel Energy Engineering
Avenue Ariane 7
B-1200 BRUSSELS

Tel: +32 (0) 2 773 82 06
Fax: +32 (0) 2 773 89 00
Eml: roger.fraikin@tractebel.be

SMETS, Werner
Tractebel Energy Engineering
Avenue Ariane 7
B-1200 BRUSSELS

Tel: +32 (0) 2 773 7468
Fax: +32 (0) 2 773 8900
Eml: werner.smets@tractebel.be

CZECH REPUBLIC

BELAC, Josef
Manager
Dept. of Theoretical Reactor Physics
Nuclear Research Institute plc
250 68 REZ U PRAHY

Tel: +420 (334) 78 3535
Fax: +420 (2) 2094 0156
Eml: belac@nri.cz

CIZEK, Jiri
Chemcomex Prague plc
Prazska 16
102 21 PRAGUE

Tel: +420 (0)2 81017329
Fax: +420 (0)2 71750456
Eml: jciz@cce.cz

KMENT, Jaroslav
NPP Dukovany
CZ-675 50 DUKOVANY

Tel: +420 618 814298
Fax: +420 618 866360
Eml: kmentj1.edu@mail.eez.cz

PECKA, Marek
Chemcomex Prague plc
Prazska 16
102 21 PRAGUE

Tel: +420 2 8101 7287
Fax: +420 2 7175 0456
Eml: mpec@cce.cz

SEDLAK, Anton
Chemcomex Prague plc
Prazska 16
102 21 PRAGUE

Tel: +420 2 810 17310
Fax: +420 2 717 50456
Eml: ased@cce.cz

FINLAND

ANTILA, Martti
Design Manager
Fortum Engineering Ltd.
P.O. Box 10
FL-00048 FORTUM
Tel: +358 10 45 3 2477
Fax: +358 10 45 3 2477
Eml: martti.antila@fortum.com

LAVI, Petri
Head of Reactor Technics
Teollisuuden Voima OY
FL-27160 OLKILUOTO
Tel: +358 2 8381 5410
Fax: +358 2 8381 5509
Eml: petri.lavi@tvo.tvo.elisa.fi

SOLALA, Mikael
Head of Reactor Physics
Teollisuuden Voima OY
FL-27160 OLKILUOTO
Tel: +358 2 838 15420
Fax: +358 2 838 15509
Eml: mikael.solala@tvo.tvo.elisa.fi

FRANCE

JANVIER, Daniel
EDF/SEPTEN
Département Théorie – Division PN
12-14 avenue Dutriévoz
F-69628 VILLEURBANNE Cedex
Tel: +33 4 72 82 73 53
Fax: +33 4 72 82 77 10
Eml: Daniel.Janvier@edf.fr

MOURLEVAT, Jean-Lucien
FRAMATOME
Tour FRAMATOME
F-92084 PARIS LA DEFENSE Cedex
Tel: +33 1 4796 3134
Fax: +33 1 4796 5048
Eml: jlmourlevat@framatome.fr

VERDIEL, Francois
SEPTEN
12-14 Avenue Dutrievoz
F-69628 VILLEURBANNE Cedex
Tel: +33 472 82 7561
Fax: +33 472 82 7705
Eml: francois.verdiel@edf.fr

GERMANY

BECZKOWIAK, Michael
SIEMENS AG
Power Generation (KWU) Dep. NBTC
Postfach 3220
Bunsenstr. 43
D-91050 ERLANGEN
Tel: +49 (9131) 18 7369/7117
Fax: +49 9131 18 4045
Eml: Michael.Beczkowiak@erl19.siemens.de

ENDRIZZI, Ivo
Senior Scientist
SIEMENS AG
Power Generation (KWU) Dep. NBTI
Postfach 3220
Bunsenstr. 43
D-91050 ERLANGEN
Tel: +49 (9131) 18 3081
Fax: +49 (9131) 18 5243
Eml: ivo.endrizzi@erl19.siemens.de

* HEIDEMANN, Peter
University of Hannover, IKPH
Elbestr. 38A
D-30419 HANNOVER

Tel: +49 511 762 9341
Fax: +49 511 762 9353
Eml: Heidemann@mbox.ikph.uni-hannover.de

LUNDBERG, Sten
Stoller Energietechnik GmbH
SEG
Humboldtsrasse 12
D-90542 ECKENTAL-FORTH

Tel: +49(0)9126 286107
Fax: +49(0)9126 286108
Eml: stenl@compuserve.com

POHLUS, Joachim
Institut fur Sicherheitstechnologie
(ISTec) GmbH
Abteilung Diagnose
Forschungsgelaende
D-85748 GARCHING, PF 1313

Tel: +49 (89) 3200 4542
Fax: +49 (89) 3200 4300
Eml: poh@istecmuc.grs.de

POTSTADA, Henning
SIEMENS AG
Power Generation (KWU) Dep. NBTC
Postfach 3220
Bunsenstr. 43
D-91050 ERLANGEN

Tel: +49 (9131) 18 7544
Fax: +49 (9131) 18 4045
Eml: Hans-Henning.Potstada@erl19.siemens.de

* RUNKEL, Joachim
Nuclear Engineering and NDT Institute
(IKPH)
University of Hannover
Elbestrasse 38a
D-30419 HANNOVER

Tel: +49 (511) 762 9356
Fax: +49 (511) 762 9353
Eml: runkel@mbox.ikph.uni-hannover.de

WENISCH, Juergen
Hamburgische Electricitaets-Werke AG
Dep. TUK
Ueberseering 12
D-22286 HAMBURG

Tel: +49 (0) 40 6396 3941
Fax: +49 (0) 40 6396 3661
Eml: jwenisch@hew.de

HUNGARY

ADORJAN, Ferenc
Senior Researcher
Hungarian Academy of Sciences
KFKI Atomic Energy Research Institute
P.O. Box 49
1525 BUDAPEST

Tel: +36 1 395 9116
Fax: +36 1 395 9293
Eml: adorjan@sunserv.kfki.hu

JAPAN

FUKAO, Akihiro
TODEN Software, Inc.
In-Core Fuel Management System
6-19-15 Sinbashi
Minato-ku
TOKYO, 105-0004

Tel: +81 (3) 3596 7680
Fax: +81 (3) 3596 7670
Eml: fukao@tsi.co.jp

INAGAKI, Köki
Engineer
Mitsubishi Heavy Industries Ltd.
1-1 Wadamisaki-Cho
1-Chome
HYOGO-KU

Tel: +81 (78) 672 5084
Fax: +81 (78) 672 3277
Eml: koki_inagaki@kind.kobe.mhi.co.jp

KAKUTA, Tsunemi
Department of Nuclear Energy System
JAERI
Shirakata Shirane 2-4
Tokai
Naka-gun, Ibaraki-ken 319-11

Tel: +81 29 282 6078
Fax: +81 29 282 6122
Eml: kakuta@stsp2a0.tokai.jaeri.go.jp

KOSAKA, Shinya
Toden Software Inc.
6-19-15 Shinbashi
Minato-ku
TOKYO

Tel: +81 3 3596 7680
Fax: +81 3 3596 7670
Eml: kosaka@tsi.co.jp

MATSUMURA, Toshiaki
Mitsubishi Electric Corporation
Wadasaki-cho
Hyogo-ku, KOBE

Tel: +81 78 682 6337
Fax: +81 78 682 6367
Eml: xwtm@pic.melco.co.jp

SAJI, Etsuro
TODEN Software, Inc.
In-Core Fuel Management System
6-19-15 Sinbashi
Minato-ku
TOKYO, 105-0004

Tel: +81 (3) 3596 7680
Fax: +81 (3) 3596 7670
Eml: saji@tsi.co.jp

SHIMAZU, Yoichiro
Division of Quantum Energy
Graduate School of Engineering
Hokkaido University
North 13, West 8, Kita-ku
SAPPORO 060-8628

Tel: +81 11 706 6676
Fax: +81 11 707 7888
Eml: shimazu@qe.eng.hokudai.ac.jp

KOREA (REPUBLIC OF)

PARK, Moon Ghu
KEPRI
Korea Electric Power Corporation
103-16 Munji-Dong, Yusung-Gu
TAEJON 305-380

Tel: +82 42 865 5571
Fax: +82 42 865 5504
Eml: mgpark@kepri.re.kr

LITHUANIA

BUBELIS, Evaldas
Lithuanian Energy Institute
3 Bresiaujos St.
3035 KAUNAS

Tel: ++370 7 351 403
Fax: ++370 7 351 271
Eml: evaldas@isag.lei.lt

URBONAVICIUS, Egidijus
Lithuanian Energy Institute
Ignalina Safety Analysis Group
Breslaujos 3
3035 KAUNAS

Tel: ++370 7 348 067
Fax: ++370 7 355 271
Eml: egis@isag.lei.lt

NORWAY

MOBERG, Lars
Director
Studsvik Scandpower AS
P.O. Box 15
2007 KJELLER

Tel: +47 (64) 84 45 34
Fax: +47 (64) 84 45 31
Eml: lm@scandpower.no

SPAIN

ALBENDEA, Manuel
Iberdrola (Nuclear Fuel)
Hermosilla 3
28001 MADRID

Tel: +34 91 577 65 00
Fax: +34 91 576 67 62
Eml: manuel.albendea@iberdrola.es

SWEDEN

ALMBERGER, Jan
Scientific Advisor
Vattenfall Fuel Co.
Joemtlandsgatan 99S
S-16287 STOCKHOLM

Tel: +46 87395444
Fax: +46 8178640
Eml: jan@fuel.vattenfall.se

ANDERSSON, Stig
Company Senior Scientist
Nuclear Fuel Division
ABB Atom AB
S-72163 VASTERAS

Tel: +46 (21) 347 153/(70)5347153 M
Fax: +46 (21) 348 299
Eml: stig.andersson@seato.mail.abb.com

ANDERSSON, Tell
Swedish State Power Board
Ringhals Nuclear Power Section
Department RBT
S-430 22 VAROBACKA

Tel: +46 (340)6670 28
Fax: +46 (340)6651 02
Eml: tean@ringhals.vattenfall.se

BEJMER, Klaes-Hakan
Vattenfall Fuel AB
Jamtlandsbatan 99
S-16287 Stockholm

Tel: +46 8 739 7384
Fax: +46 8 178 640
Eml: klaes@fuel.vattenfall.se

CASAL, Juan J.
Senior Specialist
Nuclear Fuel Division
ABB Atom AB
Box 53
S-721 63 VASTERAS

Tel: +46 21 347108
Fax: +46 21 348299
Eml: atojuca@ato.abb.com

CLAESSON, Per
OKG AB
Reactor Physics
S-572 83 OSKARSHAMN

Tel: +46 491 786 000
Fax: +46 491 785 050
Eml: Per.Claesson@okg.sydkraft.se

GARIS, Ninos
Swedish Nuclear Power Inspectorate
S-106 58 STOCKHOLM

Tel: +46 (0)8 698 8461
Fax: +46 (0)8 661 9086
Eml: ninos.garis@ski.se

JONSSON, Carl-Ake
ABB Atom AB
Nuclear Fuel Division
S-721 63 VASTERAS

Tel: +46 21 347230
Fax: +46 21 348299
Eml: carl-ake.jonsson@se.abb.com

KELFVE, Per
ABB Atom AB
Nuclear Fuel Division
Calculation Systems
S-721 63 VASTERAS

Tel: +46 (21) 34 71 21
Fax: +46 (21) 34 82 99
Eml: per.kelfve@se.abb.com

KRUNERS, Magnus
Studsvik Scandpower AB
Stenåsavägen 34
S-432 31 VARBERG

Tel: +46 (0)340 92966
Fax: +46 (0)340 92967
Eml: magnus@varberg.scoab.se

KURCYUSZ, Ewa
Vattenfall Fuel AB
Jamtlandsbatan 99
S-16287 STOCKHOLM

Tel: +46 8 739 6910
Fax: +46 8 178 640
Eml: ewa@fuel.vattenfall.se

LANSAKER, Paer
Vattenfall
Forsmarksverket
S-74203 OESTHAMMAR

Tel: +46 173 81543
Fax: +46 173 82100
Eml: p1k@forsmark.vattenfall.se

LEFVERT, Tomas
Head
Swedish Centre for Nuclear Technology
Dept. of Energy Technology
Royal Institute of Technology
S-100 44 STOCKHOLM

Tel: +46 8 790 86 40
Fax: +46 8 20 80 76
Eml: tomas@egi.kth.se

MULLER, Erwin
Nuclear Fuel Division
ABB Atom AB
S-721 63 VASTERAS

Tel: +46 21 347889
Fax: +46 21 347733
Eml: atoermu@ato.abb.com

NILSSON, Alf
Barsebäck Kraft AB
Box 524
S-246 25 LVDDEKOPINGE

Tel: +46 46 72 240 81
Fax: +46 46 77 58 48
Eml: alf.nilsson@bkab.sydkraft.se

OVRUM, Stein
Marketing Manager
ABB Atom AB
S-721 63 VASTERAS

Tel: +46 21 347 478
Fax: +46 21 182 737
Eml: stein.ovrum@se.abb.com

SANDERVAG, Oddbjörn
Head
Dept. of Reactor Technology
Swedish Nuclear Power Inspectorate (SKI)
S-106 58 STOCKHOLM

Tel: +46 (8) 698 84 63
Fax: +46 (8) 661 90 86
Eml: oddbjorn@ski.se

STEIRUD, Urban
OKG AB
Reactor Physics
S-572 83 OSKARSHAMN

Tel: +46 491 786 000
Fax: +46 491 785 050
Eml: urban.steirud@okg.sydkraft.se

SVENSSON, Hakan
Manager
Nuclear Fuel Division
ABB Atom AB
S-721 63 VASTERAS

Tel: +46 21 347 095
Fax: +46 21 348 299
Eml: hakan.n.svensson@se.abb.com

THUNMAN, Mats
Forsmark Kraftgrupp AB
FKA
S-742 03 ÖSTHAMMAR

Tel: +46 173 81969
Fax: +46 173 81697
Eml: mtn@forsmark.vattenfall.se

WIKSELL, Goeran
Reactor Physics
OKG AB
Oskarshamnsverket
S-57283 OSKARSHAMN

Tel: +46 49 491 786143
Fax: +46 49 491 787850
Eml: goran.wiksell@okg.sydkraft.se

SWITZERLAND

NOEL, Alejandro
Principal Consultant
Studsvik Scandower AB
Maderstrasse 17
CH-5400 BADEN

Tel: +41 56 221 7359
Fax: +41 56 221 7359
Eml: alejandro.noel@bluewin.ch

TVEITEN, Bengt
Fuel Regulatory Licensing
EGL/KKL
Postfach 1280
CH-8034 ZURICH

Tel: +41 1 388 2525
Fax: +41 1 388 2550
Eml: Bengt.Tveiten@egl.ch

VAN DOESBURG, Willem
Swiss Federal Nuclear Safety Inspectorate (HSK)
Hauptabteilung fuer die Sicherheit der Kernanlagen
CH-5232 VILLIGEN-HSK

Tel: +41 56 310 3862
Fax: +41 56 310 4979
Eml: willem.vandoesburg@hsk.psi.ch

ZIMMERMANN, Martin
Swiss Nuclear Society
Laboratory for Reactor Physics and Systems Behaviour
Paul Scherrer Institut
CH-5232 VILLIGEN PSI

Tel: +41 56 310 27 33
Fax: +41 56 310 23 27
Eml: martin.zimmermann@psi.ch

UNITED STATES OF AMERICA

COVINGTON, Lorne
Studsvik Scandpower, Inc.
1087 Beacon Street, Suite 301
NEWTON, MA 02159

Tel: +1 (617) 965 7450
Fax: +1 (617) 965 7549
Eml: ljc@soa.com

MOON, Hoju
Siemens Power Corporation
Nuclear Division
P.O. Box 130
RICHLAND, WA 99352-0130

Tel: +1 (509) 375 8265
Fax: +1 (509) 375 8402
Eml: hoju-moon@nfuel.com

INTERNATIONAL ORGANISATIONS

SARTORI, Enrico
OECD/NEA Data Bank
Le Seine-Saint Germain
12, boulevard des Iles
F-92130 ISSY-LES-MOULINEAUX

Tel: +33 1 45 24 10 72
Fax: +33 1 45 24 11 10
Eml: sartori@nea.fr

BERG, Öivind
Institutt for Energiteknik
OECD Halden Reactor Project
P.O. Box 173
N-1751 HALDEN

Tel: +47 (69) 21 22 00
Fax: +47 (69) 21 24 60
Eml: oivind.berg@hrp.no

TSUIKI, Makoto
Control Room Systems Division
IFE/OECD HRP
P.O. Box 173
N-1751 HALDEN

Tel: +47 (69) 21 22 00/22 34
Fax: +47 (69) 21 24 60
Eml: Makoto.Tsuiki@hrp.no

* *Regrets not having been able to attend.*

ALSO AVAILABLE

NEA Publications of General Interest

1998 Annual Report (1999) *Free: paper or Web.*

NEA Newsletter
ISSN 1016-5398 Yearly subscription: FF 240 US$ 45 DM 75 £ 26 ¥ 4 800

Radiation in Perspective – Applications, Risks and Protection (1997)
ISBN 92-64-15483-3 Price: FF 135 US$ 27 DM 40 £ 17 ¥ 2 850

Radioactive Waste Management in Perspective (1996)
ISBN 92-64-14692-X Price: FF 310 US$ 63 DM 89 £ 44

Radioactive Waste Management Programmes in OECD/NEA Member countries (1998)
ISBN 92-64-16033-7 Price: FF 195 US$ 33 DM 58 £ 20 ¥ 4 150

Nuclear Science

Ion and Slow Positron Beam Utilisation (1999)
ISBN 92-64-17025-1 Price: FF 400 US$ 72 DM 119 £ 43 ¥ 8 500

Physics and Fuel Performance of Reactor-Based Plutonium Disposition (1999)
ISBN 92-64-17050-2 Price: FF 400 US$ 70 DM 119 £ 43 ¥ 8 200

Shielding Aspects of Accelerators, Targets and Irradiation Facilities (SATIF-4)(1999)
ISBN 92-64-17044-8 Price: FF 500 US$ 88 DM 149 £ 53 ¥ 10 300

Speciation Techniques and Facilities for Radioactive Materials at Synchrotron Light Sources (1999) *Free on request.*

Prediction of Neutron Embrittlement in the Reactor Pressure Vessel: VENUS-1 and VENUS-3 Benchmarks (2000) *Free on request.*

Calculations of Different Transmutation Concepts: An International Benchmark Exercise (2000) *Free on request.*

International Evaluation Co-operation *(Free on request)*

Volume 1: *Comparison of Evaluated Data for Chromium-58, Iron-56 and Nickel-58* (1996)
Volume 2:*Generation of Covariance Files for Iron-56 and Natural Iron* (1996)
Volume 3: *Actinide Data in the Thermal Energy Range* (1996)
Volume 4: *^{238}U Capture and Inelastic Cross-Sections* (1999)
Volume 5: *Plutonium-239 Fission Cross-Section between 1 and 100 keV* (1996)
Volume 8: *Present Status of Minor Actinide Data* (1999)
Volume 12: *Nuclear Model to 200 MeV for High-Energy Data Evaluations* (1998)
Volume 13: *Intermediate Energy Data* (1998)
Volume 15: *Cross-Section Fluctuations and Shelf-Shielding Effects in the Unresolved Resonance Region* (1996)
Volume 16: *Effects of Shape Differences in the Level Densities of Three Formalisms on Calculated Cross-Sections* (1998)
Volume 17: *Status of Pseudo-Fission Product Cross-Sections for Fast Reactors* (1998)

Order form on reverse side.

ORDER FORM

OECD Nuclear Energy Agency, 12 boulevard des Iles, F-92130 Issy-les-Moulineaux, France
Tel. 33 (0)1 45 24 10 10, Fax 33 (0)1 45 24 11 10, E-mail: nea@nea.fr, Internet: http://www.nea.fr

Qty	Title	ISBN	Price	Amount
			Postage fees*	
			Total	

*European Union: FF 15 – Other countries: FF 20

❑ Payment enclosed (cheque or money order payable to OECD Publications).

Charge my credit card ❑ VISA ❑ Mastercard ❑ Eurocard ❑ American Express

(N.B.: You will be charged in French francs).

Card No.	Expiration date	Signature
Name		
Address	Country	
Telephone	Fax	
E-mail		

LES ÉDITIONS DE L'OCDE, 2, rue André-Pascal, 75775 PARIS CEDEX 16
IMPRIMÉ EN FRANCE
(66 2000 11 1 P) ISBN 92-64-17659-4 – n° 51247 2000